THE CONTROL WAR

THE CONTROL WAR

THE STRUGGLE FOR SOUTH VIETNAM, 1968–1975

Martin G. Clemis

UNIVERSITY OF OKLAHOMA PRESS : NORMAN

Publication of this book is made possible through the generosity of Edith Kinney Gaylord.

Library of Congress Cataloging-in-Publication Data

Name: Clemis, Martin G., 1967– author.

Title: The control war : the struggle for South Vietnam, 1968–1975 / Martin G. Clemis.

Description: Norman : University of Oklahoma Press, [2018] | Includes bibliographical references and index.

Identifiers: LCCN 2017042897 | ISBN 978-0-8061-6009-2 (hardcover ; alk. paper)

Subjects: LCSH: Vietnam War, 1961–1975. | Military geography—Vietnam (Republic) | Insurgency—Vietnam (Republic) | Counterinsurgency—Vietnam (Republic)

Classification: LCC DS557.7 .C55 2018 | DDC 959.704/3—dc23

LC record available at https://lccn.loc.gov/2017042897

The paper in this book meets the guidelines for permanence and durability of the Committee on Production Guidelines for Book Longevity of the Council on Library Resources, Inc. ∞

1 2 3 4 5 6 7 8 9 10

To my mother, Mary E. Clemis,
and to the memory of my father, William R. Clemis

CONTENTS

MAPS

ACKNOWLEDGMENTS

This book represents the end of a long and arduous journey—one that would have been impossible without the kind and generous support of colleagues, friends, and family. It is fitting and proper, then, to recognize those who helped me along the way and were instrumental in this achievement.

This is my first monograph and—like all landmark firsts in our lives—it holds special meaning. I would like to begin by thanking Adam Kane at the University of Oklahoma Press for his interest in my work and for taking the time and effort to shepherd this manuscript to completion. Stephanie Evans at OU Press and Bob Fullilove also deserve much thanks. I also owe a particularly large debt of gratitude to three superb scholars: Gregory A. Daddis of Chapman University, Andrew J. Birtle of the U.S. Army Center of Military History (CMH), and Ron Milam of Texas Tech University. Greg was the catalyst behind this project, providing inspiration for a dissertation topic in 2010 at the West Point Summer Seminar in Military History. Since then, he has become a mentor and a friend and I appreciate his assistance and advocacy beyond measure. Andy was incredibly generous with his time while I conducted research at CMH, and he offered wise counsel and provided detailed suggestions for refinement during the early stages of this project. Greg's and Andy's expertise and feedback were invaluable. They not only helped me navigate the exasperatingly labyrinthine world of the Second Indochina War, but also helped redefine, clarify, and strengthen many of this work's key arguments. Ron agreed to read the manuscript without hesitation, and he (along with Greg Daddis) wrote an extremely generous reader's report on my behalf. Overall, I am indebted to the assistance and expertise of these outstanding historians and am humbled by the time they spent reading, assessing, and promoting this work. Other exemplary scholars who offered encouragement and support over the past seven years and have inspired me to continue exploring the Vietnam War include Gian Gentile, Kyle Longly, Heather Stur, Rob Thompson, Eric Villard, Andy Wiest, and James Willbanks.

I would also like to extend a special thanks to my former mentors and graduate school advisers Gregory J. W. Urwin of Temple University and Christopher S. DeRosa of Monmouth University. Greg and Chris are model professionals and teachers who inspired, challenged, and expanded my capacities as a (figuratively) young scholar and historian. I cannot thank them enough for their wisdom, guidance, and sponsorship. I am also indebted to the time, insights, and advice given by the following Temple University professors: Beth Bailey, Dieu T. Nguyen, Jay Lockenour, and David Farber. Faculty and administrators at Valley Forge Military College, particularly Patrick Murray, Maureen Malone, and Nan Hood, also deserve thanks.

This work would have been impossible without generous financial and research assistance from key institutions. I would therefore like to thank the following for their support: the U.S. Army Center of Military History, the U.S. Army Heritage and Education Center, the John Votaw Endowed Research Fellowship, Temple University's Center for the Study of Force and Diplomacy, and the Society for Military History. Individuals from these institutions who deserve mention include Richard Stewart, Brian Newman, Jeff Hawks, and John Votaw. Special thanks also go out to the archivists and librarians at the National Archives and Records Administration, the CMH, the U.S. Army Military History Institute, and the Vietnam Center and Archive at Texas Tech University.

Members of the infamous "Temple Mafia" also deserve recognition for their direct and/or indirect contributions to this work. Two outstanding historians and close friends merit special mention: Earl J. Catagnus of Valley Forge Military College and Michael R. Dolski of the Defense POW/MIA Accounting Agency. Earl served as a sounding board for this project since its inception, and his scholarship and insights as both a practitioner and student of war helped focus this work's arguments and refine its discussion. Mike took the time to read the manuscript in its entirety and offer suggestions for conceptual (and cosmetic) improvement. Earl and Mike both had a profound impact on this manuscript, and I am indebted to them for their assistance and, especially, their friendship. Other close colleagues and friends who warrant thanks for their spiritual and intellectual support include Josh Wolf, Eric Klinek, Chris Golding, John Pierre Beugoms, Jason Warren, Jason Smith, Dave Ulbrich, Matt Muehlbauer, Bob Wintermute, Mike Lynch, Rich Grippaldi, and Tom Reinstein.

Finally, special thanks go out to my beloved family and friends, especially my wife, Becky, and our daughters, Lily and Ivy, who sacrificed so much so that I could complete this project. I love and cherish you all beyond measure!

ABBREVIATIONS

APC	Accelerated Pacification Campaign
ARVN	Army of the Republic of Vietnam
CIA	Central Intelligence Agency
CORDS	Civil Operations and Revolutionary Development Support
COSVN	Central Office of South Vietnam
CPDC	Central Pacification and Development Council
DAO	American Defense Attaché Office
DRV	Democratic Republic of Vietnam (North Vietnam)
DMZ	Demilitarized Zone
FWMAF	Free World Military Assistance Forces
GVN	Government of Vietnam (South Vietnam)
HES	Hamlet Evaluation System
ICCS	International Commission of Control and Supervision
JMC	Four-Party Joint Military Commission
MACV	Military Assistance Command Vietnam

MR	Military Region
NLF	National Front for the Liberation of South Vietnam
NVA	North Vietnamese Army
PAVN	People's Army of Vietnam
PLAF	People's Liberation Armed Forces
PRG	People's Revolutionary Government
RD	Revolutionary Development
RF/PF	Regional Forces/Popular Forces
RVN	Republic of Vietnam (South Vietnam)
RVNAF	Republic of Vietnam Armed Forces
VC	Vietnamese Communist, Viet Cong
VCI	Viet Cong Infrastructure
VSSG	Vietnam Special Studies Group
VWP	Vietnam Workers' Party

A NOTE ON NOMENCLATURE

The byzantine nature of the Second Indochina War makes it imperative that historians and other scholars provide clear and consistent identification of the historical actors that were involved. The conflict contained a complex array of state and non-state actors, including governments, political parties and institutions, front groups, alliances, delegations, social groups, and a wide variety of military and paramilitary organizations. In numerous instances, a single institution, organization, or entity assumed or was assigned different names. It is necessary, therefore, to provide a brief taxonomy of the primary political and military groups discussed in this study in order to avoid confusion and to help clarify the many names and acronyms used to describe the agents that actively took part in the struggle for South Vietnam.

The Second Indochina War can largely be broken down into two opposing camps—one communist and the other noncommunist—each with its associate and/or sponsoring powers. On the noncommunist side of the ledger there was the Republic of Vietnam (RVN), often called the Government of South Vietnam (GVN) or South Vietnam. Also, because the RVN capital was Saigon, the South

Vietnamese government is sometimes referred to as the "Saigon government" or the "Saigon regime." The RVN's primary political and military ally was the United States. America's predominant role in the conflict, particularly its direct intervention and use of combat forces from 1965 to 1972, should not obscure the fact that South Vietnam was supported by a broad coalition of nations. American GIs comprised the largest and most well-known contingent within the noncommunist coalition, but they were just one group of foreign nationals that fought alongside South Vietnam and its military, the Republic of Vietnam Armed Forces (RVNAF). Military forces from the Republic of Korea and Commonwealth of Australia, among others, also took part in the conflict. Collectively, those military forces that comprised the noncommunist military coalition were designated the Free World Military Assistance Forces (FWMAF).

The communist side of the ledger was more convoluted. In large part, the Second Indochina War was a product of the Vietnamese revolution and efforts by the Vietnam Workers' Party (VWP) to seize power, overthrow the government of South Vietnam, and reunite the divided country under communist rule. The chief means to accomplish these goals was revolutionary warfare or people's war—a hybrid approach that merged political organizing, diplomacy, conventional war, and, most importantly, a southern insurgency situated within the national borders of South Vietnam. This complex method produced a mélange of political and military organization and institutions. First, there was the Democratic Republic of Vietnam (DRV), or North Vietnam. The DRV was ruled by the VWP from Hanoi. Contrary to what many have argued over the years, the DRV was the primary power that drove the southern insurgency. Although there has been much debate over the years concerning the roles noncommunists and the DRV had in managing and executing the rebellion, recent scholarship, including new documentary evidence from Vietnamese archives, has confirmed what the Saigon regime and U.S. policymakers had claimed throughout the war—that despite claims to autonomy and the participation of noncommunist parties, the southern insurgency was in fact under the direct supervision and control of the VWP and its associated organs, the Central Committee and the Politburo. For this reason, the umbrella term "communists" is used widely to describe the insurgents in South Vietnam. It is not meant to ignore the role played by those groups and individuals who participated in the revolution but did not identify themselves as Marxist-Leninists. Rather, it is used to simplify the narrative and to acknowledge the hegemonic role the VWP played in planning, directing, and executing the insurgency.

On the operational level, the insurgency in South Vietnam was a composite of numerous organizations and committees, each with distinct (though often overlapping) functions. The primary organs for conducting revolutionary activities in South Vietnam were the People's Revolutionary Party (PRP), which was essentially the southern branch of the VWP; the National Front for the Liberation of South Vietnam (NLF); and the Central Office of South Vietnam (COSVN). The NLF was, as its name implied, a front group composed of communist and noncommunist entities united in a common effort to topple the government of South Vietnam. The NLF was superseded in 1969 by the People's Revolutionary Government (PRG), an entity that was created as an official communist "government" within South Vietnam and intended to influence the ongoing peace negotiations. COSVN, meanwhile, was the headquarters that directed all insurgent activities, both military and political, in the southern half of South Vietnam. Communist guerrillas and political cadres in the northern half of the country were under the direct command and control of the DRV. The VWP controlled all political and military activities in South Vietnam at every echelon, from the village/hamlet level up through the regional level. Political activities were conducted by the party. Military activities were carried out by the People' Liberation Armed Forces (PLAF), or, as they were more well known, the "Viet Cong" (VC)—a pejorative given to both communist guerrillas and political cadres by the South Vietnamese and their American ally. The PLAF were assisted in the field by an expeditionary wing of the People's Army of Vietnam (PAVN), or the North Vietnamese Army (NVA) as it was more often called. NVA units were infiltrated into South Vietnam to conduct conventional military operations in support of the PLAF's guerrilla activities. Each of these designations—with the exception of "VC," which I have tried to avoid except in quotations and other rare instances—are used when recognizing a specific entity, although the identifier "communist" is often used when discussing the politico-military activities of these groups in general terms.

THE CONTROL WAR

> As long as there were two sides—revolution and counterrevolution—and they continued to struggle to control the land and people there would be a contested area. It would disappear only when there was no longer a struggle between the two sides, i.e., when one side yielded and the other side won complete victory.
>
> —Tran Van Tra, *Concluding the 30-Years War*, 1983

> All histories are geographically specific and their making is . . . inseparable from the social production of spaces and places.
>
> —Allan Pred, *Making Histories and Constructing Human Geographies*, 1990

INTRODUCTION

THE LOCUS AND MEDIUM OF REVOLUTIONARY CONFLICT: THE OPERATIONAL ENVIRONMENT

The Second Indochina War was a competition for political power, fought largely in the South Vietnamese countryside. The 1954 partition of Vietnam gave rise to two competing regimes: one communist and determined to overthrow its adversary and unify the country under the socialist banner; the other anticommunist and resolute in defending what it considered its right to exist as a free and autonomous nation. This opposition guaranteed a bitter struggle for the power to define the future of Vietnamese society. The Geneva Conference did not establish two separate states, but rather, in the words of historian David W. P. Elliot, "two contesting parties within a single state . . . each continued after Geneva, just as each had done before, to lay claim to the whole country."[1] In 1969, after nearly a decade of political and military struggle in South Vietnam, scholar John McAlister made the following observation: "The war now raging in Vietnam is not a war being fought between two separate nations, but a revolutionary struggle within one nation. . . . It is a competition between two . . . governments, each of which wants to become the sole legitimate government of a people."[2] The Vietnamese communists and the Republic of Vietnam both claimed to be making revolution, not only for the proprietary right to rule, but for a mandate to construct a new social and political order in Southeast Asia. The ultimate objective for each was securing absolute political power to govern the people and the territory of South Vietnam and, according to historian Neil Jamieson, to set the "social thermostat" by resolving through war "their own competing and incompatible visions of what Vietnam was and what it might and should become."[3]

At its heart, the war in Vietnam was a contest to establish national political suzerainty. Achieving the power to determine the nation's future first meant establishing dominance over the villages and hamlets of rural South Vietnam. As one authority has noted, a paradigmatic shift in warfare had occurred in the decades following World War II; from an age of "interstate industrial war" to an era of "war amongst the people."[4] This new paradigm, which was fought largely in the underdeveloped countries of Asia, Africa, and Latin America, was unique because civilian populations and the places they lived were not peripheral to armed conflict; they were the sine qua non. Thanks to postcolonial ambitions and the advent of "people's war" throughout the developing world, and the infusion of Cold War antagonisms into local intrastate conflicts, indigenous populations had become both the object and the medium of revolutionary conflict. Mao Zedong famously pronounced that in revolutionary wars the population was the "water" that sustained and nurtured the insurgent "fish" that inhabit it.[5] David Galula, a French officer turned counterinsurgency expert, claimed the population was the "real terrain" of revolutionary conflicts. "This is where the real fighting takes place," he wrote, "where the insurgent challenges the counterinsurgent, who cannot but accept the challenge."[6] In the view of Henry Kissinger, national security advisor and secretary of state in the Richard Nixon administration, wars of revolution differed from those involving traditional military operations because the objective, the "key prize," was "control of the population."[7] Regardless of which metaphor one preferred—the land, the sea, or the prize—the civilian population was considered the most essential element within the context of revolutionary conflict.

Renowned Vietnam scholar Bernard Fall wrote in 1964, "What differentiates revolutionary warfare from conventional ground operations is the objective. In a conventional war, the key objective is terrain and enemy troops; in revolutionary warfare it is the *civilian environment* in which the opponent operates."[8] Fall's observation was judicious. It was not just the population that was important in revolutionary war; it was the people *and* the environment in which they lived that was central. "It's the land where the people are that counts in this conflict," U.S. Ambassador Robert W. Komer stated in 1968.[9] The civilian environment—in this case the densely populated villages and hamlets of the South Vietnamese countryside—was the cockpit of revolutionary struggle. Rural areas held profound strategic importance because they were the place where a majority of the population lived and where the material means for waging a protracted civil conflict were located. In short, they were both the incubator and the platform for social revolution. Control of the countryside and its people, therefore, was crucial to both the communists and the RVN. For the purposes of this study, control is defined as the imposition of force, whether by the government or the insurgency—the ability to exercise power and authority over territory and population via military, political, or psychological means. "Control of the people and resources of the countryside," read a study generated by the Vietnam Special Studies Group, "is vital to both contestants. . . . [T]he struggle for control of the countryside is the 'control war.' This is what gives the Vietnam conflict its insurgency character and distinguishes it from more conventional types of warfare."[10] Others shared this assessment. "The countryside represents a significant and still relatively unorganized and untapped source of political strength and support for anyone wishing to attain or maintain real political power in South Vietnam," a 1970 article published in *Asian Survey* declared.[11]

The Vietnamese themselves recognized the vital strategic importance of the countryside. According to former South Vietnamese general Tran Dinh Tho, rural areas were the "main battleground" of the war:

> To our side as well as to the enemy, the rural area of South Vietnam was to be the decisive battlefield. Without it, the enemy would lose his foothold and the opportunities to protract his war, for the rural area was his major source of subsistence and manpower. The countryside was the arena for the ideological struggle between the free world and communism. It was where the battle for the hearts and minds of the people was fought and whoever won their trust, cooperation, and support would be the final victor. Without the rural area, which in communist doctrine included not only the agricultural lands, but also the forests, swamps, and mountains, the nation could hardly survive.[12]

Tho's adversary, North Vietnamese general Vo Nguyen Giap, concurred. The South Vietnamese countryside, Giap declared, was the "firm, long-term prop and battleground" of the war. "The rural . . . areas are densely populated areas with enormous resources, which the aggressor enemy has constantly tried to occupy," he said.

> If we succeed in gaining control over the rural areas, the revolution will acquire a firm basis for mobilizing human and material resources to develop its forces for protracted combat, to insure that the more the revolutionary forces fight, the more powerful they become—thus bankrupting the enemy's policy of trying to obtain human and material resources and using Vietnamese to fight Vietnamese and war to support war—and to create favorable conditions for the rural, delta, and mountainous areas, so that the urban revolutionary movement can be supported vigorously and the enemy attacked at his nerve centers and in his lair.[13]

Giap's statement underscores the significance of the countryside to the revolution. It also reveals the strategic importance placed on South Vietnam's urban areas within communist revolutionary war doctrine.

Thus, while communists' efforts to establish control over the countryside were fundamental to their overall strategy, rural areas constituted part of a hybrid approach that emphasized a coordinated attack and seizure of political power both in the cities and in the provincial capitals—an "urban revolutionary movement." To the Democratic Republic of Vietnam and the National Liberation Front, urban areas were critical despite South Vietnam's largely rural context. In fact, an urban-rural approach was the lynchpin of the communist strategy known as the General Offensive and General Uprising, a dual politico-military plan to synchronize military operations with a popular uprising by the masses. Inspired by the 1945 August Revolution (an event where the Vietminh took advantage of the political vacuum that resulted from Japan's defeat in World War II and seized power in Vietnam through a combination of military offensives in the countryside and political uprisings in the cities, villages, and towns), the strategy was formally introduced in December 1963 at the Vietnam Workers' Party Ninth Plenum of the Central Committee by the general secretary, Le Duan. Under the first secretary's leadership, the General Offensive and General Uprising became the cornerstone of communist political and military doctrine in the war against the RVN and the United States.[14] In February 1965, as American intervention loomed, Le Duan restated his hybrid vision for toppling the Saigon regime: "As we expand our guerrilla warfare operations and shrink the size of those areas occupied by the enemy in the mountain jungles and in the rural lowlands and at the same time raise the struggle movement in the cities to new levels, we will then launch a coordinated general uprising with a general military offensive aimed straight at the enemy's heart, his center, to seize control of the government."[15]

The General Offensive and General Uprising was the most important modification the Vietnamese had made to Mao Zedong's theory of revolutionary warfare. Although the Vietnamese strategy had borrowed heavily from Mao's protracted war principles, it amended the plan for an evolutionary three-phase process (defense to equilibrium to general counteroffensive) by advocating a more precipitous approach. In this way, the strategy was tailored to the specific circumstances of Vietnam and built according to a carefully balanced blueprint for harmonizing political struggle and armed struggle. Le Duan's plan for a general offensive and uprising was not a rejection of protracted war strategy.[16] Rather, it was a modification that sought rapid, decisive victory as a means to supplement the more patient, gradual method of protracted war. The resolution passed at the Ninth Plenum declared that there was "no contradiction" in the concept of protracted war and a general offensive and uprising. "On one hand, we must thoroughly understand the guideline for protracted struggle," the resolution read, "but on the other hand we must also seize the opportunities to win victories in a not too long period of time. In a revolutionary war against aggression and a domestic war, as in [South Vietnam], we should be ready to take advantage of sudden changes in the situation to disintegrate the enemy's troops and government. . . . The more we try to build and develop our forces to gain victories in a short time, the more we will have favorable conditions to fight a protracted war."[17] As we can see, the General Offensive and General Uprising did not supplant a protracted war strategy; it merely prepared for contingencies that would facilitate the rapid seizure of power.

In 1965, Le Duan underscored the difference between the Vietnamese and Chinese model of revolutionary war. "The revolution in the South," he stated, "will not follow the path of protracted armed struggle, surrounding the cities by the countryside and advancing to the liberation of the entire country by using military forces as China did, but will follow a Vietnamese path. That means there will be partial uprisings and establishment of base areas, and there will be a guerrilla war leading to a general uprising which will primarily use political force in coordination with armed forces to grasp political power in the hands of the people."[18] Although Le Duan recognized there were "many similarities" between Mao's China and Vietnam, and acknowledged that the Chinese revolution had bequeathed "many valuable lessons" to the Vietnamese revolution—including using the peasantry to build a revolutionary army, following a protracted war strategy, building base areas, and adopting guerrilla warfare—he maintained, nonetheless, that the Vietnamese struggle was unique. "While we have studied the experiences of the fraternal parties in other countries," Le Duan declared before the VWP Central Committee, "we used a spirit of independence and self-reliance to set policies and directions that were suitably adapted to suit the realities and the circumstances of our own nation."[19]

Uprisings and civil unrest in the cities, in tandem with revolutionary efforts in the countryside, were a fundamental component to the Vietnamese communist strategy for collapsing the South Vietnamese government and achieving victory. "Along with the political and military struggle in the rural areas," read a statement issued by the VWP, "there must be a political struggle movement coordinated with military violence at a definitive level in the cities to prepare for the General Offensive and Uprising movement which is the decisive step of the South Vietnam Revolution."[20] Only through the careful coordination of military and political struggle and "a close relationship between rural areas and cities," communist strategists believed, could revolutionary forces expand their hold on South Vietnam and create favorable conditions for seizing political power. "Our protracted struggle [is] aimed at gaining control part by part as a step toward gaining control of the entire country, our method of initiating uprisings step by step and eventually advancing toward the general uprising [is] to seize total power by force."[21]

Despite the importance of South Vietnam's urban areas at the doctrinal or theoretical level, their empirical significance was eclipsed by its rural areas. Urban areas were important, particularly in theory; in some instances, they even had real-world significance, as evidenced during the 1968 Tet Offensive. But the struggle to control rural areas remained the war's most salient feature. "If we want to defeat the puppet army in the lowlands," Le Duan reemphasized in November 1965, "the most important thing for us is to firmly maintain and expand our mastery and control in the rural countryside. In order to secure and maintain our control in the rural countryside, the first thing we must do is to build and rationally deploy our battlefield armed forces in each district, province, and military region. We must strengthen the forces and improve the fighting skills of our local force troops. We must powerfully and broadly expand our guerrilla network. We must build combat villages and hamlets."[22]

As significant as populous areas were, both rural and urban, the civilian environment was but half of South Vietnam's overall *operational environment*, defined as a composite of the conditions, circumstances, and influences that shape the conflict area.[23] This amalgam included the deployment of friendly and enemy military forces; the country's social, political, and economic character; and its physical environment. "The conduct of warfare is conditioned by the physical and cultural characteristics of the conflict area," noted one group of scholars. "Military plans and operational doctrine, mission force structure, weapons, and equipment requirements vary depending on the physical and cultural conditions in which they are deployed."[24] As we will see, South Vietnam's operational environment played a central role in the control war.

The significance of geography and terrain is often dismissed or marginalized when discussing Vietnam and other revolutionary conflicts aside from their utility in assisting guerrillas at the tactical level. In fact, the belief that insurgents fight to influence the population and not to hold or occupy territory is widespread. Such an assertion, however, is incomplete; it does not provide a full or accurate picture of communist revolutionary warfare, particularly its means and ends. Although the practical realities of asymmetrical conflict dictated that communist guerrillas at times abandon territory and use stealth, speed, and mobility as a means to preserve themselves in the face of a numerically and technologically superior enemy, territory—particularly base areas, insurgent villages and hamlets, and other forms of revolutionary terrain—was extremely important. "The special characteristics of revolutionary warfare, of national liberation warfare, and of our strategic guidelines in the resistance war against the Americans demand that our armed forces and our political forces must . . . have firm footholds," Le Duan pronounced in 1965. "They must be able to stand their ground firmly on the mountain jungle battlefields as well as in the lowlands in order to fight the enemy. We cannot fight and then run away, abandoning the land and abandoning the people."[25] Le Duan's statement reveals an aspect of the war in Vietnam that is often overlooked, which is that guerrilla warfare was merely a tactic. Revolutionary warfare, as practiced by Hanoi and the NLF, was a comprehensive strategy for defeating the enemy, seizing political power, and achieving total societal transformation. As such, it required a terrestrial as well as an ideological and human foundation.

Efforts to win the war and establish hegemony in South Vietnam vis-à-vis the operational environment centered on the production and control of politico-military space—that is, physical and ideational spheres of influence that were used to build and consolidate political and military power. Conventional understandings and conceptions of space and the places people inhabit have been challenged and modified over the years. Scholars have refashioned traditional conceptions of the natural world, arguing that the environment is far more than physical phenomena defined by the absolute materialism of Cartesian geography and the scientifically quantifiable and measurable world we inhabit. It is at once an absolute (physical or real) and a relative (ideational or imagined) object that is constituted and given meaning through human agency.[26] According to sociologist Henri Lefebvre, humans create, mold and transform the natural environment. To a large degree, the landscape we inhabit is a human product—an artefact of man's activities.[27] Thus, the environment is not merely a scientific object that serves as backdrop to human action, but a manmade construction that is designed, fabricated, given purpose, and assigned meaning through social activity.[28] "We have moved from the production of things *in space*," Lefebvre declared, "to the production *of space* itself."[29] Thus, the human environment is both a mental and a material construct, an absolute and relative creation that "owes as much to the conceptual realms as to material activities."[30]

The spaces humans create are not divorced from politics or ideology but are in fact political and ideological constructs. All human activity occurs in the physical world and within a spatial context. States and interest groups within the state create, appropriate, and structure the environment for political purposes. The physical world, therefore, does not merely exist; it is produced by humans for social purposes and invested with political meaning and value. "There is a politics of space," Lefebvre once wrote, "because space is political."[31] Manmade environments are also essential to maintaining social and political power. Physical and ideational spaces can be created and ordered as a method of social control and an exercise in authority.[32] "Space is fundamental to any exercise of power," French historian and philosopher Michele Foucault once claimed.[33] Lefebvre agreed with this conception, arguing that humans create spaces as "a means of control, and hence of domination, of power."[34] Political power, then, can only be actualized and transformed from a latent to an active state through the "construction of human geographies and the social production of space."[35]

Spaces that have been produced through social and political processes are not only highly ideational, they are highly competitive. Geographer Robert Sack has shown that conflicts over ownership and control of territory are commonplace. They occur when individuals and groups work to "affect, influence, or control people, phenomena, and relationships by delimiting and asserting control over a geographic area."[36] This is known as territoriality. According to one source, territory, at its most fundamental level, is a spatial expression of power; a claim to political sovereignty over a particular space.[37] Social and political spaces, noted another authority, are "the product of competing ideas."[38] Competition over space inevitably leads to conflict; and conflict frequently leads to violent struggle, armed conflict, and war. "The problem of war," stated one authority, "is rooted in geography because military conflict is primarily undertaken to gain control of land and influence people."[39] Geographer Colin Flint has argued that political geographies "shape and are shaped by the many processes of war and peace." Geography and war, he claims, "are the products of human activity; war creates geographies of borders, states, empires, and so on, and in turn these geographic entities are the terrain over which peace is maintained or new wars are justified."[40] As these descriptions illustrate, space is produced through human action and imbued with meaning. Moreover, once spaces are made they become objects of power and are coveted by political actors who seek to control them. Space is and has been contested, often violently, throughout history as a result of social and political differences over its design and purpose. Such was the case in South Vietnam.

The Second Indochina War was in large part a competition for space, as communist insurgents and the RVN grappled for control over a specific geographical region. Political scientist Stathis N. Kalyvas has shown that in revolutionary civil wars political control has a clear territorial foundation; power and dominance, he argues, cannot be achieved without a constant and credible armed presence, one that has the power to modify behavior and produce popular collaboration through the threat or use of violence and coercion.[41] This holds true for the political dimension as well. Without an administrative or governing presence to reify political ideals and put theory into practice, a revolution is merely a will-o'-the-wisp, a spectral apparition with no corporeal foundation to legitimize, validate, or demonstrate its power.

Thus, the struggle for South Vietnam, aside from its more obvious manifestations, was a contest to control the operational environment. Like other revolutionary conflicts, the war was inherently spatial because both sides struggled to shape the physical and ideational world as a means to gain power and advance their respective visions of a more just and equitable society. Spatial contestation, territoriality, and efforts to control the landscape—both real and imagined—were a salient feature in the war for the countryside and the fight to determine the future social and political order of South Vietnam. Land and people were critically important, both to communist revolutionaries and to the RVN. Physical and human geography not only provided each belligerent with the corporeal and material requirements it needed to wage a protracted civil war in the countryside—including manpower, supply, intelligence, and logistics; these factors also fulfilled critical psychological needs as well by legitimating ontological claims to sovereignty, authority, and power. This material and ideational significance made populated areas and the surrounding countryside the locus and medium of political and ideological struggle. Much like a conventional military battlefield, where the site of combat is also part of the territory over which contending armies fight, the operational environment was both a *place* of political conflict and an *object* of political struggle.[42]

South Vietnam's operational environment was incredibly diverse. To begin with, it was populated by a fiercely independent people, predominantly peasants, who worked and lived according to ancient local customs and lifeways. Although the RVN had by the early 1960s developed a tiny economic elite, nascent middle class, and modest working class in and around urban areas, the overwhelming majority of the population were rural farmers who worked small plots of land, either as owners or tenants. Peasants were dedicated to their family, their ancestors, their village, and the land they worked and lived on. Most of them, including farmers, fishermen, and small artisans, had some property and education despite their rustic condition. Moreover, they were energetic, resourceful, and ambitious, and considered themselves culturally and economically superior to their regional neighbors. The RVN's total population was estimated to be about 16 million. Of this number, roughly 85 percent were ethnic Vietnamese. The remaining 15 percent included native Highlanders, or Montagnards as they were known; ethnic Cambodians; and nonindigenous Chinese. Approximately 80 percent of the South Vietnamese population practiced Buddhism. The remainder included Taoists, Confucians, Catholics, and an assortment of local sects. The diversity of the South Vietnamese citizenry, along with the absence of any real sense of national identity among the rural population, however, exacerbated tensions among South Vietnam's ethnic and religious groups. More importantly, it undermined efforts to forge a national political community and to build a grassroots movement that would assist the Saigon regime in its struggle against communist subversion in the countryside.[43]

The climate in Vietnam's southern half is tropical. Although generally hot and humid, the average temperature and heavy annual rainfall varies during the region's two monsoonal wet seasons. The total area of the former crescent-shaped RVN was approximately 66,000 square miles. It was some 700 miles in length and anywhere from 40 to 120 miles wide. The country was bounded by the DRV to the north, the South China Sea to the east, and Laos, Cambodia, and the Gulf of Thailand to the west. The terrain was unevenly divided into three primary geographical regions: the mountainous highlands, the coastal lowlands, and the broad, flat alluvial Mekong delta. The highlands, which cover nearly all of the northern and central portions of the country, was a sparsely populated region dominated by the Annamite Range, a series of mountains, ridges, and foothills that rise 5,000 feet in the north before gradually decreasing in elevation toward the piedmont and delta regions. The rugged elevated areas of the central highlands are interspersed with valleys, river gorges, and a large grassy plateau, roughly forty by one hundred miles, which dominates the western edge of the country along the Laotian and Cambodian borders. The vegetation of the highlands consists of double- and triple-canopy jungle, brush, elephant grass, and bamboo. A lack of roads and other lines of communication isolated the area, restricted movement, and made operations extremely difficult. Although sizable, the highlands were home to only 5 percent of the population. South Vietnam's second geographical region, the coastal lowlands, occupied roughly three-quarters of the RVN's eastern border along the South China Sea. Here, the rugged heights of the highlands give way to a gently rolling coastal landscape of broad valleys, rivers and marshes, and pockets of woodland, brush, and bamboo. Although relatively small, the area between the highlands and the sea was heavily populated and strewn with densely packed villages and hamlets, bustling provincial capitals, and rice paddies. The coastal lowlands were home to roughly 30 percent of the population, numerous agricultural and commercial centers, and a well-developed system of roads, highways, and railroads. The Mekong Delta, the country's third geographical region, is a wide low-lying alluvial plain flush with populous villages, endless rice paddies, and a complex network of irrigation dikes, streams, and canals. Although just one-third of South Vietnam's total landmass, the Mekong River Delta was home to 65 percent of the population.[44]

TERRITORIALITY AND SPATIAL CONTESTATION: COMMUNIST REVOLUTIONARY WARFARE AND PACIFICATION

Communist revolutionaries and the government of the RVN, despite their competing and incompatible visions, pursued an identical strategy albeit under different guises. Both sides fought to establish and maintain dominance and/or effective influence over population and territory while simultaneously shrinking or eliminating enemy control. The objective, as previously noted, was to control the operational environment; the means to do so were largely territorial.

The method used by Hanoi and the NLF to create politico-military space and control the operational environment was revolutionary warfare, a prescription for seizing political power through violence, subversion, and mass politics. "The seizure of power by armed force, the settlement of the issue by war," Mao Zedong once declared, "is the central task and the highest form of revolution."[45] Tran Huy Lieu, a member of the VWP, concurred: "The fundamental issue of a revolution is political power. A revolution is launched for the purpose of seizing political power from the hands of the enemy and setting up a new regime to realize the aims and principles of the revolution."[46] Inspired by Mao's famous dictum that political power "grows out of the barrel of a gun" and based on his phased strategy of protracted warfare, the communist revolution in Vietnam was a hybrid politico-military movement that employed a sophisticated blend of grassroots political mobilization, discriminate violence and terrorism, guerrilla warfare, and conventional military operations. Together, these elements formed a comprehensive yet flexible design meant to produce politico-military space, seize control of the operational environment, and topple the Saigon regime. "In the conduct of their fundamental strategy of revolutionary warfare," noted one expert, "the Vietnamese communists employ many and diverse instruments, both political and military. Designed to be mutually supporting, these are each focused on but one end: the seizure of political power in the South."[47]

Political and military spaces maintained by the communists during the Second Indochina War were known as "liberated areas." According to the NLF's official news organ, the Liberation Press Agency, "The construction, consolidation, and development of the liberated zone is a fundamental and urgent task of the revolution."[48] Liberated areas had been an essential component within communist revolutionary warfare since the late 1930s, when Mao had first emphasized the critical relationship between building base areas and developing guerrilla warfare within enemy territory.[49] In South Vietnam, zones of communist control were diverse in form and function. Besides supporting the battlefield, they were a place where communist revolutionaries could enact deep and meaningful social change. In the eyes of many, such areas were places where socialism was implemented and the tenets of Marxist-Leninist revolution reified. "We are building socialism!" a North Vietnamese editorial declared. The topic of discussion was daily life within the liberated areas of South Vietnam.

Establishing liberated areas was merely the first phase within a larger process of steadily growing the insurgent presence and its influence throughout the countryside. The overall objective was to collapse the South Vietnamese government by establishing nodes of communist-controlled territory that would, over time, metastasize and conjoin to create a physical and ideational antistate—an "insurgent state"—within the national territory of the RVN.[50] Le Duan described this process in 1965. "On the rural battlefield . . . [w]e must have a plan to tightly coordinate the combat operations of various units and forces and of the different local areas. We must always fully understand and follow our formula of attaining mastery and control to destroy the enemy . . . in order to gain increasingly firm control," he wrote. "We must use every possible means . . . to secure and defend each village and hamlet, and that we gradually expand our area of control and shrink the size of the enemy's area of control."[51] According to North Vietnamese general Hoang Van Thai, the process was "to gradually transform the contested areas into liberated areas and areas under enemy control into contested areas."[52] In this way, Hoang, Le Duan, and others reasoned, the forces of national liberation would develop and grow while the power of the Saigon government atrophied. Over time, the growth of liberated areas would strengthen the revolution and create the necessary conditions for launching a popular uprising and general offensive in the final push toward victory.

The method used by the RVN and the United States to build politico-military space and control the operational environment was rural pacification, an umbrella term for a variety of political and military programs implemented over the years under different names. Pacification was a civil-military endeavor developed to counter the corrosive effects of insurgency and assist nation-building efforts in Southeast Asia. Implemented predominantly by the RVN with technical, administrative, and military support from U.S. armed forces and American civilian agencies, rural pacification was managed and executed jointly by both powers. General William Westmoreland, commander of the Military Assistance Command, Vietnam (MACV) from 1964 to 1968, characterized pacification as "the ultimate goal of both the Americans and the South Vietnamese government. . . . [I]ts aim was to achieve an economically and politically viable society."[53] Former undersecretary of state U. Alexis Johnson claimed pacification was the extension of governmental control and presence into rural areas, including such services as police, education, health care, and agricultural production.[54] According to U.S. Ambassador William Colby, the civilian head of Civil Operations and Revolutionary Development Support (CORDS) from 1968 to 1972, pacification was nothing less than "the allied answer to the communists' tactics of people's war."[55]

The ends postulated by Westmoreland and Johnson were pursued through two primary means: security and development. The former involved both large-scale conventional combat actions and small-unit counterguerrilla operations and policing activities at the village and hamlet level. The intent was to destroy communist mainforces and eliminate local guerrillas and political cadres. The latter included grassroots democratic reform and a variety of local socioeconomic programs to improve the lives and livelihoods of the peasantry. Together, security and development were meant to dissolve the insurgency and expand and/or consolidate the government's presence and influence in the countryside. Thus, pacification worked to gradually eliminate the insurgency as it created a safe environment and built civil institutions. Rooted in counterinsurgency theory and practice and the precept that control of the population was the key to defeating internal subversion and communist "wars of national liberation," pacification was the reciprocal of communist insurgency despite its countervailing purpose. Much like the phenomenon it was meant to arrest, pacification used a combination of hard and soft power, and employed every element of modern war—military, political, economic, social, and psychological.

The politico-military spaces produced and conserved by the RVN were known as "pacified" or "government-controlled" areas. Much like communist "liberated" zones, geographical areas under Saigon's control served material as well as ideational purposes. Government-controlled areas not only provided Saigon the manpower, combat support, supplies, and intelligence necessary for waging war among the people (thereby denying these resources to the enemy), they also served as a symbol of government power and authority, a vehicle for nation-building efforts, and the means for constructing an economically productive, democratically elected, noncommunist society.

The spatial strategy pursued by Saigon and Washington was identical to that of the communists, only in reverse. Ultimately, the goal was to transform communist-controlled areas into government-controlled areas. According to MACV, the overall objective of pacification was "the transformation of all areas now under [communist] control, in whole or in part, into areas firmly controlled by the [RVN]."[56] Through its two primary mechanisms—military force to eject insurgent forces from populated areas and prevent their return; and social and political programs to stimulate rural development and win popular support—pacification worked to seize control of the operational environment by breaking the communists' hold over rural areas and extending government authority in the countryside. Although tailored to meet specific contingencies at the provincial, district, and hamlet level and implemented variably in response to a wide variety of local conditions and circumstances, the program's main objective was to achieve government control throughout all of South Vietnam. "The pacification of the entire nation and the reimposition of GVN control," read one report, "constitutes the end goal of our effort here."[57]

The relative nature of control was a product of the spatial strategies, territorial commitments, and force-structure requirements of each side. Political suzerainty in the villages and hamlets and the presence of a sympathetic or (at minimum) neutral rural population were common objectives for both communist insurgents and the RVN. Nonetheless, the practical application of control was quite dissimilar.

For Saigon and its allies, security and control were predicated on the corporeal presence of friendly military and administrative forces and the concomitant absence of enemy units and political cadres. In short, control meant physically occupying territory. The RVN required both military strength (local security forces) and a functioning political and administrative apparatus (hamlet chief, village council, and civil administrators) to achieve control over populated territory and elicit civilian compliance. Moreover, as an independent and sovereign state fighting in defense of self-determination, the RVN had to establish and maintain an authoritative presence everywhere. Ceding any portion of the national territory and its population to insurgent control not only granted tacit political recognition to the NLF, it legitimated communist claims to power and authority as well. Such political concessions were obviously unacceptable to the RVN, a fledgling but incumbent government that was desperately trying to prove it was the sole legitimate authority in South Vietnam. Existential credibility, therefore, spurred the RVN to defend "everywhere" in an attempt to protect every inch of national territory. What emerged was an inelastic, area-based defense posture that lacked strategic flexibility and tied down the bulk of Saigon's ground forces in territorial defense. The RVN was able to live with this arrangement during the years of U.S. combat assistance and robust military and economic aid from the American government. After 1972, however, when the RVN became increasingly isolated and its military units exhausted, underequipped, and overextended, the strategy of territorial defense proved fatal.

Communist control, meanwhile, was not predicated exclusively on the ongoing physical presence of armed units. Hanoi and the NLF could achieve control with a strong political organization alone owing to its systemic, covert nature and strategic use of terrorism. The mere threat of violence by local guerrillas or political cadres extended insurgent influence and control over populated areas by spreading fear, modifying behavior, and coercing compliance. The communists simply did not need to physically extend their forces everywhere or tie up a significant portion of their manpower policing the population. Their network of base areas and safe havens, hybrid protracted war strategy, and skillful clandestine campaign that assassinated or killed government officials and planted Front members in RVN territorial forces and local administrative bodies, enabled the communists to appear omnipresent and project their authority over a much larger area than they could physically occupy. These economy-of-force and spatial strategies also provided a double advantage in that they freed up manpower and allowed communist guerrillas and main-forces to concentrate on those areas where the government was most vulnerable.[58]

Efforts by both sides to produce politico-military space and expand control over South Vietnam's operational environment did not involve merely creative processes; destruction also played a central role. Like the Hindu god Shiva, who embodied at once the roles of creator and destroyer, insurgency and pacification were processes that involved the simultaneous erasure *and* production of politico-military space.

As each side created and expanded areas under their control, they concurrently reduced and/or eliminated territory controlled by the enemy. Thus, the contest between revolutionary warfare and pacification was an ever-shifting process of "creative destruction." In this manner, the control war was absolute: any increase in space for one side was a "double gain" in that it denied the enemy use of the same territory and access to the same population. "The struggle in Vietnam," wrote Vietnam expert Douglas Pike, "is what the academic gamesters call a zero-sum game, like poker, in which what one player wins the others must lose."[59]

THE OBJECT OF REVOLUTIONARY CONFLICT: CONTROL OF THE OPERATIONAL ENVIRONMENT

For both sides of the conflict, attaining the power to determine Vietnam's social and political future first meant establishing dominance over the operational environment. Therefore, despite earnest desires for winning broad-based political support, the more immediate objective was seizing and exercising power and control over territory and population—not winning the hearts and minds of the South Vietnamese people. Over the years, the revolutionary struggle in Vietnam has largely been cast as a fight for political legitimacy or popular support. This characterization, however, is only marginally accurate, and applies predominantly to the period before direct American military intervention, when the popularity and strength of the communist revolution were ascendant, and when emphasis on the military dimension of the conflict had not yet eclipsed the political dimension.

As the war progressed, a growing number of individuals had recognized that the key to winning the war in South Vietnam was population control, not political support. The question of which side of the conflict the people of South Vietnam genuinely supported, many argued, was largely irrelevant. "No government or political grouping has been able to win widespread popular support—or seems likely to do so," political science professor Samuel Huntington wrote. "The most one can realistically speak of is the relative ability of the government and the [revolutionary movement] to exercise authority and to control population."[60] Huntington's colleague Gil Alroy agreed. "The question is not whether governments are popular," Alroy wrote that same year, "but whether they can exercise effective control over their peoples. The point is whether they can influence behavior rather than produce loyalty."[61] A number of recent quantitative "micro-studies" that examine political violence in South Vietnam have reached similar conclusions.[62] Research conducted by Yale University professors Stathis Kalyvas and Matthew Kocher shows that irregular wars, contrary to widespread perception, are not "mere contests" for the hearts and minds of the population but are rather "a competition for territorial control, where violence is used to challenge and to create order." According to Kalyvas and Kocher, both sides in revolutionary conflicts must generate popular "collaboration" and deter "defection." Ultimately in such contests, the extent of collaboration each side achieves "hinges largely on the degree of control they are able to exercise."[63] As Kalyvas and Kocher have observed, the struggle for control of contested space in South Vietnam was not a popularity contest or a competition for popular support and political affirmation. It was instead, as Douglas Pike indicated, a bitter and violent "zero-sum" contest to establish and maintain political dominion over the civilian environment.

The fight for contested politico-military spaces and the battle for influence and control over land and population had always been a critical component in the war for South Vietnam. These factors assumed unprecedented urgency, however, in the years following the 1968 Tet Offensive—a period when the political shockwaves generated by this seminal event profoundly altered the trajectory of the war and fundamentally impacted its conduct, its character, and the strategy pursued by both sides. The offensive, called *Tet Mau Than* (Tet, year of the monkey) by the Vietnamese, was a watershed event because it brought about the sudden and dramatic emergence of a new set of politico-military variables that redefined the war in Southeast Asia. Although each phase of the offensive—one in February, one in May, and one in August—failed to achieve their primary objectives (collapsing the Saigon government and forcing a precipitous American withdrawal), and were unable to persuade the combatants to abandon their grand strategic aims and the pursuit of political objectives through military force, *Tet Mau Than* demonstrated that the war was essentially stalemated in 1968, and that a favorable balance of power could only be achieved if both sides reappraised their efforts and pursued new initiatives.

For the communists, *Tet Mau Than* was an immense military disappointment, one that failed to deliver victory and forced politico-military retrenchment and temporary reversion back to low-level protracted warfare occasionally punctuated by conventional operations and large-scale offensives. The crushing defeat and staggering losses suffered by communist guerrillas and political cadres during 1968 also compelled Hanoi and the NLF, if only as a temporizing measure, to enter into formal negotiations to discuss a potential diplomatic resolution to the war. Rather than seriously entertain diplomacy and a brokered peace that would have denied the communists their ultimate goal of toppling the South Vietnamese government and unifying the country under communist rule, however, negotiations were used as another "front" in the war and pursued merely to buy time for rebuilding guerrilla forces and a shattered political administrative apparatus in preparation for future offensives. Nonetheless, the presence of representatives from the DRV and NLF at the peace talks in Paris, along with the temporary return to predominantly low-level protracted guerrilla war in the South, represented a profound shift in communist policy.

For the RVN and the United States, *Tet Mau Than* was a tactical victory that opened up new possibilities even as it closed others. To begin with, the initial shock and state of emergency brought on by the first round of attacks quickly faded. By the summer of 1968, the Saigon regime was moving rapidly toward rebuilding South Vietnam's shattered towns and cities and overcoming the defensive-mindedness that had paralyzed its armed forces and its population since the series of offensives began on January 30. The sudden shock and brutality of the attacks also dissolved the war-induced lethargy that had for years gripped the South Vietnamese population, sapped its morale, and produced chronic "fence-sitting" among much of the population. Most importantly, allied forces began pushing back into rural areas to regain control and fill the political and military vacuum that had been created in the countryside when communist forces launched their attacks against major cities and provincial capitals. On 1 November, the RVN and the United States began the largest and most ambitious pacification campaign of the war, the Accelerated Pacification Campaign (APC), in an effort to wrest control of villages and hamlets from the communists and rapidly expand Saigon's presence and authority over rural areas. The offensive initiated a progressive expansion of security and government control over rural areas.

Despite the galvanizing developments brought on by *Tet Mau Than*, however, the political fallout generated by the attacks was largely negative for the noncommunists. This was particularly true in the United States, where antiwar elements in Congress and the public at large—after being lulled into an uneasy quiescence by months of rosy reports of progress on the ground in South Vietnam—were suddenly reawakened and electrified by the suddenness and audacity of the offensive. The shock of Tet, along with growing frustration and increased pessimism among policymakers and the public, initiated a series of sweeping political and strategic changes. These included new leadership in Washington, termination of the strategic bombing campaign against the DRV, a greater emphasis on pacification and population security, and the commencement of formal peace negotiations that opened up the possibility of a ceasefire and a negotiated settlement to end the war. Most notably, the communist offensive forced the United States to close what had been up to that point an open-ended commitment to the defense of the RVN, and begin a process of deescalation and transfer of military stewardship back to the Republic of

Vietnam Armed Forces. Known as "Vietnamization," the policy paired the gradual withdrawal of U.S. combat forces with military and political programs designed to build military self-sufficiency and allow the RVN, under the leadership of President Nguyen Van Thieu, to assume the burden of its own defense. According to historian James Willbanks, the "ultimate objective" of the Vietnamization program was "to strengthen the armed forces of South Vietnam and bolster the Thieu government to make the South Vietnamese capable of standing alone against their communist opponents."[64] A 1970 report submitted to the House Committee on Appropriations by Secretary of the Army Stanley R. Resor claimed Vietnamization was more than a military program meant to enable the RVNAF to take over combat operations from their American counterparts; it was, Resor argued, a program that involved "helping the Vietnamese people to build their own future—to assure effective internal security forces, a stronger economy, and a cohesive society."[65]

The profound political and military changes brought about by the Tet Offensive made control of population and territory and the expansion of power and influence over South Vietnam's operational environment a top priority, both for the RVN and for the Vietnamese communists. Determined to gain the upper hand at the negotiations in Paris as well as on the ground in South Vietnam, each side worked feverishly to maximize its social, political, and military footprint and establish some degree of physical presence in as much of the countryside as possible. In 1970, a report submitted to the U.S. House Committee on Foreign Relations succinctly captured the essence of the war after Tet. "The struggle today is to control," it read. "To be on top when the ceasefire finally comes and the half-million Americans go home. The name of the game on both sides is to get your people into places of power, to win the allegiance of the countryside and its rice-roots leadership for the future, and, conversely, to get the enemy's people out of corresponding positions."[66] According to another report, "each side wanted to be able to maneuver itself into a position of political strength should the peace talks result in either a ceasefire or a broader political settlement of the war." In the years following *Tet Mau Than*, it continued, "each side appeared to be jockeying for a better position in South Vietnam in order to increase its leverage at the bargaining table in Paris."[67] As these statements illustrate, establishing control over South Vietnam's operational environment had become the overriding objective in Southeast Asia in light of the systematic reduction of American combat forces and the possibility of a negotiated settlement that would freeze opposing political and military forces in place, grant each contender authority over the population and territory it occupied at the time of ceasefire, and heighten rather than diminish the struggle for South Vietnam.

The control war and efforts to establish dominion over territory and population also largely defined the conflict in the years that followed the Paris Peace Accords and American withdrawal from Southeast Asia. Much like the fighting that occurred between *Tet Mau Than* and the signing of the Paris treaty, the period between January 1973 and the fall of Saigon in April 1975 was marked by incessant low- to mid-intensity "land grabbing" and "nibbling" operations, as the communists and the RVN each struggled to expand its political and military footprint inside of South Vietnam and defend the areas it possessed from enemy encroachment. Still unable to militarily crush its adversary and constrained to some degree by the restrictions put in place by the Paris agreement, each side probed the other for possible weakness and attempted to expand its control over the operational environment. The struggle for land and people was largely inconclusive until the autumn of 1974, when Washington's decision to dramatically curtail aid to the Saigon regime tipped the balance of forces in the communists' favor. By May 1975, Hanoi had triumphed and the RVN fell into history. In large part, communist victory in Vietnam can be attributed to the communists' success in expanding dominion over the landscape and constructing an insurgent antistate within the national boundaries of South Vietnam.

SCOPE OF THIS WORK

This study examines the mediation of power in Southeast Asia within a spatial and environmental context. It does so by exploring the strategic use of geography and the operational environment, including both the natural and the human-made world. It argues that the war in South Vietnam after 1968 was primarily a competition to control contested political spaces—a "control war" in which the competing factions vied for political supremacy over the people, territory, and resources of South Vietnam. Efforts to control the operational environment were a salient feature in the struggle to establish political hegemony and determine the future social and political order of South Vietnam. Moreover, these efforts shaped the war in Indochina and affected its outcome. This approach is intended to take a fresh look at the "other war" in Vietnam. It is also meant to underscore the critical and often overlooked relationship between war and its environs. Most studies of revolution and counterrevolution in Southeast Asia have used traditional lines of analysis. Political and military histories of the Vietnam War are legion. This study tries to break away from conventional methodologies by presenting a hybrid approach that integrates the methods and insights from a number of disciplines, including military history, diplomatic history, environmental history, and political geography. Rather than apply a predominantly political and/or military analysis, this study examines politics and war through the lens of geography and the operational environment, paying particular attention to the ways in which communist revolutionaries, on the one hand, and the RVN and its American ally, on the other, conceptualized space and harnessed the physical and ideational landscape to serve social, political, military, and ideological objectives.

The sheer complexity of the Second Indochina War has complicated our ability to understand, to discuss, and to write about this conflict. This stems in part from the fact that what transpired in South Vietnam between 1954 and 1975 was a multifaceted "hybrid" war that involved a complex and ever-evolving mixture of conventional war, guerrilla warfare, and mass politics. Another reason for our difficulty is that the war for the countryside lacked clearly defined front lines and zones of political and military control. The war in South Vietnam, like all armed conflicts, possessed a unique spatiality. Although often called a "war without fronts," the reality is that the Second Indochina War was a war with innumerable fronts, as insurgents and counterinsurgents feverishly wrestled to win political power and control of an operational environment that encompassed 44 provinces, 236 districts, and roughly 12,000 hamlets. The struggle for South Vietnam was not one geographical war, but many; it was a highly complex conflict that fragmented space and atomized the battlefield along countless divergent points of conflict.

This study examines revolutionary war from the perspective of politico-military control rather than politico-ideological legitimacy. The intent, however, is not to dismiss the importance of popular support, or to advance a false dichotomy that casts political legitimacy and political control as mutually exclusive phenomenon—in the eyes of the Vietnamese communists and the Saigon regime, they were not. Both sides of the conflict considered political legitimacy an indispensable element in fulfilling long-term political and strategic objectives in Southeast Asia. Moreover, communists and noncommunists alike saw legitimacy and control not as separate and distinct but rather as integrated and interdependent phenomena within the pursuit of political and military objectives. The issue for communists and the RVN, therefore, was not a false choice, a decision of selecting control *or* legitimacy; it was a matter of emphasis and timing. Each side recognized that the question of which side the people of South Vietnam favored was largely irrelevant *in the short term*. Influencing behavior and getting the population—whether willingly or unwillingly—to assist in defeating the enemy and establishing authority was the primary consideration; winning political legitimacy was subsidiary. In 1971, American scholar Allan E. Goodman noted that control was an expedience, a necessary precursor to legitimacy: "At the present time the military struggle between the two sides is dominated by a preoccupation with achieving security, the extension of an administrative apparatus over people and territory, and the extraction of resources from the population under their control. In the future, the GVN and the Viet Cong will face each other in a political struggle with essentially the same institutions that now exist, but, presumably, the goals of each side will change from ones which stress the necessity of controlling the population to ones which stress the need to mobilize the population in political terms."[68] As Goodman indicates, each side shrewdly pursued the immediate short-term goal of politico-military control in the war's final phases in an effort to achieve the gradual long-term goal of politico-ideological legitimacy.

This work also dispels some of the historical fallacies and misconceptions concerning communist revolutionary warfare, pacification, and the role these factors played in Southeast Asia. The first is the idea that the struggle for South Vietnam hinged on political legitimacy and the support of the people—concepts that are subsumed within the popular catchphrase "winning hearts and minds." Initially given credence in contemporary and early histories of the conflict, this interpretation has resonated widely among scholars in the decades that followed.[69] Proponents of the "hearts and minds" school have argued that because the Vietnamese communists possessed what the Saigon government did not—a blueprint for social mobility and genuine national independence, the right to rule under the Mandate of Heaven, and a winning political philosophy and method that resonated strongly with the hopes and aspirations of a majority of the South Vietnamese population—Hanoi and the NLF were able to harness the energies of a disaffected people and transform the population into an inspired, unified, and well-organized movement, one that destroyed Vietnam's antiquated "semifeudal" social order and erected a "modern" socialist Vietnam in its place. Conversely, because the RVN lacked true nationalist credentials and legitimacy in the eyes of its constituency, stayed wedded to the social inequalities and repressive practices of the country's colonial past, remained fundamentally disconnected from the needs and desires of the South Vietnamese peasantry, and was inept and hopelessly corrupt, the Saigon regime alienated a majority of the rural population and either drove them into the waiting arms of the communist insurgency or produced a widespread apathy and indifference that in turn generated public inertia and a social and political malaise that proved fatal to its existence. In this interpretation, then, it was political legitimacy, or rather the rural population's recognition and acceptance of the communists' claim to power, that was key to the war in Vietnam. As noted by historian Gary Hess, proponents of this school of thought believe that Hanoi and the National Liberation Front ultimately triumphed because the communists had, unlike the faux revolutionary movement advanced by the Saigon regime, "preached and delivered" true social revolution.[70]

This interpretation and the general belief that popular sovereignty was the key to the struggle for South Vietnam are in many ways inaccurate. Although there is some truth as to the vitality of the revolution, particularly its ability to maintain a degree of enthusiasm among the South Vietnamese population, and to the widespread and debilitating sense of apathy found within the body politic, control of the civilian environment rather than the winning of hearts and minds was the more pressing and relevant objective of both sides throughout most of the war. Moreover, it was dominion over territory and population—not popular support or political legitimacy—that shaped the trajectory of the war and led to the collapse of the RVN and triumph of the communist revolution in April 1975. Although important, support and legitimacy were subordinated to the more pressing and critical short-term objective of expanding influence and control. Moreover, they were an incidental rather than primary factor in the arc and the outcome of the conflict.

A second myth to that has held sway over the years is the belief that the RVN and the United States had "won" the war in South Vietnam by 1971 thanks to a successful pacification program that was implemented in the years following the Tet Offensive. Advanced by revisionist scholars and key participants in the war, purveyors of this thesis believe not only that pacification and Vietnamization had defeated the political and military aims of the communist revolutionary movement, but that that this success was then "thrown away" when the United States reneged on its commitments to the RVN and "abandoned" its longtime ally in 1974–75. Moreover, this view contends that the post-Tet period constituted a "better war," where the conduct, character, and trajectory of the allied military effort took a positive and constructive turn, thanks to new leadership and a new strategy that broke sharply with the destructive and counterproductive policies used prior to *Tet Mau Than*.[71]

Contrary to what proponents of this "lost victory" thesis have claimed, counter-insurgency, as it was applied after *Tet Mau Than*, did not attain the political or the military goals it set out to achieve. The failure to fully "pacify" South Vietnam or control the operational environment fatally weakened the Saigon government and precipitated the tragic consequences that unfolded in 1975. Furthermore, the post-Tet period did not constitute a "better war" marked by a smarter, more beneficent military policy. Despite an unprecedented emphasis on pacification and a substantial reduction in the number of large-scale (battalion size and above) conventional operations beginning in the fall of 1968, most of the policies implemented during the post-Tet period represented, in actuality, a continuation rather than a divergence from those that had been in place since the United States first introduced combat forces into Vietnam.

In challenging some of the major tenets of the "orthodox" and "revisionist" schools of Vietnam War historiography, this study positions itself somewhere in between the two. Thus, it fits comfortably within a growing trend in the field that can be considered postrevisionist. The scope of recent scholarship on the Second Indochina War defies easy categorization due to its sheer diversity in method and focus. Nonetheless, there has been a discernible trend in the literature that not only complicates and challenges both the orthodox and the revisionist interpretations that have dominated the field, but that restores the Vietnamese people and their struggles to their rightful place at the center of the Second Indochina War. This monograph looks to continue these trends.[72]

The control war was a fundamental aspect of the conflict in Southeast Asia. Rather than a sideshow or corollary to the "main-force" or conventional war, the fight for contested politico-military spaces, often referred to as the "other war" in Vietnam, was an integral part of a complex and multifaceted conflict. For many, the revolutionary fight for political dominance and authority over the operational environment was the principal arena in which the war took place. "Central to success, both in ending the war and in winning the peace, is the pacification program," Secretary of Defense Robert McNamara informed President Lyndon Johnson in September 1966.[73] The U.S. Army and the Central Intelligence Agency agreed. According to a study titled *A Program for the Pacification and Long-Term Development of South Vietnam*, "the critical actions are those that occur at the village, district, and provincial levels. This is where the war must be fought; this is where the war and the object which lies beyond it must be won."[74] In 1969, the CIA declared "the struggle in Vietnam is in essence a struggle for political domination over those Vietnamese who now live in that portion of Vietnam lying south of the 17th parallel. . . . [T]he primary issue is control."[75] The conventional or "main-force" war between regular military forces remained paramount during the post-Tet period. Moreover, the contest between conventional armies proved to be the deciding factor in the final collapse of the RVN and the communist victory in 1975. Nonetheless, it was the balance of forces in the countryside and the relative strengths and weaknesses of both sides as they wrestled to expand and consolidate their hold over the operational environment that largely determined the war's trajectory and helped set the stage for its outcome.

This work consists of two parts. Part I explores the theoretical underpinnings of the control war through four thematic chapters. Chapter 1 examines communist revolutionary warfare and the specific ways in which Hanoi and the National Liberation Front conceptualized space and shaped the operational environment to serve political, military, and strategic objectives. Like other revolutionaries, the Vietnamese communists worked to produce, expand, and control space by developing a variety of insurgent terrain, including base areas, war zones, and rural villages and hamlets. The purpose of this spatial approach was to create and grow an insurgent "antistate" within the national boundaries of South Vietnam, one that would "liberate" South Vietnam through the gradual attrition of RVN government control over the operational environment. Chapter 2 takes a look at the theoretical foundations of the pacification program and its civilian and military aspects aimed at defeating the communist insurgency. It illustrates that the means and ends of pacification were similar to those of communist revolutionary warfare in that they were intended to gain control of land and people and expand friendly presence and authority throughout the countryside. The government of South Vietnam and U.S. policymakers had made the production of political space and control over the operational environment a top priority in the war against communist insurgency and internal subversion. In fact, seizing control of rural areas and extending Saigon's hold over territory and population was the essence of the pacification program.

Chapter 3 describes the ideational dimension of the control war and the ways in which Vietnamese communists and the government of South Vietnam each attempted to shape the civilian environment to reflect its own national and political worldview. The significance of South Vietnam's rural areas, which were of immense importance militarily, transcended the procurement of manpower and material resources for war. The countryside was also a laboratory for social and political revolution, both communist and noncommunist. The civilian environment was more than a means of marshaling resources and a place where population and territory merged; it was the medium through which both sides of the conflict strove to build a unified, independent, and modern state. The occupation and administration of the civilian environment was an exercise in political power and a vehicle for putting ideology into practice; civilian areas were, in essence, testing grounds for the respective ideologies of the contestants. As such, occupied areas were fashioned in ways that revealed contrasting conceptions of how Vietnamese society should be organized. Chapter 4 explores how violence and armed force was used to establish and extend influence and control over the operational environment. The control war was in essence an all-or-nothing struggle for political power and hegemony. As such, the modality most commonly applied by both sides was force. The largely political means of winning popular support through positive economic development and social reform were marginalized by the prevalent use of violence, force, and coercion. Although the war in South Vietnam remained a contest between competing social and political systems, and support of the population continued to be of long-term political and strategic significance, positive actions that addressed the hopes and aspirations of the rural population were subordinated to the short-term objective of securing physical and psychological control over land and people. Rather than devote the bulk of their efforts to winning the enthusiastic support and affections of the populace through social, political, and economic policy, both sides instead relied on armed force and other forms of coercive power to influence and control the population—the logic being that even if the peasantry could not be won over or made to eagerly support either the insurgency or the government, they could at least be made compliant and cooperative.

Part II includes four chronological chapters that trace the trajectory of the control war from 1968 through 1975. Chapter 5 examines the profound political and strategic changes that came about as a result of *Tet Mau Than*. The chapter's primary focus is the Paris peace talks and the fundamental negotiating issues that defined and shaped the deliberations, including American withdrawal, regime change in South Vietnam, and the possibility of an in-place ceasefire and territorial accommodation. The jarring impact of Tet and subsequent diplomatic efforts in Paris had a major effect on the control war, transforming the conflict from a largely conventional war dominated by attempts to annihilate the enemy on the battlefield to a hybrid conventional-irregular war driven by efforts to expand control over territory and population. This chapter is intended to highlight these changes and situate the struggle to control the operational environment within the larger political currents that were unfolding in Paris, Saigon, Hanoi, and Washington. Chapter 6 examines the "talk-fight" strategy pursued by both sides after *Tet Mau Than* as well as continued efforts to expand influence and control over the operational environment vis-à-vis the ongoing peace talks in Paris. During this period a combination of communist retrenchment and "fast-and-thin" pacification helped the RVN and the United States push the communist revolution to the brink of defeat. The confluence of a weakened insurgency and newfound emphasis on pacification allowed the RVN to rapidly expand its presence into the countryside and achieve unprecedented levels of security in rural areas. Consequently, this period marked the peak years of pacification progress and the allied war effort, and conversely, the nadir of the communist insurgency and the revolution's bleakest and most desperate hour.

Chapter 7 examines the control war from January 1969 through September 1972. The pacification program, notwithstanding genuine progress, failed to destroy the communist revolution or achieve sustainable control over territory and population during these years. Despite the severe attrition and near extinction of communist guerrillas and political cadres in the South as a result of *Tet Mau Thanh* and subsequent military operations, the communist revolution was able to hold on and survive thanks to the tenacity of its political organization, the infiltration of men and matériel into South Vietnam, and the maintenance of liberated areas throughout Indochina. Together, these allowed Hanoi and the NLF to retain enough influence and control over the operational environment to continue a protracted war in the countryside and prepare for future offensives. The situation was made worse for the allies in 1972, when the Nguyen Hue, or "Easter," Offensive expanded the insurgent state and communist control of territory to peak levels. This development set the stage for Hanoi's victory in the spring of 1975. The failure of the RVN and the United States to eliminate the territory under communist control by the end of 1972 along with their inability to either negotiate the withdrawal of NVA units from South Vietnam or push them out by force before the Paris Peace Accords went into effect proved fatal in the years that followed. Chapter 8 examines the final chapter of the Second Indochina War. It follows the control war from the aftermath of the 1972 Easter Offensive through the fall of Saigon in April 1975. The pacification program's inability to defeat the revolutionary movement or blunt its capacity to metastasize and expand influence and control over land and population fatally undermined the RVN's ability to control and defend the national territory. Faced with a situation where it was required to retain and defend nearly every inch of ground without American combat assistance and with diminishing military resources, the Saigon regime and its armed forces were denied strategic flexibility and were stretched well beyond their capacity to effectively defend the nation against pervasive communist attack and subversion. In the end, the communists' ability to foil pacification and expand their control over territory and population south of the seventeenth parallel brought down the Saigon regime and ensured victory for the communist revolution.

PROLOGUE

THE ROAD TO *TET MAU THAN*

The year 1954 was pivotal in the history of Vietnam. The fall of Dien Bien Phu in May and the subsequent peace settlement signaled the termination of the First Indochina War, the end of French colonialism, and the beginning of a new era defined by national division, civil war, and state building. The Geneva Accords, put in place after France's defeat, partitioned Vietnam at the seventeenth parallel and created two rival factions. In the north lay the Democratic Republic of Vietnam (DRV), a communist state headed by revolutionary leader Ho Chi Minh and the Vietnam Workers' Party, or VWP. Ho and the VWP were committed to toppling the southern regime and reunifying Vietnam under a communist government. South of the dividing line lay the Republic of Vietnam (RVN) led by Ngo Dinh Diem, a fiercely nationalistic anticommunist and Catholic who had returned to Vietnam after a long hiatus with his own designs for the future of his country. In essence, the war in Vietnam was a civil war driven by competing nationalisms. The ideological divisions of the Cold War along with the high geostrategic value of Indochina, however, drew foreign powers into the conflict. Hanoi and Saigon both had powerful sponsors who provided military, economic, and psychological assistance. The DRV was supported by both the People's Republic of China and the Soviet Union. Saigon's primary benefactor, meanwhile, was the United States. Other regional actors participated in one form or another, including Laos, Cambodia, Thailand, Australia, and the Republic of Korea (South Korea). Thus, the war in Vietnam was a national, regional, and international conflict whose dynamics and general parameters were shaped by civil conflict and the ideological divisions of the Cold War. When combined, the 1954 national partition and the Cold War antagonisms of the "two Vietnams" and their sponsors set off twenty-one years of bloodshed and suffering.

After Geneva, Hanoi and Saigon concentrated on modernization and nation building. Ho Chi Minh and the VWP worked to consolidate power, build socialism, strengthen the economy, and modernize the armed forces. Despite this internal focus, however, Hanoi looked to the south in anticipation of overthrowing the RVN and reuniting the country under communist rule. When it became evident that Saigon would not hold national elections, as stipulated at Geneva, and the possibility of national reunification by political means became unlikely, Hanoi began building a clandestine political organization and forming local military units throughout the South Vietnamese countryside in preparation for armed struggle and revolution.

Diem also set his sights on modernization and nation building at this time. During the latter half of the 1950s, the United States initiated an economic and civic aid program in South Vietnam under the auspices of the Agency for International Development. It also began a military advisory mission headed by the Military Assistance Advisory Group (MAAG). Under MAAG, U.S. military personnel helped organize, equip, and train the fledgling South Vietnamese army. With military and economic assistance flowing into South Vietnam from Washington, Diem worked toward fashioning a conventional military, promoting economic development, and consolidating political power. The fractious nature of South Vietnamese society, lack of a unifying national identity, and the growing revolutionary movement in the countryside, however, undermined Diem's efforts to build a national political community and strengthen Saigon's authority. In response, Diem turned to repression and authoritarian rule to solidify government control and crush dissent. His hard-line policies made some headway in these areas, but they provoked a violent backlash among segments of the population and, more importantly, spurred Hanoi to pursue insurgency and regime change in the South. In January 1959, the VWP passed Resolution 15, which made revolutionary violence and the destruction of the

Republic of Vietnam official policy. The following year, southern revolutionaries created the National Front for the Liberation of South Vietnam, or NLF, to lead efforts at subversion against the Diem government. Together, these two developments precipitated the Second Indochina War.

Between 1961 and 1963, Hanoi and the NLF intensified and expanded guerrilla war in the South. As Diem worked to crush the rural insurgency, extend government control into the villages, and build bridges between Saigon and the peasantry, the NLF sowed disorder and expanded its political base in the countryside. In January 1963, an NLF main-force battalion aided by local guerrilla units achieved a stunning victory against South Vietnamese forces near the village of Ap Bac. This victory, along with the rapid growth and striking success of the communist insurgency, was a matter of profound and growing concern in Saigon and Washington. Upon taking office in January 1961, President John Kennedy dramatically increased aid to South Vietnam and made "special warfare"—that is, counterinsurgency—a cornerstone of U.S. Vietnam policy. In this spirit, Diem launched the "strategic hamlet program" in 1962 to quell the burgeoning insurgency and "pacify" the countryside. Intended to separate NLF guerrillas from the rural population, extend government presence and control into the countryside, and foster rural development through the construction of fortified villages and local defense initiatives, the strategic hamlet program was widely unpopular and became a primary target for revolutionary forces. By 1963, the communist insurgency had effectively wrecked the program and undermined Diem's efforts to crush the insurgency, restore order in the countryside, and strengthen popular support for the government. The failure of pacification and the strategic hamlet program, along with the NLF's victory at Ap Bac, raised serious doubts about Diem's leadership. This uncertainty was felt both in South Vietnam and in the United States. Near the end of 1963, the crisis in faith over Diem's leadership provoked a tragic turn of events inside South Vietnam. In early November, Diem was overthrown by a military coup and executed. Kennedy himself fell to an assassin's bullet weeks later. Responsibility for the deteriorating situation in South Vietnam fell to Kennedy's successor, Lyndon Johnson, and a revolving door of subpar leadership in South Vietnam. These changes at the top led to further escalation and a profound shift in the war's trajectory.

Within months of the November 1963 coup, the situation in the countryside had deteriorated to an alarming degree. Pacification had collapsed in many areas, and South Vietnam's regular armed forces were in a state of decline as communist guerrillas and political cadres pressed forward in rural areas. This, in large part, was a result of political instability and poor leadership stemming from Diem's demise. The worsening situation in the Republic of Vietnam prompted Hanoi to reevaluate efforts to defeat America's "special war" and topple the South Vietnamese government through guerrilla war and subversion. In December, at the Ninth Plenum of the Central Committee, the Politburo agreed to capitalize on Saigon's weakened condition and attempt a total victory over the Republic of Vietnam in 1964 before the United States could intervene. The resolution passed at the Ninth Plenum codified Hanoi's decision to escalate the war and attempt a knockout blow against the RVN. The increase in communist insurgent activity that followed the passage of the resolution prompted the Johnson administration to take another step up the escalation ladder by providing additional assistance to the South Vietnamese government. America's enlarged military footprint in Indochina invited attack. In August, communist torpedo boats allegedly fired on two American destroyers, the USS *Maddox* and USS *Turner Joy*, in the Gulf of Tonkin. In October, communist forces attacked a U.S. Special Forces camp at Plei Me in the central highlands. In response to these and other incidents, some of which had resulted in American deaths, the Johnson administration authorized punitive airstrikes against the Democratic Republic of Vietnam. Over time, these limited sorties grew into Operation Rolling Thunder, a graduated but sustained strategic bombing campaign. The purpose of Rolling Thunder was twofold: to cripple Hanoi's ability to wage war and to shatter the Politburo's political will.

Rolling Thunder was another critical step toward escalation and Americanization of the war in South Vietnam. The advent of strategic bombing against Hanoi led to the construction of American air bases inside the RVN. When these bases were inevitably targeted and attacked by communist units, the Johnson administration introduced U.S. ground forces to protect them. Safeguarding American air bases, however, required more than static defense around a fixed perimeter; sustained offensive patrolling to locate and eject communist guerrillas and main-force units from the hills and jungle surrounding these bases required substantial numbers of men and matériel. Moreover, as Rolling Thunder and support operations ramped up in the spring of 1965, the situation in South Vietnam became increasingly dire and the Saigon regime appeared to be on the ropes. Political instability in the cities along with insurgent success throughout the countryside convinced the Johnson administration that American combat forces must be introduced to save the South from certain collapse. Thus, American units were introduced under the U.S. Military Assistance Command, Vietnam (MACV), which was established in February 1962 to assist the American advisory effort. In 1964, MACV's original commander, General Paul Harkins, was replaced by General William Westmoreland at a time when direct military intervention by the United States was becoming more likely. In March 1965, U.S. combat forces began arriving in South Vietnam. By the end of the year, the Americans had arrested Saigon's declining strategic position, blunted the momentum built up by communist forces during the first half of the year, and stabilized the situation in RVN.

In 1966, U.S. combat forces continued to grow in size and number, as the Pentagon augmented existing marine and army units. During the U.S. buildup, the South Vietnamese government and its armed forces worked to reestablish and expand government control in the countryside through a revitalized pacification effort. Although much of the year was spent managing the massive influx of troops and equipment, American combat units still managed to take the war to the enemy and frustrate communist efforts to regain the initiative. This took place as MACV grew its forces and established an expanding network of bases, airfields, and other military installations. The year 1966 witnessed several large-scale operations along with dozens more smaller engagements. These operations mauled communist forces as the American buildup progressed. By December, the government of South Vietnam was relatively stable, and most of the personnel and logistical support for the large American units were in place. U.S. combat forces were poised to go on the offensive.

The next year, 1967, has accurately been called the "year of big battles." After more than a year and a half building up American forces and setting the stage for large-scale efforts to sweep communist forces from South Vietnam, MACV had made massive, sometimes months-long, main-force operations a cornerstone of the American war in Vietnam. In January, U.S. and ARVN units launched Operation Cedar Falls a multidivisional search-and-destroy mission conducted in III Corps Tactical Zone (CTZ). It was the first corps-sized operation of the war and included 30,000 allied troops, including 16,000 Americans and 14,000 South Vietnamese. The objective was to locate and eliminate the headquarters for the communists' Military Region 4 and to clear the Iron Triangle of enemy forces. The following month, the U.S. 1st Cavalry and 25th Infantry Divisions along with South Vietnamese and Republic of Korea forces conducted Operation Pershing a joint search-and-destroy/clearing mission that lasted more than a year and was intended to assist pacification efforts in Quang Ngai and Binh Dinh Provinces. Also in February, U.S. forces launched Operation Junction City in War Zones C and D, located northwest of Saigon in III CTZ. The largest operation in South Vietnam to date, the mission included four South Vietnamese battalions and twenty-two American battalions. The operation was one of the largest airmobile assaults in history. Its purpose was to locate and take out the headquarters of the Central Office of South Vietnam (COSVN), the command-and-control organization that directed and managed the insurgency in the South, and to clear enemy forces from War Zone C, a longtime communist redoubt and infiltration route located along the Cambodian border in Tay Ninh Province. The overall objective of the operation was to pacify the province by driving communist forces away from populated areas and out into the open where they could be destroyed more effectively with superior firepower and without incurring heavy civilian casualties. By December 1967, it was evident that while allied combat operations had not achieved everything allied planners had hoped, they were having a detrimental impact on communist forces. Heavy communist losses not only impeded Hanoi's and the NLF's ability to wage protracted war in the southern countryside, they helped gradually expand Saigon's hold over population and territory.

The purpose of U.S. combat forces from 1965 to 1967 was to use their traditional strengths in firepower and mobility to locate and destroy communist main-forces, base areas, and supply caches, and to interdict the southward flow of men and matériel down the Ho Chi Minh Trail into South Vietnam via Laos and Cambodia. In this manner, the Americans served as a "shield" for pacification and rural development—programs that were implemented largely by the South Vietnamese with U.S. logistical and advisory support. Search-and-destroy operations—a hallmark of Westmoreland's main-force war—were intended to find and kill communist main-force soldiers and guerrillas at a rate neither Hanoi nor the National Liberation Front could sustain. Paired with the strategic bombing campaign against the North, Westmoreland's big-unit "war of attrition" was designed to break the communists' political will by inflicting so much punishment on southern guerrillas, northern regulars, and North Vietnamese civilians that Hanoi and the NLF would sue for peace. After two years of main-force combat, however, Washington and Saigon were no closer to this goal than they had been when it began. Although the timely introduction of U.S. combat forces had saved the RVN, brought it back from the brink of extinction, and paved the way for increased government control in the countryside, MACV failed to break either the military stalemate in South Vietnam, which persisted despite the high-intensity warfare and bloodletting, or the communists' political will.

By the fall of 1967, Hanoi clearly recognized the damage it had suffered since the arrival of U.S. combat forces. Yet despite high numbers of casualties suffered by communist local and main forces and the North Vietnamese civilian population, Hanoi and the NLF were not irreparably damaged. Nor were they prepared to throw in the towel or surrender their goal of conquering South Vietnam and reunifying the country. In fact, revolutionary leaders were planning an audacious and unprecedented attack that they believed would precipitously collapse the Saigon regime, drive the Americans from Indochina, and fulfill the Vietnamese revolution. In Hanoi, Le Duan held firm control over the VWP leadership. Despite dissent in the Politburo and some strong opposition to the first secretary's preference for main-force warfare and fealty to the general offensive and uprising as the primary vehicle for achieving the primary goals of the revolution, Le Duan strong-armed critics and put in place his plan for the 1968 winter–spring campaign. The 1968 Tet Offensive—*Tet Mau Than*—was in motion.

PART I
Theoretical Dimensions of Revolutionary Conflict and the Operational Environment

> Control of a geographic part of the state is a manifesto proclaiming: "We have arrived. We are ready to replace the existing government."
>
> —Robert McColl, *The Insurgent State*, 1969

1 COMMUNIST REVOLUTIONARY WARFARE AND INSURGENT GEOGRAPHIES

Communist revolutionary warfare relied heavily on the strategic use of geography and the environment. In Southeast Asia, Hanoi and the National Liberation Front incorporated lessons from the communists' struggle for power in China as well as from Vietnam's long history of war and resistance against foreign invaders into their own method of revolution. The result was a unique strategy that conceptualized space and shaped the operational environment to serve political, military, and strategic objectives. During the Second Indochina War, the Vietnamese communists pursued a territorial imperative to produce, expand, and ultimately control politico-military space and the operational environment through the development of base areas, war zones, communist-dominated villages and hamlets, and other forms of insurgent terrain. The purpose of this spatial approach was to create and grow an insurgent "antistate" within the national boundaries of the RVN, one that would "liberate" South Vietnam by gradually expanding communist influence and control over territory and population while simultaneously shrinking Saigon's hold on the countryside. The overall objective was to overthrow the existing regime by steadily eroding government control over the state.

THE INTANGIBLE FACTORS OF SUCCESS: SPACE, TIME, AND WILL

Control of the operational environment and the places where physical and human geography merge had been a central tenet of communist revolutionary warfare since its inception. In the 1930s and 1940s, as the Chinese communists struggled against both the Kuomintang (Nationalist Party) and the Japanese, Mao Zedong developed an ad hoc theory of revolutionary war founded on mass politics and the peasantry. This theory hinged on three sequential stages that progressed upward from grassroots political organizing to insurgency to conventional war. Mao's theory was unique not only because of its synthesis of political and military action and fusion of Marxism-Leninism, guerrilla warfare, and peasant-based rural revolution, but because the successful execution of an agrarian "people's" war rested on intangible as well as material factors. The production of politico-military space and control of the environment lay at the core of these intangibles. "When the invader pierces deep into the heart of the weaker country and occupies her territory in a cruel and oppressive manner," Mao wrote, "there is no doubt that conditions of terrain, climate, and society in general offer obstacles to his progress and may be used to advantage by those who oppose him. In guerrilla warfare, we turn these advantages to the purpose of resisting and defeating the enemy."[1]

Mao believed revolutionary guerrilla warfare was the key to enabling a relatively weak and undeveloped power the means to defeat a much stronger and well-equipped adversary. "Guerrilla warfare," he wrote during China's War of Resistance against Japan, "is a weapon that a nation inferior in arms and military equipment may employ against a more powerful aggressor nation."[2] Because revolutionary or anti-colonial wars of national liberation were by nature asymmetrical, Mao emphasized spatial and immaterial factors in addition to measurable factors such as weapons systems, supply, and manpower. As Harvard University professor Edward L. Katzenbach noted in 1956, Mao constructed his blueprint for revolutionary guerrilla war on three primary intangibles: space, time, and will.[3]

The first of these, *space*, involved the development of guerrilla warfare behind enemy lines and the establishment of a network of guerrilla bases throughout enemy territory. The purpose of these two elements, Mao argued, was to "dominate the spaces"—that is, to geographically grow guerrilla territory while shrinking areas occupied by the enemy. According to Mao, base areas were an indispensable and fundamental part of revolutionary war. To begin with, they allowed insurgent forces to survive and carry out protracted warfare against a well-armed and entrenched occupying force. Base areas, he wrote, "are the strategic bases on which the guerrilla forces rely in performing their strategic tasks and achieving the object of preserving and expanding themselves and destroying and driving out the enemy. Without strategic bases, there will be nothing to depend on in carrying out any of our strategic tasks or achieving the aim of the war. . . . Guerrilla warfare could not last long or grow without base areas."[4] A system of interlocking rural base areas also opened up venues for attacking the enemy by allowing insurgent activity to develop throughout enemy-occupied territory. Such a system transformed the enemy's "rear" into a fighting "front" and forced him to fight ceaselessly throughout the territory he controlled.[5] Moreover, it preserved guerrilla forces while simultaneously overextending and exhausting the enemy through incessant attack and harassment at multiple geographical points. "The establishment of innumerable . . . bases behind the enemy's lines," Mao declared, "will force him to fight unceasingly in many places at once, both to his front and his rear. He thus endlessly spends his resources. . . . We must make war everywhere and cause dispersal of his forces and dissipation of his strength."[6] Aside from protecting revolutionary forces and providing a platform for launching offensive military operations, rural base areas and other terrain provided the means to marshal resources and grow the revolution. The ability to control China's operational environment and build base areas provided Mao and the communists the manpower and material resources necessary to raise an army and fight the Kuomintang; it also established an appropriate space to build a mass political movement throughout the countryside, one based on the tenets of Marxism-Leninism and built on the widespread support and participation of the peasantry.

The spatial techniques developed by Mao had become fundamental tenets of revolutionary warfare during the "long twilight struggle" of the Cold War. As Robert Taber famously noted in 1965, "Analogically, the guerrilla fights the war of the flea, and his military enemy suffers the dog's disadvantages: too much to defend; too small, ubiquitous, and agile an enemy to come to grips with. If the war continues long enough . . . the dog succumbs to exhaustion and anemia without ever having found anything on which to close his jaws or to rake with his claws."[7] Base areas and insurgent-controlled terrain formed the terrestrial foundation on which the "war of the flea" was fought, and the means for guerrillas to exploit the "dog's disadvantages" and exhaust their enemies through incessant attack at innumerable geographical points. This held true both for Mao and for those that adhered to his theory of revolutionary warfare in the decades that followed.

The second intangible, *time*, rested on space. The construction of an interlocking and expanding system of rural base areas, along with the establishment of guerrilla warfare throughout enemy-occupied territory, allowed the revolution to survive and grow despite its material deficiencies and in spite of concerted counterrevolutionary efforts to destroy it. The perseverance of the revolutionary movement stretched the conflict out across time and created the conditions for what Mao coined "protracted war." Protraction of the conflict, in turn, facilitated the development or erosion of political *will*, Mao's third intangible factor. As the war turned into a prolonged struggle, and territory under the control of the revolution expanded, popular support for the movement, along with the will of the people to resist and fight the enemy, would grow stronger. In this way, the revolution gathered strength and propagated. Conversely, incessant guerrilla attacks throughout enemy-occupied territory overextended the enemy's forces, weakened his strategic posture, and eroded his will to fight. Thus, while communist guerrillas were materially weak, at least initially, they were able to overcome this handicap and grow stronger as the conflict progressed. The presence and political will of their adversary, meanwhile, atrophied. In China, the ability to control both the operational environment and the tempo of war proved key to the Chinese Communist Party's transition from a small and relatively weak assemblage of guerrillas and political cadres to a well-organized, well-equipped, and viable politico-military movement with a large conventional military. Ultimately, the ability to control these two factors created a favorable set of conditions that allowed the Chinese revolution to grow in strength, build the Red Army, transition from a guerrilla to a conventional war, defeat its enemies (both foreign and domestic), and seize political power through armed force.

Space, *time*, and *will* were profoundly important to the Vietnamese communists and their revolution as well. As pointed out by former British officer turned counter-insurgency pundit Sir Robert Thompson, Mao's three intangibles were fundamental to the VWP's approach to people's war. Building revolutionary *space* through a network of base areas and developing an insurgency throughout South Vietnam, Thompson argued, laid the foundation for prolonging the war and subsequently breaking the political will of the RVN and its American ally. The spatial objective of Hanoi and the NLF, Thompson noted, was to expand the communists' presence and/or influence "throughout the length and breadth of Vietnam and into every element of its society so that the government [was] threatened in every quarter."[8] Similarly, communist pronouncements argued that the objective of revolutionary forces was to spread insurgency everywhere. "The only way to cope with an enemy who has a large number of troops," read a 1969 resolution issued by the Central Office of South

Vietnam (COSVN), the organ in charge of the southern insurgency, "is to wage guerrilla warfare throughout the three strategic areas [mountains, lowlands, cities], to wear down and destroy the enemy on a continuing basis; to compel the more than one million enemy troops to stretch thin, to reduce the enemy to a situation where his troops are thinly dispersed, his forces and war facilities are decimated, his fighting stamina is worn out, and his morale and organization collapse."[9] This statement not only supported Thompson's claim that the territorial aim of the insurgency was to extend itself everywhere; it also validated Mao's dictum that the key to revolutionary guerrilla warfare was exhausting and disintegrating enemy forces through constant pressure over a wide geographical area.

Base areas and other revolutionary terrain provided a foundation for prolonging the war and stretching it out over time. During the Second Indochina War, the protection and concealment they provided allowed the NLF to survive despite its material deficiencies. Survival, in turn, allowed them to sustain the revolution. Time also allowed for a measure of strategic flexibility by enabling the techniques of guerrilla warfare. When an opportunity presented itself, the Vietnamese communists advanced. When the strategic situation was unfavorable, they took one step back. American analyst Thomas C. Thayer argued that the communist insurgency survived a grueling war of attrition against two militarily superior powers because it followed this formula. In the end, the ability to control when, where, and under what conditions combat would occur enabled the communists to control their losses and endure. "They held the initiative in this respect, not the allies," he wrote. The "vastly superior" military forces of the RVN and the United States, he argued, "found it impossible to pin down and defeat an enemy who chose to avoid combat."[10] The Vietnamese communists' capacity to wage protracted guerrilla warfare and survive depended on their use and control of the operational environment. Hanoi and the NLF's extensive and frequent use of base areas, sanctuaries, and other forms of insurgent terrain was a key aspect of their ability to avoid destruction and sustain the revolution.

As in other revolutionary conflicts, the political will of the powers fighting in Vietnam relied on time and protraction of the war. The longer the fighting stretched out in Southeast Asia and a decisive outcome remained elusive, the greater the levels of impatience, frustration, and war-weariness, both in South Vietnam and in the United States. During the Second Indochina War, political will was synonymous with patience, and as the war dragged on, it was the United States that increasingly lost this virtue. In the end, the American decision to withdraw its military forces from Southeast Asia and leave the RVN to assume the burden of its own defense was the result of a loss of political will in Washington and among the American people owing to a long, costly, and inconclusive conflict. As noted by former RAND analyst turned historian David W. P. Elliott, the aim of the communists during the Second Indochina War, particularly after U.S. intervention, was not to shatter the enemy's military forces in an "apocalyptic counteroffensive," but to bring about a psychological defeat that would prompt the United States, first, to avoid deeper involvement in the war; and second, to withdraw its forces in the face of a strategic and political stalemate. "The fundamental objective of the strategic model used by Hanoi in the Second Indochina War," Elliott wrote, "was to understand the options open to the United States for maintaining its interests in South Vietnam and formulate ways of eliminating these options. The emphasis was not on a military defeat for the United States, but, rather, on exhausting the strategic possibilities open to it."[11] According to another authority, Vietnamese communist strategy was not aimed at "the positive objective of inflicting a decisive defeat on its enemies," but rather on "the negative one of undermining their enemy's morale to such a point that he gives up the struggle as much because of the unpopularity of the war on the civilian front as of war weariness in the field."[12] Although the overthrow of the Saigon regime remained the single primary objective of the war, Hanoi and the NLF devoted much of their strategic attention on the United States. This was because they knew that the fulfilment of the revolution and destruction of the RVN could only be achieved after American forces had been withdrawn from Southeast Asia, and after the "limited war" that had arisen due to the active but restricted intervention of U.S. combat forces reverted back to a direct confrontation between the contending Vietnamese factions. The strategic use of space to buy time and erode the political will of their enemies was the key to the communists' victory in Vietnam.

INSURGENT TERRAIN—THE "LIBERATED AREA"

The revolutions in China and in Vietnam demonstrated that intangibles, particularly protraction of the war and the gain or loss of political will, could prove decisive. They also revealed that time and will were founded on space, or to be more specific, that the development of base areas and other insurgent terrain provided the foundation for exhausting the enemy through a long and drawn-out guerrilla war. Lin Biao, vice chairman of the Central Committee of the Communist Party of China and minister of national defense under Mao, argued that the party's ability to produce politico-military space and control the operational environment in China had played a quintessential role in its ascension to power. "The base areas established by our

Party became the center of gravity in the Chinese people's struggle to resist Japan and save the country," Lin wrote in 1966. "Relying on these bases, our Party expanded and strengthened the people's revolutionary forces, persevered in the protracted war and eventually won the War of Resistance against Japan. . . . Revolutionary base areas established in the War of Resistance later became the springboards for the People's War of Liberation, in which the Chinese people defeated the Kuomintang reactionaries. In the War of Liberation we continued the policy of first encircling the cities from the countryside and then capturing the cities, and thus won nationwide victory."[13] In the eyes of Lin and others, the use of rural base areas to foster revolutionary guerrilla warfare and surround and eventually capture the enemy's urban power base possessed a utility that transcended the communist experience in China. The ability to defeat one's adversary by controlling space and the environment, he asserted, was "of outstanding and universal practical importance" for other revolutionary movements, particularly those that were unfolding in Asia, Africa, and Latin America during the 1960s against "imperialism and its lackeys."[14]

Mao's blueprint for a successful revolution built on rural base areas and protracted warfare did in fact prove applicable elsewhere, particularly in Southeast Asia. Between 1946 and 1975, Vietnamese communists successfully blended these two elements, along with Mao's three intangibles, and used them to great result against three opponents: France, the United States, and the RVN. Although much smaller and less populous than China, Vietnam and its largely agrarian society provided the environmental means for successfully waging a "people's war" in Indochina. The Vietminh and the NLF both adopted Mao's prescription for building guerrilla bases and growing "liberated" areas throughout the countryside. "During the . . . long resistance war [against the French], we had vast free zones and a strongly organized rear for the armed struggle, besides the guerrilla bases and guerrilla zones in the enemy's rear," Vo Nguyen Giap wrote in 1964. "In the liberation war now waged by our countrymen in the South, as a result of the political struggle of the masses combined with vigorous and extensive guerrilla warfare, liberated zones have come into being and are expanding, playing an increasingly important role in bringing about victory."[15]

To the Vietnamese communists, liberated areas were, first and foremost, a means of supporting the battlefield. According to one VWP member, "Revolutionary bases . . . ensure our safety when we encounter difficulty, they enable us to build up our forces, to extend the . . . movement, to carry out the armed struggle . . . [and] they provide our [administrative] apparatus with a stable position from which to operate and provide the facilities for training and feeding elements of the Liberation

Army."[16] Liberated areas were also a means for procuring the manpower and resources necessary for waging revolutionary warfare in the jungles, mountains, and densely populated lowlands of rural South Vietnam. In the eyes of communist strategists, liberated areas were the "broad and firm foundations" for waging a protracted war of liberation because they secured access to vital strategic resources, particularly manpower, supply, and intelligence.[17]

Liberated areas came in a variety of forms, including populous villages and hamlets, hidden sanctuaries and safe havens, remote base areas that housed conventional military forces, and specially designated war zones. Besides supplying manpower and material resources for local and main-force guerrilla units, liberated areas served as platforms for offensive military operations. They also provided safe havens for resting, refitting, and restoring the physical and psychological capacities of combat forces. Liberated areas were located throughout Indochina. The DRV was often called the "great rear area," and liberated areas in Laos and Cambodia were used for the Ho Chi Minh Trail—an incredibly complex logistical system of tunnels, trails, roads, and depots that sustained the southern insurgency and provided a conduit for infiltrating men and matériel from the DRV.

Base areas were not independent, fortresslike zones but rather mutually supporting regions located inside South Vietnam and along the eastern portions of Laos and Cambodia. They served three principal functions: transference of personnel and supplies, logistical resupply, and staging areas for political and military operations.[18] Base areas contained installations, defensive fortifications, and other physical structures, and they were places where communist units could evade allied forces and/or initiate offensive operations. These areas also functioned as sites for resting, refitting, and retraining battered and exhausted NLF/NVA units. In addition, base areas were used as permanent or temporary locations for political, military, or logistical headquarters as well as sites for the storage and distribution of food, medicine, petroleum, oil-based lubricants, ordnance, and other war materials. Unlike war zones, which had relatively fixed territorial borders, base areas were frequently reconfigured and expanded or contracted to meet changing circumstances and requirements.[19]

The network of communist base areas found throughout Indochina was an indispensable element within communist strategy. Moreover, it was central to the General Offensive and General Uprising plan put forth by Le Duan. "Securing base areas is one of the vital factors to obtaining victory," the Central Committee's Ninth Plenum resolution declared in December 1963. "When the war is developed to a certain degree and when the main-force units are strong, it is necessary to have large and secure base areas. Only when we expand and consolidate the base areas, can we accelerate the capability of the main-force units to launch large scale attacks and cause a basic change in the present situation in our favor."[20] Le Duan's loyal opposition within the Politburo agreed. "In compliance with the rule of governing the development of the revolutionary war's rear base, and to exploit, mobilize, and fully develop the total forces of our country," Giap wrote in 1969, "we have advocated combining the building of local base areas everywhere, along with the common rear base of our entire country [DRV]. Our experience has clearly shown that we must not only build common strongholds and rear bases for the entire country but also build strongholds and rear bases everywhere, on every battlefield and in each locality."[21] Although there was a great deal of disagreement in Hanoi over communist strategy during the Second Indochina War, Giap, Le Duan, and others agreed that the constellation of liberated areas maintained throughout Indochina was indispensable to the revolution's success.

War zones consisted of relatively large sparsely populated areas that were situated on or near principal communist infiltration routes and formed around natural terrain that lent itself to defense. As a result, these regions were not easily accessible to RVNAF and American military forces. The primary function of war zones was to marshal manpower and matériel for the battlefield. They too were used to rest and recuperate combat forces and as staging areas for military operations. These areas often contained high-level headquarters and served as the location for important functions and installations, including combat support services, training, and hospitals. Primary war zones included War Zone C, War Zone D, and the Iron Triangle, each located north of Saigon in Military Region 3.[22] Other war zones, including the Do Xa, Dong Thap Muoi, and U Minh, along with a fair number of lesser war zones and safe havens, were sprinkled throughout South Vietnam's other military regions.[23]

Aside from their material and martial benefits, liberated areas fulfilled ideational and existential needs. Base areas, communist-occupied villages and hamlets, and other forms of insurgent terrain provided tangible evidence of the revolution's existence and served as a symbol of the VWP's cause as well as its viability as a military, economic, and sociopolitical movement. Moreover, liberated areas provided an appropriate space for implementing socialism and reifying the tenets of Marxism-Leninism. As will be discussed in detail later, the significance of liberated areas transcended the procurement of manpower and material support for the battlefield; to the Vietnamese communists, areas under their control were a demonstration of legitimacy, an exercise in authority, and a vehicle for putting ideology into practice. Once communist guerrillas and/or main-forces gained control of an area, they

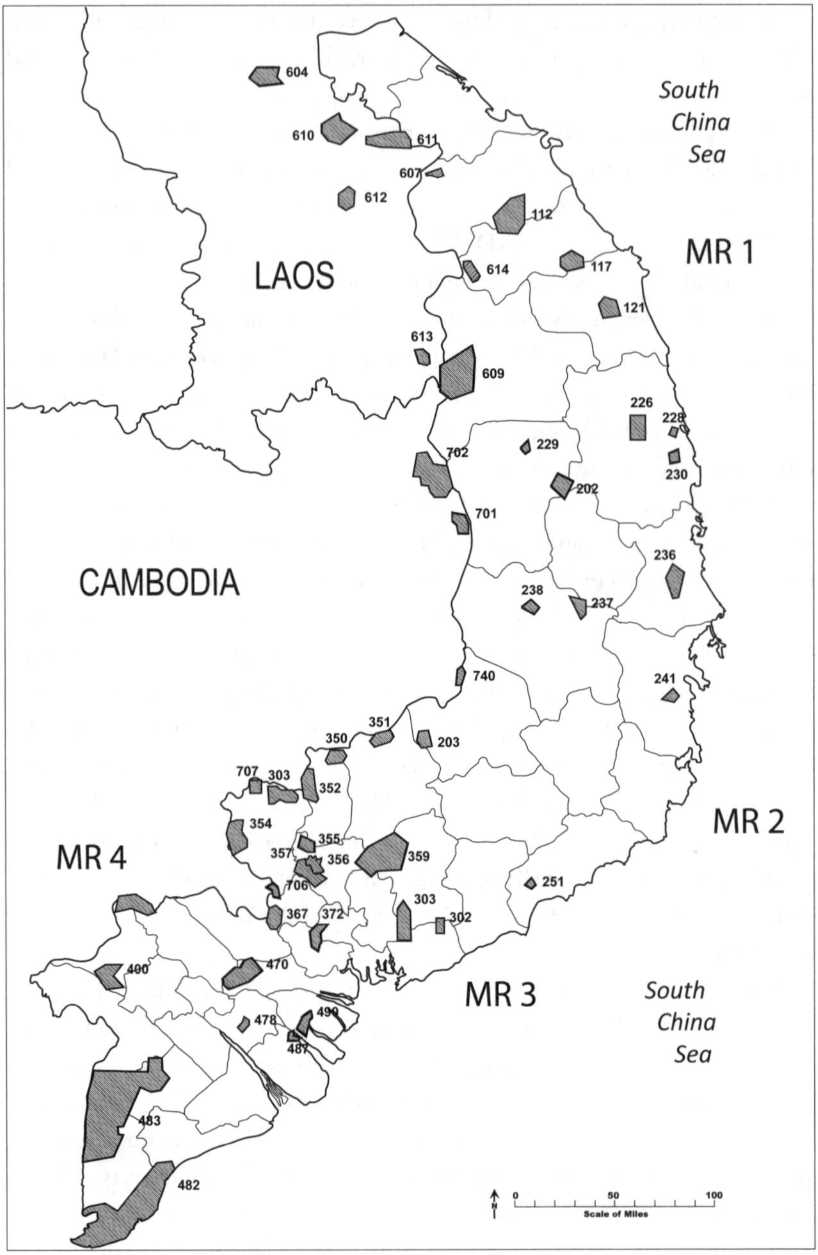

Viet Cong/North Vietnamese Army Base Areas as of 1 July 1969.

Map by the author. Copyright © 2018 by the University of Oklahoma Press.

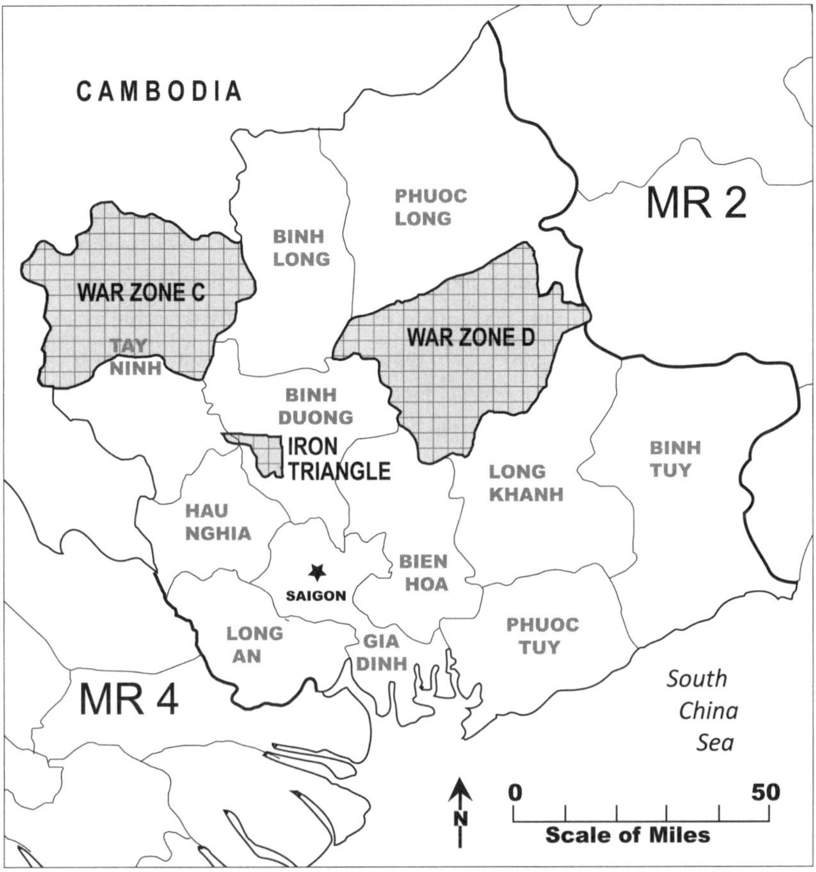

War Zones C, D, and the Iron Triangle.

Map by the author. Copyright © 2018 by the University of Oklahoma Press.

worked to establish "state" functions and actualize the VWP's blueprint for society by putting into place programs and policies that reflected their social, political, and economic worldview. Although Hanoi and the NLF treated areas under their control primarily as sources of manpower, rice, and money for carrying out the war, they used liberated areas to "build socialism" and create a new society in South Vietnam, one that was intended to close the book on Vietnam's colonial past and fashion a modern state built on land reform, "ideological remolding" of the population, and the creation of a new leadership class.[24]

Establishing liberated areas was merely the first step in a larger process designed to expand and consolidate the communist presence and authority over the countryside.

Expansion and consolidation were the two major elements for territorial control. Without a consolidated base area, further territorial expansion would have been difficult. Without territorial expansion, the existing base areas would have been endangered.[25] The overall objective was to maximize control over the operational environment. "The area of South Vietnam is not extensive nor its population very large," noted one party cadre. "For this reason, in fighting this war our primary strategic objective would be to defend and expand the liberated zones—our direct rear—to strengthen and develop the ownership right of the people, and to destroy enemy forces. We would pursue this objective at all costs. By the same token, the enemy himself considers it vital to gain the people over to his side, to capture land, to expand his own rear line, and to reduce our rear line."[26] Within the context of a war to control the operational environment and win absolute political power in Southeast Asia, enlarging liberated areas and strengthening the insurgency's grip on the countryside was strategically vital. In fact, Hanoi and the NLF's territorial objective was to create a physical and ideational antistate within the national territory of South Vietnam, one that would, over time, metastasize and grow until it consumed its host.

INSURGENT STATE BUILDING AND THE STRATEGIC ENVIRONMENT OF SOUTH VIETNAM

For some observers, geography played a critical role in revolutionary conflicts and wars of national liberation in the twentieth century. In 1969, university professor Robert McColl claimed space and the environment were the keys to understanding both the means and ends of modern insurgencies. The fundamental objective of Cold War–era insurgencies, McColl pronounced, was to achieve total control of population and territory by overthrowing and abolishing an existing regime through the "gradual attrition of government control over the state." The means to do this was geographical. Revolutions in China, Indonesia, Cuba, and Indochina, he argued, had a "territorial imperative" to capture and control space in order to establish base areas, liberated zones, and other forms of politico-military terrain: "Modern national revolutions have accepted as a basic tactic the creation of a territorially based anti-state (insurgent state) within the state," McColl asserted. "This is no mere 'shadow government' involving only political leaders. The national revolution is aimed at directly involving the general population as well as national real estate in the revolution. The mechanism is the creation of territorial units complete with all the attributes of any legitimate state, namely a raison d'être, control of territory and population and, particularly, the creation of its own core areas and administrative units as well as a power base [for] its guerrilla army."[27] Lin Biao supported McColl's assertion. "Our base areas were in fact a state in miniature," he declared, referencing China's War of Resistance against Japan. Within areas controlled by the Chinese communists, Lin claimed, "we built the Party, ran the organs of state power, built the people's armed forces and set up mass organizations; we engaged in industry and agriculture and operated cultural, educational, and all other undertakings necessary for the independent existence of a separate region."[28] McColl's and Lin's descriptions of liberated areas support Bernard Fall's assertion that the communists' revolutionary war objective was to control not just the population but the societal context as well—the civilian environment.

The "territorial imperative" to create an insurgent state within a host nation was directly linked to the overall goal of socialist revolutions as pursued in Asia, Africa, and Latin America: fundamental social transformation. People's war, in contrast to other guerrilla conflicts throughout history, was not simply aimed at achieving limited political objectives such as the removal of an occupying power, national independence, a change of government, or redress of grievances. Nor was it, as many at the time had suggested, merely a response to social and economic inequality or an unjust or repressive government. It was, instead, a blueprint for the radical reorganization of society, and the forced imposition of a new order in places that had only recently been emancipated from the yoke of Western imperialism. "Revolutionary war," Douglas Pike argued, "seeks a totally new social order, and it levies on the participant the demand of total involvement, total immersion. It touches all persons of the society at all points of their existence."[29] Achieving the "total immersion" of the population in the revolution first meant shaping the operational environment. This explains why communist revolutionaries went to such great lengths to control the geography of South Vietnam and create a physical and ideational world that would advance revolutionary aims and fulfill political, military, and social objectives.

To facilitate armed struggle and political control over South Vietnam's complex operational environment, Hanoi and the NLF subdivided South Vietnam into administrative and military regions. By 1961, they had apportioned South Vietnam into six regional military divisions under the command of COSVN.[30] These, in turn, comprised provinces, districts, villages, and hamlets. Each military region had a military headquarters, and each village had a village headquarters.[31] Five years later, as the tempo of the war quickened and the southern conflict became more complex due to the intervention of American and North Vietnamese combat forces, the communists reconfigured the politico-military geography to meet the changing circumstances. Military Region 5, which had originally extended from Quang Tri Province on the DMZ down through central South Vietnam, was divided into three separate commands: the Tri-Thien-Hue–Central Laos Military Region, which included Quang Tri and Thua Thien Provinces; a new Military Region 5, made up of Quang Da, Quang Nam, Quang Ngai, Binh Dinh, Phu Yen, and Khanh Hoa Provinces; and the B-3 Front or Western Highlands Front Command, which consisted of the communist provinces of Kontum, Gia Lai, and Dac Lac.[32] The B-2 Front comprised the rest of South Vietnam and included the area from Quang Duc, Tuyen Duc, and Ninh Thuan Provinces down to the Ca Mau Peninsula.[33] By 1965, Hanoi and COSVN had reorganized the country into nine administrative/military regions and a handful of specially designated zones. These areas included thirty-three provinces (in contrast to the RVN's forty-four provinces) and approximately 230 districts, each controlled by a Province Party Committee. Like their predecessors, these regions varied in size and mission. Moreover, each had its own political and military headquarters. Except for some minor geographical alterations made in 1966, 1968, and 1970–71, periods when new zones were carved out or existing ones were consolidated or modified, the administrative/military regions reconfigured by the communists in 1965 remained intact throughout the conflict.[34]

Besides establishing zones of military and administrative jurisdiction, the Vietnamese communists divided South Vietnam into three major geostrategic regions: mountains, rural lowlands, and urban areas. Although each was distinct and called for a carefully prescribed strategic approach, Hanoi and the NLF used all three areas to attack and weaken the Saigon regime. Le Duan insisted revolutionary forces must fully understand the special characteristics of each geostrategic region if they were to coordinate military operations and assign appropriate missions and responsibilities to each. "The lowlands battlefield is responsible for inflicting attrition on the enemy, for killing enemy troops, for pinning down enemy troops, and for providing personnel and material resources to our forces. The mountain jungle battlefield is responsible for annihilating enemy troops and pinning down enemy forces, with the primary focus being on U.S. and puppet regular army units. At the same time, it is also responsible for improving and expanding our base areas," he declared in a November 1965 letter to COSVN.[35] Giap agreed. "To attack the enemy in all three strategic areas—mountains, lowlands, and cities—with adequate tactics is a very important point in our method for carrying out our revolutionary war," he asserted in 1968.[36] Although Le Duan and Giap had serious disagreements on strategy, both men concurred that a geographically diversified approach was essential. Thus, they paid close attention to South Vietnam's operational environment and

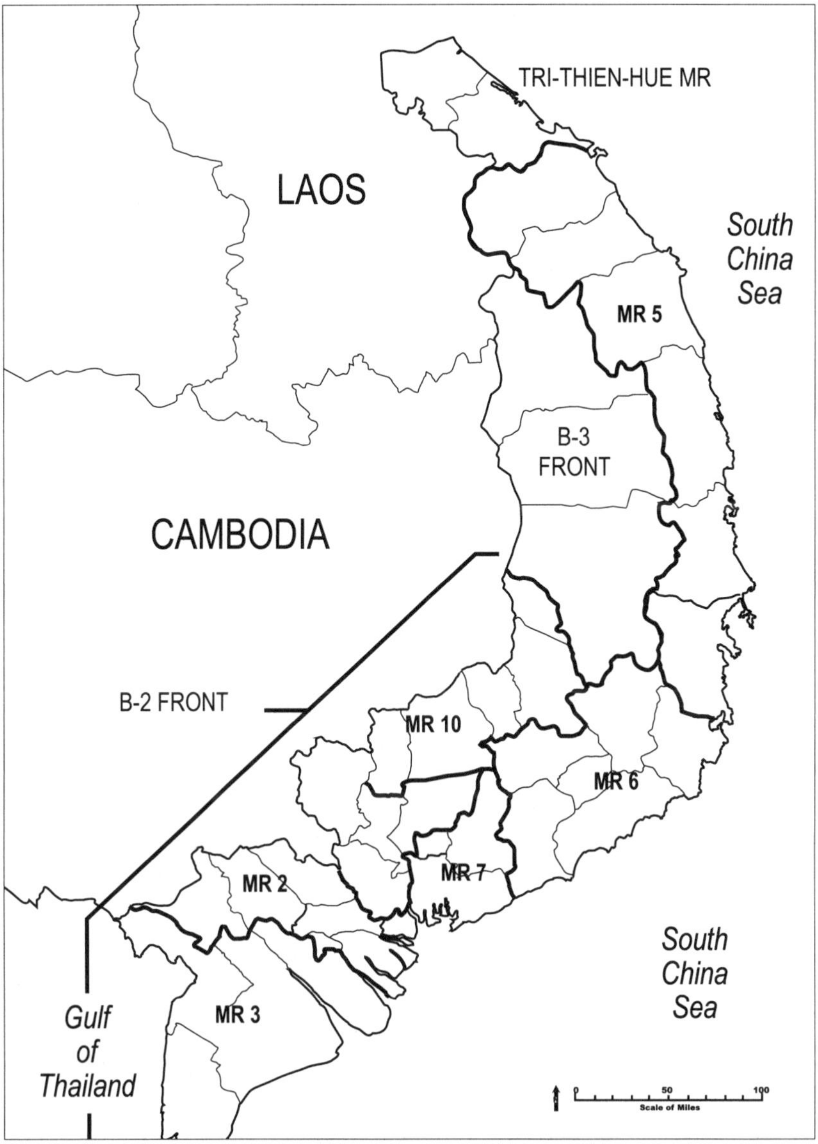

Viet Cong Political Geography.

Map by the author. Copyright © 2018 by the University of Oklahoma Press.

used both the natural and human-made world in their efforts to overthrow the Saigon regime. For Le Duan, Giap, and others, South Vietnam's landscape was more than just a backdrop for war; it was critical to the success of revolutionary warfare and, ultimately, the revolution itself.

Mountainous areas were important because they provided favorable conditions for waging protracted war. Comprised of the sparsely populated highlands, which stretched from the DMZ to an area fifty miles north of Saigon, South Vietnam's mountains not only provided territory for base areas and other places where communist main-force units could assemble and prepare for combat; they also afforded suitable areas for launching large-scale conventional military attacks against RVN troops. Mountainous areas were also significant because they were used as "stepping stones" for launching military offensives and spreading political activity into the heavily populated coastal lowlands and interior portions of the B-2 Front. "In case the enemy expands the war to a larger scale," the VWP declared in 1963 in explaining its decision to escalate the conflict in the South, "the mountainous area together with the lowland will enable us to fight a protracted war against him. We should make every effort to control the mountainous areas and have the determination to build these areas into a solid base area."[37] Le Duan repeated this argument in 1965 in response to the arrival of U.S. combat forces. "The mountain jungles are our bases," he informed COSVN. "These are the places where our military units have been built into powerful forces, and these are the battlefields that we know like the back of our hand and where we have a powerful battlefield posture that we can use to destroy both American as well as puppet troops. If we want to defeat the American army in the mountain jungle, the first thing we must do is to gain control of, gain mastery over, the mountain jungles."[38]

South Vietnam's rural areas included the densely populated coastal plain, the Mekong Delta, and the piedmont provinces surrounding Saigon. Hanoi and the NLF considered rural areas the key to victory for two reasons. First, Saigon lacked presence and firm control of the countryside. According to Le Duan, this made the RVN vulnerable to attack and subversion. "In South Vietnam . . . the aggressors and their henchmen ruling in urban centers cannot establish a strict control over the rural areas. That is why, when conditions are ripe for revolution, the villages constitute the weakest spot where the puppet administration becomes shaky and sinks into a crisis, hence the possibility of most rapidly starting local insurrections and of destroying the enemy's power apparatus considerably."[39] Second, rural areas were perfect for constructing base areas to build up large-scale military forces and served as a foundation for launching attacks against the strategically vital populated regions along the coast.[40] "Relying on these [rural] very steadfast bases of the revolution and revolutionary war," Giap editorialized, "our people can maintain, build, and develop their forces and persist in a protracted struggle. . . . These bases can serve as a steadfast springboard from which we can expand our struggle to the plains."[41] In the eyes of Vietnamese strategists, the population- and resource-rich rural lowlands were the epicenter of the conflict, the place where the war to control the civilian environment was most vital. "There [the rural lowlands] are the areas where the enemy is striving to scoop up human and material resources for the purpose of using Vietnamese to fight Vietnamese, using war to feed war; these are also the areas where we are determined to dispute with the enemy for control of human and material resources with which to fight a protracted war."[42]

Urban areas, the third and final geostrategic region, were important because that was where Saigon's political power was concentrated and the organs of the central government were located. Cities and district capitals provided opportunities for a host of clandestine subversive activities, including political agitation and proselytizing, propaganda campaigns, and recruiting from among the civilian population, civil servants, and the military.[43] Urban areas were also significant because an increasingly large portion of the population lived there. Although South Vietnam remained a predominantly rural society, increased levels of urbanization (mostly due to the ravages of war and widespread destruction throughout the countryside) swelled South Vietnam's cities and district capitals during the conflict. According to Samuel Huntington, urbanization was the "most dramatic and far-reaching" product of the war in South Vietnam. The principal reason for an enlarged urban population was the intensification and growing destructiveness of the war beginning in 1965, when American combat forces were first introduced. Writing for *Foreign Affairs* in 1969, Huntington pointed out that the urban population had more than doubled, perhaps even tripled, within a few years. He estimated that 6.8 million out of a total population of 17.2 million, about 40 percent, lived in cities of 20,000 or more people. This made South Vietnam not only the most urban nation in Southeast Asia, but more urbanized than Sweden, Canada, Poland, Austria, Switzerland, Italy, and the Soviet Union. "The image of South Vietnam as a country composed largely of landlords and peasants," he wrote, "an image still prevalent among many Vietnamese intellectuals who continue to quote the 85 percent rural figure, has little relationship to reality."[44] According to a newspaper article published that same year, "South Vietnam is now an urban place."[45] Although the actual percentage of the population living in cities and district capitals fluctuated over time in response to major military offensives, the overall intensity of combat in rural areas, and progress or regression in government-sponsored refugee resettlement programs, urban areas remained an important target for communist revolutionaries in the fight to control the operational environment. "General uprising," declared one communist directive, "means a concerted insurrection or a series of successive insurrections conducted by the revolutionary people . . . to seize political power for the people, in all areas under control of the puppet authorities, and particularly, in big cities which are the enemy important rear bases, where key agencies of the puppet government are assembled."[46] The use of military and political action to attack cities and provincial capitals was an important element in communist revolutionary warfare. This explains the communists' emphasis on major offensives aimed at South Vietnam's cities and provincial capitals beginning in 1968.

THE STRATEGY, TACTICS, AND ORGANIZATION OF INSURGENT STATE BUILDING

The General Offensive and General Uprising was predicated on a dual rural-urban attack and the strategic use of South Vietnam's diverse operational environment. In fact, the close coordination of political struggle and armed struggle and the strategic use of geography was the foundation of Le Duan's strategy. This geocentric hybrid approach, he and others believed, would destroy the RVN's armed forces, disintegrate its political infrastructure, and provoke popular uprisings in rural areas and cities across South Vietnam. The communists would then seize control of the operational environment and overthrow the "reactionary" government in South Vietnam.[47]

This dual urban-rural method along with South Vietnam's complex operational environment demanded a carefully calibrated approach, one that utilized an appropriate mixture of political and/or military means within each geostrategic area. The sparsely populated mountainous and heavily forested areas, for instance, were well suited for large-scale military operations and a predominantly military approach. The densely populated villages and hamlets of the rural lowlands, meanwhile, called for a blend of low- to mid-level insurgency and political action. Lastly, South Vietnam's urban areas, its cities and provincial capitals, required a predominantly political attack due to the overwhelming and entrenched government position in these areas.[48] Variations in topography, population density, and other complex sociopolitical factors prompted Hanoi and the NLF to adopt a strategy that not only used a variety of forces, including conventional military forces, guerrillas, and political operatives, but that applied a varied politico-military method tailored to meet specific and often shifting local and regional circumstances.

Hanoi and the NLF responded to South Vietnam's spectrum of environmental and situational contexts by developing a multifaceted politico-military strategy that was built on a diverse yet interdependent mixture of military and political forces. Douglas Pike argued that the politico-military strategy conceived, developed, and fielded by the Vietnamese communists was a "dimensional new method for making war"—one that was unique to modern warfare and had no proven counterstrategy. According to Pike, the genius of communist revolutionary warfare lay in its carefully calibrated application of military and political means.[49] General William Westmoreland's onetime deputy, Phillip B. Davidson, summarized this "new method" of warfare as follows: "The strategy of Revolutionary War has as its objective the seizure of political power in a nation-state. In operation, it totally integrated two principal forms of force, armed conflict and political conflict, which the North Vietnamese call military *dau tranh* (struggle) and political *dau tranh*. Their integration creates a kind of war heretofore unseen: a war waged on several . . . programmatic fronts, all conducted by one authority, all carefully meshed to serve one end—the seizure of national power."[50] Thus, revolutionary war as practiced by the Vietnamese communists rested on a unique dual strategic approach that fused political and military action and engaged both guerrilla and conventional warfare in the pursuit of strategic and political objectives.

The first half of this approach, military *dau tranh*, Pike called the "revolutionary violence program." It included military operations and "other forms of bloodletting" not usually associated with the activities of formal armed forces. Military *dau tranh* utilized military and quasi-military activities, including conventional combat, guerrilla raids and ambushes, direct and indirect attacks by fire, sapper team activities, subversive action and demolition, assassinations, and kidnapping. Although initially based on Mao's three-stage blueprint for revolutionary war and its progression from grassroots political organizing up through conventionally armed mobile warfare, military *dau tranh* was refashioned and altered during the Second Indochina War in response to shifting politico-military circumstances. After the introduction of American combat forces and other allied ground troops in 1965, Hanoi and the NLF applied a flexible, nonlinear approach that shifted back and forth between the different levels of revolutionary warfare. Thus, a flexible and adaptive approach, one marked by periodic main-force offensives followed by extended periods of low-level protracted insurgency, became an important part of Vietnamese revolutionary strategy. For instance, the communists abandoned large-scale conventional military operations and returned to low-intensity guerrilla warfare and insurgency in 1968, 1972, and January 1973, after the Paris Accords had been signed. These reversions to low-level guerrilla warfare were themselves overturned when Hanoi reverted back to high-intensity main-force warfare during the 1972 and 1975 spring offensives. Taken together, these shifts in strategic approach reveal a more flexible and nuanced application of armed struggle than a singular progression from guerrilla warfare to conventional war. In this way, communist planners did not conceive military *dau tranh* as "a single generalized doctrine," but rather as "a series of finite, precise, strategic-tactical variants."[51]

Military *dau tranh* was only half of a holistic approach that blended armed force and political action. Violence, Pike argued, was always cast in a political context; it was never evaluated "in simple military terms."[52] Others confirmed Pike's contention that the Vietnamese communists sought to strike a symbiotic balance between military and political means. According to David Elliott, communist guerrillas and political cadres operating in the Mekong Delta had employed "a sophisticated political-military control system with the end objective of establishing control over both territory and people." Political and military efforts, he argued, were balanced to achieve this end. "Political activity," Elliott wrote at the time, "lays the foundation for military operations; military operations are only undertaken for political reasons."[53]

The other side of the coin, political *dau tranh*, involved a broad spectrum of "systematic coercive activity," including social organization, mass mobilization, and political agitation in the form of recruitment and motivation, indoctrination, and propaganda. Political *dau tranh* included three distinct "action" programs: (1) nonmilitary activities conducted among the civilian population in enemy-occupied areas; (2) nonmilitary activities conducted among enemy troops; and (3) administrative activities in "liberated" areas. The first was meant to undermine the loyalty and morale of the civilian population in government-controlled areas, predominantly through political agitation and propaganda campaigns. The second action program was a military proselytizing campaign aimed at inducing desertion and/or defection among the government's armed forces as well as its civil service. The third program consisted of administrative and motivational activities within areas that were under the control of the NLF or the Provisional Revolutionary Government (PRG), a quasi-governmental apparatus established in 1969 as a "legal" communist government in South Vietnam. The PRG was a critical component of political *dau tranh* after 1968. The organization claimed to represent the "true aspirations" of the South Vietnamese people, an assertion that was immediately echoed by Hanoi and other socialist governments, which stamped the PRG with the imprimatur of legality. Together, these three political *dau tranh* programs were synced with the military *dau tranh*; the goal was to destabilize South Vietnamese society and weaken the government's hold on the operational environment by undermining Saigon's claims to legitimacy and eroding popular support for the RVN's political/military institutions.

Pike likened the dual use of military and political *dau tranh* to "the jaws of [two] pincers used to attack the enemy." These approaches had to work together. Strategic and political objectives could only be achieved when the two were combined and there was a "marriage" of violence to politics. "The basic objective in *dau tranh* strategy," Pike wrote, "is to put armed conflict into the context of political dissidence. Thus, while armed and political *dau tranh* may designate separate clusters of activity, conceptually they cannot be separated."[54]

The hybrid politico-military approach of the *dau tranh* strategy, in tandem with the sheer complexity of South Vietnam's operational environment, prompted Hanoi and the NLF to develop a variety of political and military forces and a style of warfare that synthesized political subversion, guerrilla warfare, and conventional military operations. According to MACV, the communist revolution operated as a "system" comprised of several distinct yet codependent politico-military "sub-systems," each with prescribed functions and responsibilities.[55]

The political component of the system, in accordance with nonmilitary *dau tranh*, was a network of clandestine NLF agents and sympathizers collectively called "infrastructure." The Viet Cong Infrastructure (VCI) or "VCI sub-system" was a hierarchically arranged political administrative organization that directed and supported the communist insurgency from the regional down to hamlet level. "The political infrastructure," read one report, "is the 'shadow government' of the [Vietnamese communists] that directs the insurgency in the South and competes with the legal [RVN] government for control of the people."[56] Although classified as a single entity, the VCI was a composite of many organizations and committees, each with its own responsibilities and specific functions for supporting the total insurgent effort. It included members from the People's Revolutionary Party (the southern branch of the VWP), the NLF, COSVN, and numerous other organizations and agencies.[57] Overall, the VCI subsystem was charged with establishing front organizations, recruiting and training insurgent personnel, planning operations, drafting soldiers and laborers, directing guerrilla forces, providing logistical support for combat units, collecting taxes, coordinating terror campaigns and assassinations of government officials and other "reactionary" elements, disseminating propaganda, and indoctrinating the civilian population. These activities, along with the various organizational and administrative elements of the VCI, formed an important part of a politico-military movement whose sole purpose was to gain control of the operational environment and seize political power.[58]

The VCI was an important part of the control war and efforts to master the operational environment. According to the Vietnam Special Studies Group, an ad hoc assessment body assembled during the Nixon administration, the Vietnamese communist shadow government was a formidable threat because it endangered Saigon's position in rural areas. "As long as the VCI is largely intact," read one report, "the enemy retains a substantial capability to resist further [government] control gains, to erode [government] control covertly from within, and to quickly reassert [Vietnamese communist] control should the military position and trends in the countryside be reversed."[59]

The "military sub-system," which embodied armed *dau tranh*, encompassed all communist military forces. These were broken down into three distinct classes: main-forces, local forces, and guerrillas. Main-forces, which included North Vietnamese Army regulars as well as larger units from the military wing of the National Liberation Front, the People's Liberation Armed Forces (PLAF)—or "Viet Cong" (VC) as they were pejoratively called by the Americans and South Vietnamese—were assigned to regional- and province-level commands and were responsible for conducting large conventional military operations against allied main-forces, particularly the Army of Vietnam (ARVN). Communist main-forces were organized into battalions, regiments, and divisions and operated at the regional level, primarily in South Vietnam's mountainous and sparsely populated hinterlands. Local forces were full-time fighters organized into battalions, companies, and platoons and assigned to districts and provinces. These units predominantly conducted combat operations against enemy territorial forces. Part-time guerrillas and other paramilitary units, including local self-defense forces, rounded out the communists' military triumvirate. Guerrillas were organized into squads and platoons and charged with conducting small-scale operations at the village and hamlet levels. Typical missions for guerrillas included terrorist and sabotage activities, intelligence gathering, and the provision of local security. When necessary, guerrilla units assisted local and main-force units. Both local forces and guerrillas operated in South Vietnam's densely populated lowland regions, including coastal regions and the Mekong Delta.[60]

In order to extend the insurgent "system" into new areas, in accordance with the overall objective of building an insurgent antistate within the national territory, Hanoi and the NLF relied on "a balanced interrelationship" among main-forces, local forces, and guerrillas as well as on the "dual use" of political and military methods. Although village guerrillas, provincial local forces, and regional main-forces were separate entities, their actions were interconnected. It was the responsibility of each type of force to shield and support the others along with the political components of the system. According to one study conducted by MACV in Kien Hoa Province, "The interrelationship among the three force levels of the military subsystem has been of crucial importance to overall communist strategy." This strategy, it continued, "essentially calls for the guerrilla and local force units under the direction of the VCI to seize and hold control of both the geography and population . . . while the main-force units provide the needed backup and tie down the bulk of allied forces, thus rendering them incapable of aiding allied pacification efforts."[61] As indicated, the Vietnamese communists closely coordinated the use of three types of military forces to expand control over territory and population.

Other province-level analyses revealed the need for close coordination. The ability of communist main-force battalions to penetrate into new areas, they found, depended on the achievement of local force battalions and companies. The success of local forces in penetrating contested or government-controlled areas, in turn, was contingent on the ability of local guerrillas and VCI to establish firm footholds in these areas. According to one report, the use of three types of military forces to expand territorial control was meant to work like a wedge. Guerrilla units backed by main-forces were charged with the initial penetration into an area in order to break enemy control and prepare the way for larger forces. Once control was established and consolidated and a base area established, main-force units were then moved in to assist guerrillas in defending the area against allied incursions, and to launch offensive attacks into surrounding areas to further expand their control. This was known as the "springboard" concept. The first step was to break enemy control and establish a liberated area using armed force. The second step was to consolidate communist control through political organization and administration so that the area could be used as a jump-off point for future offensives into adjacent areas.[62] A wartime broadcast by Hanoi's Liberation Radio confirmed this process. "The creation of many large, firm interdependent liberated areas," it stated, "not only contributes toward splitting the enemy's strategic posture, but also is of great strategic significance in that these liberated areas serve as springboards for extending our offensive and uprising thrusts more deeply and extensively to areas that have heretofore been under the enemy's tight grip and strong domination."[63]

REVOLUTION AS PROCESS: THE ATTRITION OF STATE CONTROL OVER TERRITORY AND POPULATION

The objective of communist revolutionary warfare and its hybrid politico-military approach was to expand friendly control over the operational environment while simultaneously shrinking the enemy's hold over land and people. The means to do so was to transform government-controlled areas into contested territory (a state where both sides shared some degree of presence and authority) and contested areas into insurgent-controlled territory. This was the blueprint for building an insurgent antistate within the national borders of the host nation. Thus, revolution was a process, a means of social and political transformation; a method for seizing political power through the gradual attrition of state control over its territory and its population. "The aim of the insurgent," French officer turned counterinsurgency expert David Galula noted in 1964, "is to change the occupied areas into guerrilla areas, guerrilla areas into guerrilla bases, and these into regular bases."[64] Three decades earlier, during China's struggle against Japan, Mao himself had identified distinct zones of control along with a method of transference:

> In guerrilla warfare behind the enemy lines, there is a difference between guerrilla zones and base areas. . . . [Guerrilla zones] are areas which are held by the guerrillas when they are there and by the puppet regime when they are gone, and are therefore not yet guerrilla bases. . . . Such guerrilla zones will be transformed into base areas when they have gone through the necessary processes of guerrilla warfare, that is, when large numbers of enemy troops have been annihilated or defeated there, the puppet regime has been destroyed, the masses have been roused to activity . . . mass organizations have been formed, people's local armed forces have been developed, and . . . political power has been established. By the expansion of our base areas we mean the addition of areas such as these to the bases already established."[65]

In their struggle against the French, Vietnamese revolutionaries also adhered to this process. "We . . . advocated the wide development of guerrilla warfare, transforming the rear into our front line," Giap declared shortly after the Vietminh had successfully driven the French from Indochina. "Our units operated in small pockets, with independent companies penetrating deeply into the enemy-controlled zone to launch guerrilla warfare, establish bases and protect local people's power," he continued. "We gradually formed a network of guerrilla bases. On the map showing the theater of operations, besides the free zone, 'red zones,' which ceaselessly spread and multiplied, began to appear right in the heart of the occupied areas. The soil of the fatherland was being freed inch by inch right in the enemy's areas."[66]

Mao's and Giap's testimony reveals the unique spatiality produced during irregular conflicts and wars of insurgency/counterinsurgency. As their descriptions reveal, the presence of base areas, use of guerrilla tactics within occupied areas, and intermingling of friendly- and enemy-controlled territory fragmented the operational environment and produced a unique patchwork quilt of territorial control and spatial contestation. This prompted Galula to characterize the spatial pattern produced in revolutionary conflicts as "an accidental mosaic, a patchwork of pieces with one well pacified, next to it another one not so pacified or perhaps even under the effective insurgent's control."[67] Practitioners and observers provided other colorful metaphors. According to Mao, the spatial patterns produced by revolutionary warfare resembled the interlocking pieces of a jigsaw puzzle. The Vietnamese, meanwhile, believed the unique mixture of friendly and enemy enclaves within a single geographical area resembled the spots on a leopard. Thus, during the Second Indochina War, the spatial formation created by pockets of communist control in South Vietnam was often referred to as a "leopard spot" or "leopard skin" configuration.

The fragmentation of the battlefield and intermixture of oppositional zones of control described above led many to describe revolutionary guerrilla conflicts as wars without front lines. Mao himself noted in 1936, "In a revolutionary civil war, there cannot be fixed battle lines. . . . [R]elatively fixed battle lines are often to be found in the general run of war."[68] A decade later, VWP member Truong Chinh claimed that the "interlocking and disorderly" nature of the Vietminh struggle against the French made it "a war without battle fronts."[69] According to Giap, "there was no clearly defined front" in the war against the French. "It was wherever the enemy was. The front was nowhere, it was everywhere."[70] Thomas C. Thayer, an American systems analyst and director of the Southeast Asia Intelligence Division during the Second Indochina War, made a similar assessment. "The Vietnam wars," he wrote, "were highly atomized struggles to influence the population in hundreds of different districts and there were no front lines as we know them."[71]

The unique spatial configuration produced by communist revolutionary warfare, including a lack of identifiable front lines and interspersed zones of control, was more than a product of communist base areas and the development of guerrilla warfare behind enemy lines; it was an artifact of the complex operational matrix used by Hanoi and the NLF during the First and Second Indochina Wars. Both were "hybrid" wars—conflicts that involved a complex and transmuting mixture of conventional forces, guerrillas, and clandestine political cadres working in tandem. As a result, they were ever-evolving conflicts fought by a variety of forces of different size and composition, each applying a multitude of tactics at different times and in different regions. Stealing a page from David Galula, William Westmoreland's onetime deputy Philip B. Davidson labeled this phenomenon, as it applied to the war in South Vietnam, a "mosaic war." This was because the Second Indochina

War witnessed an incessant series of back-and-forth strategic and tactical shifts between Mao's three phases of protracted war along with an almost infinite variety in the mixture and emphasis of military and/or political means. At any given time, the fighting in one area of South Vietnam could have been in the third phase conventional war involving communist and allied main-force units. In another it could have been in the first phase, marked predominantly by low-level insurgency and political struggle in localities. Meanwhile in yet another area, the war could have resembled stage two, which involved a midlevel mixture of conventional and guerrilla warfare. "There were forty-four provinces in South Vietnam," Maxwell Taylor wrote in 1972, "and the situation in one province was quite different from its neighbors. Thus two reliable and conscientious observers in two different provinces could file completely contradictory reports and both be right."[72]

The means to attain the VWP's primary objectives—independence, unification, and socialism—were largely spatial. Although political organization and the development of an administrative apparatus that mimicked Saigon's governmental structure at every level (something the French called *hiérarchies parallèles*) have often been touted as the key to the revolution's success,[73] the ability to control territory and population was a critical first step in the development and preservation of such an organizational structure. The Vietnamese variant of people's war—one that carefully choreographed military and political attack, guerrilla and conventional war, and military and paramilitary forces within a spatial strategy that targeted the enemy at multiple geographical points in three different geostrategic areas—was predicated on control of the operational environment and the development of insurgent-controlled areas. The spatial objective of the Vietnamese communists was the same as Mao's: to transform enemy territory into friendly terrain through the expansion of revolutionary space and the gradual attrition of government control over the operational environment. COSVN Resolution 14, issued in 1969, articulated this process. The expansion of liberated areas, it explained, enabled friendly forces to not only establish "penetration bases" throughout South Vietnam, but "destroy large chunks of the enemy control apparatus in many villages and turn them into liberated areas."[74] The aim of revolutionary forces, declared another source, was "to transform the guerrilla zone into a guerrilla base area and the enemy-occupied zone into a guerrilla zone, in order to enlarge our liberated zone and narrow down the enemy occupied zone. In this way, guerrilla warfare does not cease to develop and advance to ever higher forms, creating conditions favorable to the promotion of regular warfare to win final victory."[75] The linchpin of communist strategy and its hybrid approach was the transformation of government-controlled areas to insurgent-controlled areas and the construction of a territorially based insurgent antistate within the national territory of the RVN. The endgame was to overthrow the South Vietnamese government and replace it with a new regime, one that would "transition to the socialist revolution" and reify the tenets of Marxism-Leninism through a total transformation of Vietnamese society. "The Party's ultimate goal," Le Duan wrote in 1970, "is to achieve communism."[76]

Hanoi and the NLF were in fact able to build and consolidate politico-military space throughout Indochina, in the form of base areas, war zones, and insurgent-controlled villages and hamlets. This, in accordance with Maoist principles, forced the RVN and its allies to fight everywhere, stretched out Saigon's armed forces, and exhausted the government's ability to check communist advances and protect the national territory. Over time, the ability of Hanoi and the National Liberation Front to develop and maintain revolutionary terrain and apply guerrilla warfare over a wide geographical area forced more than a million South Vietnamese troops to thin out, to be encircled and divided, and to be attacked from all sides. "Immersed in the sea of people's war," Le Duan declared in 1970, "the enemy has become even more divided and weakened. . . . As for the [revolutionary] armed forces and people, they can constantly encircle the enemy, unceasingly attack him along with concerted uprisings, constantly maintain their offensive initiative, harass the enemy in the front and in the rear, and fight him anytime, anyplace."[77] The spatial strategy developed by Mao and used by the Vietnamese communists—particularly efforts to create politico-military space and retain territorial control—enabled the revolution to survive, maintain a foothold inside South Vietnam, gather its strength, and prepare for the offensive that toppled the RVN in April 1975 via a conventional military coup de main.

> An insurgent movement is a war for the people. It stands to reason that government measures must be directed to restoring government authority and law and order throughout the country, so that control over the population can be regained and its support won.
>
> —Sir Robert Thompson, *Defeating Communist Insurgency*, 1966

2 PACIFICATION AND COUNTERINSURGENT STATE BUILDING

From its earliest days, the Republic of Vietnam, in conjunction with the United States, had made the production of politico-military space and control of the operational environment a top priority in its war against communist insurgency and internal subversion. Seizing control of rural areas and extending governmental presence and authority into the countryside was the essence of the pacification program. Beginning with initiatives implemented by South Vietnamese president Ngo Dinh Diem and his brother, Ngo Dinh Nhu, during the late 1950s and early 1960s, and continuing throughout the conflict until the final collapse of the Nguyen Van Thieu regime in April 1975, the pacification program focused its efforts on eliminating the communist presence in rural areas and on establishing, maintaining, and increasing government control over territory and population. According to former ARVN officer Tran Dinh Tho, the purpose of the pacification program was to "establish the GVN presence in less secure, contested areas with a view of controlling the nation's manpower and resources and denying them to the enemy."[1]

COUNTERINSURGENT STATE BUILDING AND THE STRATEGIC ENVIRONMENT OF SOUTH VIETNAM

Pacification was implemented over the years through an assortment of programs, each using different nomenclature. Whether it was called pacification, Revolutionary Development, Rural Reconstruction, or New Life Development, however, the aim was identical: to "liberate the people from communist control, assist them in choosing their own government, and help them carry out local projects that would give them a better and more prosperous life."[2] Pacification was essentially armed nation building within the context of an agrarian civil war. In the eyes of South Vietnamese and American officials, the war in Vietnam was more than just a military conflict to eradicate communist infiltration and subversion; it was a struggle to create a sense of peoplehood and nation. "[Vietnam] is not only a war of opposing military forces," stated U.S. Ambassador Robert W. Komer in 1968, "but a war for the allegiance of the people of South Vietnam. Hence it is not enough to defeat the enemy's forces. . . . It is also essential in this kind of war to provide protection to the people of the countryside, to help meet their aspirations for a better life. This is what pacification is all about."[3]

Pacification was both a program and a *process*—a military, political, and socioeconomic method designed to systematically eliminate communist insurgency, establish government control over territory and population, and, ultimately, develop a national political community. The 1966 Honolulu Declaration proclaimed: "The war for the hearts of the people is more than a military tactic. It is a moral principle. For this we shall strive as we fight to bring about a true social revolution."[4] According to a concept report generated for Civil Operations and Revolutionary Development Support (CORDS), a civil-military organization created in 1967 to manage the pacification program, it was "an integrated military and civil process to restore, consolidate, and expand government control so that nation-building can progress throughout the Republic of Vietnam." Pacification, it continued, was comprised of a number of coordinated military and civil actions that were meant to "liberate the people from VC control; restore public security; initiate political, economic, and social development; extend effective GVN authority; and win the willing support of the people toward these ends."[5] MACV defined pacification as "all civilian, military, and police actions to eliminate organized VC military activity, detect and eliminate the overt and covert VC political apparatus, and nurture economic, political, and social development of a viable economy." These "systematic and accelerated activities," it continued, served as "the foundation for nation-building activities."[6] Sometime after 1967, the following definition was given: "Pacification is the military, political, economic, and social process of establishing or reestablishing local government responsive to and involving the participation of the people. It includes the provision of sustained, credible territorial security, the destruction of the enemy's underground government, the assertion or re-assertion of political control and involvement of the people in government, and the initiation of economic and social activity capable of self-sustenance and expansion."[7] As these descriptions indicate, pacification used a variety of civil and military methods to meet three broad objectives: continuous territorial and internal security against communist attack and subversion; an effective political and administrative structure; and self-sustaining economic activity. When combined, these elements were intended to eliminate the internal communist threat, reassert national sovereignty over the territory of South Vietnam, and build a unified and viable nation supported by the people.

At the national level, pacification was a government-sponsored program that provided planning, guidance, management, and material supply. At the local level, it was a process carried out largely by the people themselves. Pacification's two main thrusts—providing physical security against insurgent activities and implementing social, economic, and political action—were to begin at the grass roots and extend upward to the village, district, provincial, and national levels. This meant that efforts were focused on contested rural areas rather than on secure urban areas or rural areas that were already under firm government control. The countryside became the locus of pacification activities and state formation not only because it was the place where communist insurgency had propagated and where the government's hold was most tenuous, but because rural areas, thanks to their immense human and material abundance, were the key to attaining and maintaining real political power in South Vietnam. This explains why the densely populated rural environment was the epicenter of revolutionary struggle and the focal point for pacification. According to South Vietnamese generals Cao Van Vien and Dong Van Khuyen, "Villages and hamlets are the solid foundation of the nation. . . . The nation-building process, therefore, should begin in rural areas and should be founded on them, and all national efforts should be devoted to strengthening this foundation."[8]

The active participation and support of the rural population were considered essential to pacification. Responsible citizenship, the program's architects realized, was the basis of effective nation building. "Before there can be a sense of nation," declared a government pamphlet, "there must be a sense of community. Before people can be expected to become patriotic, they must become involved."[9] U.S. Ambassador William Colby, the civilian head of CORDS from 1968 to 1972, also believed peasant participation to be crucial to pacification and allied state formation.

The population, he informed the Senate Committee on Foreign Relations in 1970, "must all be integrated into one overall national effort. . . . The key to it is the active involvement and participation of the people."[10] Through its assortment of political, economic, and paramilitary programs, pacification was intended to overcome the "parochialism" of the peasantry and forge a new nationalism, one that not only required citizens to identify themselves as Vietnamese first, before all other associations (including religious or ethnic), but that asked them to transcend local affiliations and loyalties by developing a close working relationship with the national government. According to the RVN's National Council of Revolutionary Development, local defense, community development, and self-help programs were intended to "mold a proud, courageous, and individualistic people into a modern cohesive and free nation. This meant, among other things, generating cooperative public attitude toward a central government in a people who have been fragmented in opposing religions and ethnic groupings, loyal to family while apathetic to all government, and who only recently had won independence from their colonial masters."[11] As this quote indicates, the purpose of pacification was to unify the population and fashion a national consciousness, one that would build bridges between the Vietnamese village and the central government and upend the old Vietnamese proverb that the emperor's writ stopped at the village gates. It was the program's task, the National Council noted, "to cut through that bamboo hedge, to force links between the hamlet and the . . . government—not to insure that the people will obey 'the Emperor's law,' but to create a true partnership between people and government."[12]

Efforts to build a national political community and expand government control over territory and population meshed with the environment. In this way, pacification closely resembled communist revolutionary warfare. Like Hanoi and the National Liberation Front, Saigon and the United States paid a great deal of attention to South Vietnam's physical and demographic landscape and made every effort to use them effectively to achieve political, military, and social objectives. To facilitate pacification, allied planners apportioned South Vietnam into a number of primary geostrategic regions known as National Priority Areas or Geographic Areas of Precedence. These target areas were chosen based on population density as well as on their proximity to lines of communication and areas of political and economic importance. Although each of these criteria was critical to the pacification effort, population density was considered the most important because the program's primary objective was to gain control of the civilian environment. "The population is a fundamental factor in the current war," read one report. "To have the people on one's side is equivalent to having everything needed to organize, nourish, and develop the war."[13] Allied efforts, therefore, concentrated on those areas where the majority of the population lived. In 1966, the South Vietnamese Joint General Staff (JGS) and MACV identified four National Priority Areas. These were Quang Nam, Binh Dinh, and An Giang Provinces and the peripheral areas surrounding Saigon.[14] By 1968, the number of National Priority Areas jumped to twenty-eight and included provinces scattered throughout the country, from the Demilitarized Zone along South Vietnam's northern border to the Ca Mau Peninsula. According to the 1969 Pacification and Development Plan, about 58 percent of the population was estimated to be living in national areas of precedence and 19 percent in regions that were classified as provincial areas of precedence. An additional 13 percent lived in areas that were designated relatively secure or in remote regions that were sparsely populated. The remainder lived in areas dominated or controlled by communist forces.[15]

The RVN also divided South Vietnam into four Military Regions (MR) to facilitate military command and control. MR1 contained the five northernmost provinces, running from Quang Tri Province at the DMZ down through Quang Ngai Province. MR2, which included twelve provinces located along the central highlands and coastal lowlands, ran from Binh Dinh and Kontum Provinces in the north to Quang Duc, Lam Dong, and Binh Thuan Provinces in the south. MR3 included the ten provinces that surrounded Saigon and the Capital Military District, comprised of Saigon City and Gia Dinh Province. MR4 was South Vietnam's southernmost region and included all sixteen provinces of the Mekong Delta. Military Regions were placed under the control of an ARVN army corps and received combat and pacification assistance from MACV/CORDS. Moreover, they were subdivided into a number of Division Tactical Areas (DTA), with subordinate infantry divisions responsible for each. Military commanders within each MR and DTA were charged with conducting mobile strike operations and territorial defense within their tactical areas of responsibility. The system of territorial defense put in place for South Vietnamese ground forces, however, turned out to be highly problematic. ARVN divisions were assigned to fixed, geographically based military regions, which meant they were essentially locked in place and tied to the defense of a particular territory. Used in this manner, RVN ground forces lacked strategic mobility.[16] An attempt to remedy this situation was made in 1970. In response to the phased withdrawal of U.S. combat forces, DTAs were disbanded and ARVN divisions ostensibly freed from their territorial responsibilities. This change was intended to increase the strategic flexibility of South Vietnamese ground forces and enhance their ability to conduct mobile operations outside of their assigned territories.[17] Increased flexibility and mobility, however, never transpired. For a number of social and political reasons that will be discussed in Part II, the dissolution of DTAs had no impact on the RVN's defense posture, which remained rooted in static territorial defense well after the withdrawal of American combat forces. The ARVN was never unmoored from the defense of particular territories or transformed into the type of mobile strike force necessary to defend the nation; this deficiency—when paired with the RVNAF's lack of strategic reserve forces—yielded tragic results for the RVN and its people in the spring of 1975.

Administratively, South Vietnam was divided into forty-four provinces, 236 districts, and five autonomous municipalities (Saigon, Da Nang, Dalat, Cam Ranh, and Vung Tau). Provinces were under the command of a province chief, usually a military officer, and divided into districts. Districts, in turn, were subdivided into villages, which were themselves divided into hamlets. Districts contained anywhere from eight to twelve villages, and villages contained from three to twelve hamlets.[18] The village was the basic unit for rural administration. By 1970, most villages had elected governments, which included a village chief, village council, and village committee.[19] The village council served as the basis of local government and was assigned the major role in planning, leading, and managing the rural population.[20]

Although villages varied greatly in size, terrain, levels of security and development, and in some cases religious and ethnic composition, they all had surveyed boundaries and formed contiguous coverage of the area within each district. Hamlets were subunits recognized for the purpose of internal village administration. Like villages, they too varied physically and demographically according to geographical location. Some were identified according to surveyed boundaries, while others had no clearly delineated borders and were recognized only as an area where population massed.[21] In its simplest and most common structure, a hamlet consisted of an aggregate cluster of houses and families, most of which were in some way interrelated. They were the organizational base of all rural activities, and each possessed its own political, economic, and social hierarchy and traditions.[22] As the basic building blocks of Vietnamese society, hamlets were organic communities constructed in rapport with the local environment and its physical characteristics. For instance, many were arranged in ordered "clustered settlements," while others stretched out naturally along roads and waterways in "string settlements."[23] In 1971, MACV estimated that there were anywhere from 2,100 to 2,552 villages and between 10,000 and 12,000 hamlets in the Republic of Vietnam. Of these, 2,094 villages and 10,187 hamlets boasted elected governments. Officials appointed by Saigon administered the remaining villages and hamlets under government control.[24]

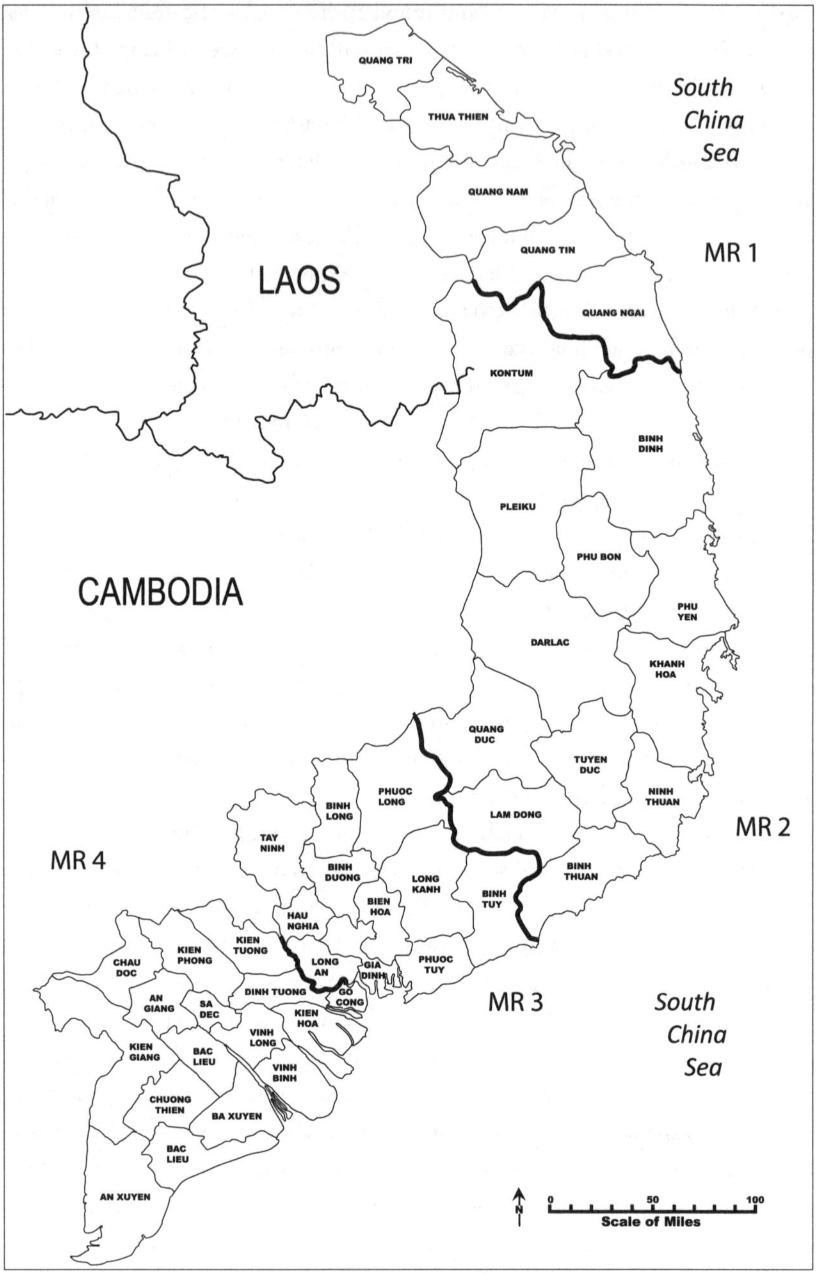

The Republic of Vietnam (South Vietnam) Political Geography.

Map by Erin Greb Cartography. Copyright © 2018 by the University of Oklahoma Press.

As discussed in the previous chapter, control and influence by both sides existed in varying degrees throughout South Vietnam's forty-four provinces, ranging from overt political and military domination in some places to low-level political struggle in others. These discrepancies in the level of control over population and territory resulted in a great deal of politico-military variation across different environments. In villages and hamlets under predominant or total government control, rural development and nation building were actively implemented. In areas under predominant or total insurgent control, combat operations dominated. Areas controlled by neither side but contested by both experienced an ad hoc mixture of military operations and civil development.[25] It was here in contested villages and hamlets that the war witnessed what one American province adviser described as "a violent see-saw military contest with various districts and villages under the dominance of first one side, then the other, as the opposing sides alternately gained or lost military dominance in [a] particular area."[26] The diversity of the conflict and its hybrid politico-military approach produced two unique circumstances. The first was a mixture of political and military mechanisms and methods applied variably throughout the country. The second was a highly irregular and fragmented collage of influence and control from the national to the local level. The mercurial and exceedingly fluid nature of the war in South Vietnam fragmented the battle space and created a unique mosaic of territorial control and spatial contestation.

The patchwork control situation that existed in Vietnam was a product of the spatiality and sequencing inherent within modern insurgency and counterinsurgency. Counterinsurgency by definition is a reaction to a social and political process (i.e., the propagation of a revolutionary movement and the development of an insurgent political and military apparatus throughout an area) that had already been put into motion and taken hold unevenly within a host nation. This situation was endemic throughout South Vietnam. For example, in December 1972, the Mekong Delta contained three of the most secure as well as three of the most insecure provinces in the country.[27] A study conducted three years earlier in Kien Hoa Province revealed that only 50 percent of the province's populated areas were under government control, predominantly its eastern districts. The western districts, meanwhile, were largely under the control or influence of communist forces and political cadres.[28] Variations in levels of government or insurgent control occurred not only at the district or province level; individual villages and hamlets were also frequently divided. For instance, a study generated by the Pacification Studies Group in 1968 revealed that two hamlets located in Phu Yen Province possessed two radically different security situations despite the fact that they were located side by side within the same village. While the hamlet of Phoc My enjoyed a high degree of government control and was considered "secure" against communist infiltration, My Le hamlet was under heavy communist influence and suffered regular penetration by guerrillas, who collected taxes, propagandized, abducted hamlet residents, and assassinated government officials with impunity.[29] That same year a field report generated by the RAND Corporation revealed that a single hamlet in Long An Province was under mixed control by both communist and government forces.[30]

General Westmoreland, in his final report as head of MACV, presented some firsthand perspective concerning the problematic nature of irregular war assessment. "The aspects of pacification most easy to measure," he wrote, "are often not the crucial aspects—and conversely, the less tangible aspects are not easy to perceive, let alone measure."[31] In order to assess what was a highly fragmented and atomized struggle, and to assess the contest between communist revolutionary warfare and pacification, U.S. planners created the Hamlet Evaluation System (HES), a computer-assisted reporting system used to measure the progress of the pacification program. Designed to evaluate the security and development of virtually all populated areas (communist-controlled hamlets were not evaluated, only cataloged), the program's overall purpose was to determine security levels and gauge the state of social, political, and economic development in regions where there was a government presence. According to Thomas Thayer, the system "was designed to yield comprehensive, quantifiable data on the security and development of every hamlet in South Vietnam under some degree of GVN control and to identify hamlets under VC/NVA control."[32] HES measured pacification progress by analyzing various programs put in place by the RVN and the United States to restore and maintain security, extend firm administrative and political control over the civilian environment, and advance the social and economic well-being of the Vietnamese people. The system generated information that could be analyzed and compared at every level, from individual hamlet, village, district, province, and region to South Vietnam as a whole. First implemented in January 1967, the initial HES program was replaced in January 1970 by an updated and revised version in an effort to overcome some of the known subjective biases and distortions that were inherent in the original system. Called HES/70, the new version required district senior advisers (DSAs) to supply a summary based on objective criteria rather than their own subjective evaluations of each hamlet. One year later, HES/70 was replaced by yet another version, HES/71. Virtually identical to its immediate predecessor, the new version refined the earlier system by adding greater weight to a number of political factors such as the presence of communist political infrastructure and levels of insurgent

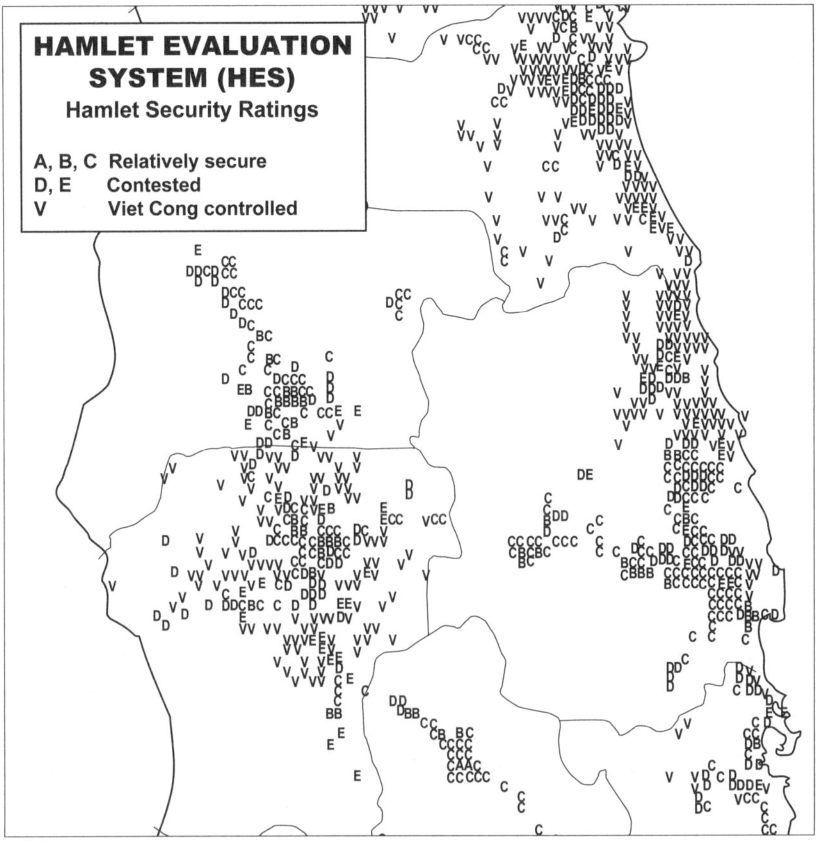

Hamlet Evaluation System (HES), May 1968, showing detail of Military Region 2.

Map by the author. Copyright © 2018 by the University of Oklahoma Press.

terrorism. Still in use at the end of 1972, HES/71 was the version handed over to the RVN after the U.S. withdrawal that was used to evaluate levels of control during the period after the signing of the Paris Peace Accords.[33]

HES required senior advisers from each district to submit a detailed monthly evaluation of pacification on a hamlet-by-hamlet basis. This was done using a standard format questionnaire known as a Hamlet Evaluation Worksheet. The original HES worksheet included a matrix of eighteen specific but equally weighted indicators—nine for security and nine for development—that covered local military, political, health, education, welfare, and economic conditions.[34] HES/70 used what many considered to be a more detailed and objective question set that included four security-related questions for each village and twenty-one security/political-related questions for each hamlet. These were to be submitted every month. An additional fifty-six hamlet-level and fifty-eight village-level questions covering all areas of the pacification program were submitted quarterly.[35] When completing their Hamlet Evaluation Worksheets, DSAs selected a response for each indicator. These ranged from A (best) to E (worst). The aggregate average of this simplified scoring system produced five possible letter ratings, one for each hamlet: A, B, C, D, or E. Hamlets under communist control were designated V hamlets and assigned a hamlet score of 0.00.

Some policymakers and analysts found HES to be a useful tool for tracking broad trends in security and development. Others, however, found the system lacking in this respect. In 1969, the Vietnam Special Studies Group (VSSG) determined that HES did not adequately or accurately measure government control in the countryside. VSSG observed that because the HES security indicator was an amalgam of security-related factors, hamlets could receive high security ratings and still contain enough communist presence (guerrillas and/or VCI) to preclude effective government control. To remedy this and other analytic deficiencies, the VSSG devised a "control" indicator. Based on raw HES data, the VSSG control indicator was used to determine the level of influence or control each side held over population and territory. As its name implied, the new indicator reflected the idea that, to some, area and population control—not development—were the critical metrics in assessing progress. Moreover, it underscored the reality that the primary objective after *Tet Mau Than* was to control the operational environment.[36] "Control of the people and resources of the countryside is vital to both contestants to this war," the VSSG declared in 1970.

> The enemy uses the countryside to generate manpower, combat support, supplies, and intelligence, and as a platform from which to launch attacks against GVN controlled areas and bases. If he achieves dominant control over the countryside, he could defeat the GVN by military action or by precipitating a GVN political or economic collapse in the face of a desperate military situation. . . . Control of the countryside is also vital to the GVN, so that it can deny the enemy the benefits just cited and gain them for itself. This control becomes even more vital to the GVN as U.S. troops are redeployed and it assumes a greater share of the war effort.[37]

As we can see, allied policymakers believed politico-military control was *the* central factor in the struggle for the countryside after 1968. Although pacification officials recognized the importance of popular support and social and economic factors, and acknowledged that these things would have a critical long-term effect on the durability of Saigon's hold on the country, they still believed, nonetheless, that political preferences and development were of secondary importance to political and military control. This was particularly true under current conditions, when the ongoing peace talks in Paris, the drawdown of American combat forces, and the possibility of a negotiated settlement and in-place ceasefire had placed a premium on expanding to the maximum extent possible dominion over the operational environment. In their estimation, popular support and political participation tended to follow rather than lead changes in control. It was vital, therefore, that politico-military control be established first so that nation building and political development could follow.[38] The Nixon administration also recognized that control had a profound effect on the attitudes and behaviors of the rural population. According to National Security Study Memorandum One, issued in March 1969, "The greatest single factor determining popular support in any area is the presence of US/Government of Vietnam or North Vietnamese Army/Viet Cong Forces. The populace generally tends to support whichever side is in military control."[39]

The VSSG used two distinct control indicators to classify hamlets: "RVN Control" and "VC Control." These were assigned to areas in which either the RVN or communist insurgents clearly demonstrated sufficient military and/or political strength to govern effectively and prevent the other side from establishing its own authority. Each control indicator contained a military and political component. A hamlet received an RVN control rating if it was protected by "adequate" security forces as determined by a U.S. district senior adviser, and possessed a fully functioning administrative apparatus in residency twenty-four hours a day. A hamlet was rated under "VC" control either if village guerrillas were combat effective and area defenses largely intact, or if it was subject to communist attack from within the village or from a location not more than two hours distant. In addition, the VCI had to be either in control of the hamlet or able to seriously inhibit government activities in the area. Within the context of the war in South Vietnam, control was a relative, not an absolute, state. Control of a village or hamlet meant that one side possessed a comparative superiority; it did not mean that the other side suffered from a complete lack of overt or covert influence. To address this condition, the VSSG devised a way to designate the level of influence each side exercised on a given area. For instance, hamlets were designated "Influence by Both" if RVN and communist organizations were of about equal strength. Meanwhile, if the political and military organization of either the RVN or the communists was relatively

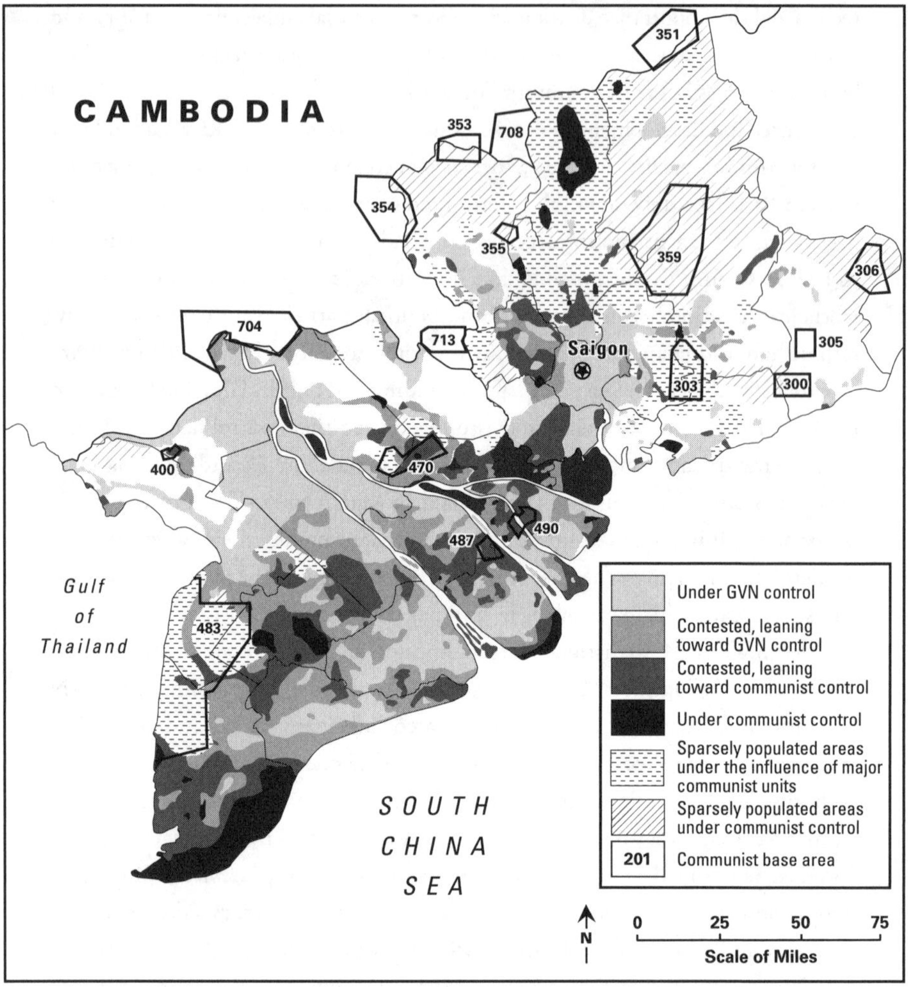

Military Regions 3 and 4, showing control by differing HES ratings.
Map by Erin Greb Cartography. Copyright © 2018 by the University of Oklahoma Press.

stronger but still too weak to assume effective control, then the hamlet was rated as either under "GVN Influence" or "VC Influence," respectively.[40]

Admittedly, HES ratings and their groupings were problematic. To begin with, generalizations such as "relatively secure" or "contested" were highly subjective and imprecise. Moreover, these classifications presented a distorted view of the on-the-ground reality in rural South Vietnam and suggested a more favorable set of circumstances than what actually existed. "The effect if not the purpose of their use," Congressman John V. Tunney stated in 1968, "is to mislead Americans about existing conditions in Vietnam."[41] Others argued that HES classifications did not adequately reflect the actual levels of presence and control by either side. According to one analyst, C-rated hamlets should have been designated "contested" rather than placed in the "relatively secure" category given the virtual parity in government and communist control. D- and E-rated hamlets were equally problematic as they were not so much contested as under varying stages of communist domination. A more realistic taxonomy, some argued, would have ranked A and B hamlets as secure, C hamlets as contested, and D,E, and V hamlets as under increasing degrees of communist control.[42] Such criticism was warranted; pacification efforts would have been better served by nomenclature that provided a more accurate and honest picture of on-the-ground developments in a revolutionary civil war fought to varying degrees in thousands of villages and hamlets.

Another significant flaw of the HES was that quantitative data and analysis simply could not evaluate accurately a process as complex as pacification. Gabriel Kolko argues that Revolutionary Development was never a matter of statistical measure. The struggle for the South Vietnamese countryside, he maintains, "was a monumental human, political, and social event whose complex effects required nuanced analysis."[43] The most vital measure of success in South Vietnam, a 1969 congressional report argued, was the degree of political support for the Saigon regime among the rural population: "Any direct measure of the extent of Saigon's political support would, of course, require criteria which are substantially different from those now being used in the Hamlet Evaluation System."[44] Historian Gregory Daddis agrees, noting that the sheer complexity of pacification precluded clear measurement. U.S. policymakers, Daddis argues, "expected HES to quantify elements of South Vietnamese society that simply could not be enumerated on a matrix—control, security, and conversion of the people to the GVN side. Advisors might assess a hamlet as VC- or GVN-controlled, but their evaluations could never identify let alone quantify the causal relationships between allied operations and population control. . . . Nor could the HES measure adequately attitudes, aspirations, and loyalties."[45] Even the most ardent supporters of HES recognized this key deficiency. "There are obviously many limitations on the overall utility of pacification measurement system data," former pacification chief Robert Komer declared in 1970. "Perhaps most significant, they do not provide other than indirect references as to what the population of the countryside really thinks—about the GVN, the VC, security, and the like."[46]

The most damning critique of HES was that its scores simply could not be trusted because they were often inflated by province and district advisers to demonstrate pacification progress. Studies conducted both during and after the war found that a significant number of DSAs inflated HES input data due to command pressure from MACV and Washington. In 1968, for example, a report generated by the House Committee on Foreign Affairs discovered that DSAs frequently produced overly optimistic reports and avoided downgrading HES scores for fear of reprimand.[47] Two U.S. Army colonels reached the same conclusion in 1975. Overall, HES scores were inflated, they concluded. Of the factors linked to inflation, the most pervasive was U.S. command influence. "This influence," they concluded, "could be categorized as a positive or negative incentive depending on whether the DSA perceived his HES evaluations as earning merits or demerits with his superiors."[48] In his recollection of time served as a Marine Corps officer in Vietnam, William R. Corson made the following observation: "What happens in Vietnam is that the U.S. Army advisor who wants to show how well his district is doing understates regression and overstates progress or improvement."[49] Much like the pressure brought to bear on combatant commands to rack up high numbers of enemy killed in action, high expectations for pacification progress among senior officers elicited pressure on DSAs to demonstrate progress in population and area control. "As with body counts in field units, officers serving as district advisors confronted pressures to demonstrate tangible progress."[50]

Civilian policymakers were also complicit in manipulating HES scores to show progress. In 1967, hamlets that were rated A, B, or C were labeled "GVN controlled." D- and E-rated hamlets were classified "contested." In July 1968, the A-B-C category was renamed "relatively secure." In 1969, the "relatively secure" category was again redesignated, this time to simply "ABC."[51] That same year, President Thieu made a bid to modify HES classifications even further. Thieu attempted to alter public perceptions concerning the war by fostering an image of pacification progress that—despite genuine forward momentum—exceeded actual gains. According to Colby, the president directed that A-B-C-D-E hamlets be designated "areas controlled by the government" and V hamlets as "areas not yet fully controlled by the government."[52] In February 1969, Colby informed General Creighton Abrams, who succeeded Westmoreland as head of MACV in July 1968, that "there is a little problem on the statistics thing here, because the president has stated that the GVN will not be in a position of admitting that the VC have control of any part of the country, and there he has divided the total category of the country into two general classes: the one, areas in which the government has control; and the second, areas in which the government doesn't quite yet have complete control."[53] Thieu intended his directive to influence the ongoing peace talks in Paris by making Saigon's hold on the civilian environment (and by extension its negotiating hand) appear much stronger than it actually was. In this manner, Saigon officials mimicked Hanoi and COSVN, who regularly incorporated propagandistic language such as "liberated areas" and "areas temporarily controlled by the enemy" into their directives and proclamations.[54]

Despite its flaws and the fact that HES data was often manipulated to exaggerate progress and was unable to measure nonquantifiable variables such as rural attitudes and the true allegiances and political preferences of the rural population, the program provided a reasonably accurate picture of the degree of control each side exercised over population and territory. "HES is admittedly an imperfect instrument," proclaimed a U.S. State Department memorandum. "It cannot determine absolutes. It cannot evaluate such human elements as popular attitudes, social awareness, and the 'hearts and minds' of the peasants. Still, HES provides useful insights into trends of pacification."[55] In 1974, Thayer repeated this claim, arguing that comparisons between HES data, captured enemy documents, and South Vietnamese voter registration statistics revealed the same movement in all tree data sets: increased population and area control for the RVN between 1966 and 1967. While such evidence was by no means conclusive, he argued, it did suggest that the progress portrayed in HES scores was not as inaccurate as many claimed. "The 'secure' population is probably overstated somewhat, and VC/NVA control was almost certainly greater than the data show, but the fundamental trends appear to be reasonably sound."[56] An in-depth study of HES submitted to the Naval War College Center for Advanced Research in 1975 reached the same conclusion. Despite inflation in reporting, it concluded, HES served as a useful management tool in South Vietnam, one that showed trends—both upward and downward—in levels of security and development. "When coupled with the capability of HES to portray what was going on in the Vietnamese countryside on a monthly basis," the report concluded, "it becomes quite evident that HES was a significantly useful tool in pacification management."[57] Many MACV officers agreed. Postwar questionnaires filled out by more than one hundred general officers who had served in Vietnam reveal that an overwhelming majority (77 percent) gave the HES fairly high marks.[58]

The controversy over HES and the reliability of its figures has continued unabated for decades. Deep skepticism concerning inflated scores, the reliability of HES indicators, and the ability of quantitative analysis to accurately gauge pacification—a complex, human, and often indecipherable sociopolitical process—is well founded. However, critics often overlook several key points. First, there is little evidence that *aggregate* HES figures were inflated so far out of bounds that they created an alternate reality. As one study indicates, HES was a useful tool for both pacification management and analysis because these two areas required relative, not absolute, values. HES provided these relative values.[59] In addition, MACV revised HES several times in an effort to improve accuracy and eliminate subjective, biased reporting. The program's evolution also produced an offspring—the VSSG control indicator. Moreover, hamlets were frequently downgraded. Although many DSAs inflated hamlet ratings and were reluctant to downgrade areas under their command for fear of censure, data show that district and province advisers did in fact downgrade hamlet ratings when the situation warranted. HES figures reveal a great deal of fluctuation when they are tracked over time. In short, hamlet ratings did not always show progress; they showed regression as well. Second, HES proved to be a reasonably reliable technique for measuring the key aspects of pacification that were in fact measurable: population, security, and development. The program and its proponents never claimed it could measure intangibles or grasp in toto the complex social and political processes that comprised pacification. "I think the statistical dope we got from the HES (if one is willing to accept it for what it is and not criticize it for what it isn't)," Komer asserted in 1969, "was a fairly impressive though highly crude way of measuring the physical security and development situation in the countryside."[60] The following year, Komer delivered a similar statement to the American Political Science Association. "Much ill-informed criticism has been levied at HES in particular," he explained, "but most seems to charge HES for what it doesn't even claim to be—a measuring of popular attitudes—rather than analyzing it for what it is. . . . In general, periodic physical status indicators are the chief output, first because these are the easiest to measure, and second because of the indispensable need for simplified, standardized procedures if the whole village/hamlet spectrum were to be covered."[61] Lastly, as previously discussed, HES successfully reported trends in pacification—trends that not only were supported with evidence from other areas, but that showed periods of regression (e.g., *Tet Mau Than* and the 1972 Easter Offensive) as well as progress. It also tracked what pacification officials called "churning," where fluctuations in security and development levels occurred over time.

THE DUAL NATURE OF PACIFICATION–MILITARY AND POLITICAL PROCESS

As previously noted, pacification was both a military and a political process. Nation building and polity construction in a time of revolutionary civil war and insurgency demanded a dual emphasis on eradicating internal threats and developing social, political, and economic institutions. Pacification, according to one contemporary source, "is aimed at transforming hamlets not under government control into secured and developed hamlets where a new life will be developed for the people within a secure environment and with improved standards of living for all. The ultimate goal . . . is to secure and develop sufficient hamlet and village areas throughout South Vietnam so that communist influences no longer can thrive or pose a threat to the central government's efforts to build a viable, free nation."[62] To fulfill this vision, the RVN and the United States implemented a program that consisted of two separate yet interdependent and mutually supporting platforms. The first was military force to eliminate communist military forces and political infrastructure. This included offensive combat operations to expel communist units and sustained combat action by permanent local security forces to protect the population and prevent insurgent reentry. The second was social, economic, and political reform in rural areas. Their purpose was to create a viable countrywide political administration, foster community development, neutralize the effectiveness and appeal of the communist political apparatus, and bind the local populace to the government by engaging their sympathies and loyalties. Together, these methods were meant to systematically diminish communist control and strengthen the government's hold on the countryside.

The RVN and the United States pursued the goal of disintegrating insurgent-controlled areas and expanding control over the operational environment by attempting to progressively break the communists' hold on rural areas and expand government presence and authority into the countryside. The program's overall objective was to gradually reassert sovereignty over the national territory by transforming all areas under communist control into zones firmly controlled by the RVN. "The ultimate objective," MACV declared, "is security up to the borders of South Vietnam. . . . This is a 'win' strategy: The enemy will not be able to exercise political control . . . [and] his intervention forces will not be supported locally."[63] In 1970, Vietnam scholar Gerald Hickey stated: "the government and its allies are trying to crush the NLF military effort and establish government control over the whole country through the pacification efforts. Essentially this means replacing NLF presence where it exists by government presence."[64] Thus, the goal of allied efforts was to extend pacification over the entire nation and reimpose RVN control everywhere within the national borders. According to the 1969 MACV Objectives Plan, the means to achieve this end involved a combined civil-military approach with three primary objectives: (1) the destruction or expulsion of NLF/NVA forces; (2) the establishment of a secure environment to extend, consolidate, and sustain RVN control; and (3) improved national development by means of an integrated host of security, political, economic, social, and psychological programs. The authors of the plan claimed that unlike 1965 to 1968, a period when allied strategy overemphasized the first objective at the expense of the other two, the aims laid out above would be pursued synchronously and with equal weight. This integrated effort became known, in the words of U.S. Ambassador Ellsworth Bunker, as the "one war" concept. According to the theory, the allied militaries would pursue a unified effort rather than distinct spheres of activity. "Currently, US, GVN, and FWMAF forces are operating within a 'one war' concept which does not recognize a separate war of big battalions, war of pacification, or war of territorial security. Under this concept, friendly forces are to carry the battle to the enemy simultaneously, in all areas of conflict."[65]

The binary civil-military nature of pacification prompted the RVN and the United States to synthesize armed force and political action and thus implement an approach identical to that of communist revolutionary warfare—only in reverse. In this manner, pacification mirrored the holistic approach of the Vietnamese communists' *dau tranh* strategy. Much like their adversaries, the Saigon regime and its American ally blended persuasion and coercion—that is, a calibrated carrot-and-stick mixture–to influence actors and shape the politico-military environment.[66] The fundamental purpose of such an approach was to liquidate enemies and generate civilian compliance. Thus, the program was designed around two primary components: armed security operations and Revolutionary Development. The former, which encompassed everything from multidivisional search-and-destroy combat operations against communist main-forces and base areas to local policing against the VCI, was meant to destroy or forcibly eject the communist presence from South Vietnam—thus creating a secure environment where pacification and nation-building activities could take hold. The latter, which involved a host of political and socioeconomic programs initiated at the village and hamlet level, was intended to improve the lives and livelihoods of the peasantry and cement their loyalty to the RVN. It was reasoned that government-sponsored local community-development programs would win the political allegiance of the peasantry and build popular support for the Saigon regime. Together, the "stick" of military force and the "carrot" of economic development and political reform were meant to influence political actors and shape the civilian environment by expanding government control and purging the communist presence.

THE STICK: MILITARY SUPPORT OF PACIFICATION AND THE AREA SECURITY CONCEPT

Saigon and Washington believed that before a national political community could be forged in South Vietnam, the malignancy of insurgency and the communist antistate had to be removed. The military objective of pacification, therefore, was to exorcize the communist presence from the countryside and expand government influence and authority over populated areas through force. "The ultimate aim," read a MACV directive issued in September 1965, "is to pacify the Republic of Vietnam by destroying the VC—his forces, organization, terrorists, agents, and propagandists—while at the same time reestablishing the government apparatus, strengthening RVN military forces, rebuilding the administrative machinery, and re-instituting the services of the government."[67] Four years later, these same objectives were in place. According to the 1969 Combined Campaign Plan, the mission of allied military forces was "to defeat VC/NVA units and assist the government to extend control throughout the Republic of Vietnam."[68] Military support of pacification was meant to eliminate communist military forces and political infrastructure and prevent their entry or reentry into populated areas. Moreover, it was intended to blunt the warmaking capacity of Hanoi and the NLF and collapse their political administrative capabilities. It was reasoned that once local guerrillas and NLF/NVA main-forces were purged from populated areas via combat operations, and the VCI diminished through paramilitary and police action, the primary goals of the program, including social and political reform, community development, and economic activities, could advance. In this capacity, military action and the security it brought provided the foundation for state-building activities. "The role of military forces in support of pacification," claimed MACV, "is to attain a requisite level of security in and around selected hamlet and village areas so that revolutionary development, civil activities, and subsequently, nation-building can proceed."[69] As this passage indicates, sustained territorial security was considered the key to successful pacification and efforts to establish and maintain government authority at the local and national level. Hence, the use of combat operations to expand and consolidate Saigon's authority was a fundamental rather than an incidental feature of pacification.

To fulfill military objectives and advance pacification, allied units conducted sustained, coordinated, and combined offensive combat operations against NLF/NVA forces, base areas, and logistics systems. The goal was to destroy enemy manpower and disrupt safe havens within South Vietnam; detect and destroy enemy incursions into the country along the Laotian and Cambodian borders; maintain security and prevent enemy attack and harassment in the areas surrounding towns, provincial capitals, and cities; and maintain and extend security in the countryside to purge local guerrillas and VCI and facilitate "uninterrupted progress" in political, economic, social, and psychological operations.[70] To fulfill these objectives and neutralize the hybrid threat of communist revolutionary warfare, the RVN and the United States developed a structured security apparatus that applied an array of hierarchically arranged security forces and a politico-military division of labor based on environmental factors.

To combat the flexible and varied approach of the communist system, the allies implemented a similarly complex mixture of military and political forces and techniques. Allied military forces in South Vietnam consisted of two prime entities. The first was the RVNAF, which encompassed conventional units from the ARVN and the territorial forces, known as the Regional Forces (RF) and Popular Forces (PF). The second was the FWMAF, composed of units from the United States, South Korea, and other allied nations. Paramilitary and armed political forces included the National Police Field Force (NPFF), Revolutionary Development (RD) cadres, and People's Self Defense Forces (PSDF). Although each was unique, and differed in size, composition, mission, and area assignment, these forces were all part of an interdependent and integrated security system that covered every geographical region in South Vietnam and extended into every strata of society, from the hamlet up through the village, district, province, and region.

In the remote, sparsely populated areas that surrounded cities, district capitals, and the densely populated villages and hamlets of the coastal lowlands and Mekong Delta, ARVN and FWMAF units conducted large-scale combat operations against NLF/NVA forces and communist base areas and logistics systems. The primary missions of allied main-forces were to conduct offensive operations to locate and annihilate enemy units and safe havens; provide territorial security for selected areas set to undergo pacification; and maintain security once enemy forces had been cleared from an area.[71] In this manner, conventional combat forces paved the way for pacification and nation-building activities. "Without initial military security operations to establish the essential secure environment," MACV declared, "the civil aspects of pacification cannot progress."[72] In 1968 Deputy Secretary of Defense Cyrus Vance underscored this point. Conventional combat operations, he informed Congress, were crucial to pacification because the superior mobility and firepower of allied forces provided a "shield" behind which the program could go forward. Without such a shield, he argued, pacification "could never get started—let alone flourish."[73] William Colby later repeated this claim, informing the Senate Foreign Relations Committee in 1970 that pacification "operates behind the shield furnished by another aspect of our efforts in Vietnam: the military operations of the Vietnamese and allied armies. However bold, however well-conceived, however logical this program, it has been amply proven that it cannot be effective unless hostile regiments and divisions are kept away."[74] General Abrams agreed. "The accelerated pacification program, which we feel is progressing quite favorably," he stated in 1969, "is made possible largely by friendly military initiative which keeps the enemy from concentrating his forces against our pacification program."[75]

The tactics and techniques used by allied main-forces to support pacification were identical to those used in other operations. "Military support of pacification is a unit mission, not a special and identifiable military maneuver," MACV asserted in a 1968 handbook for combatant commands. "In order to accomplish this mission, units will employ standard tactics and techniques adapted as necessary to meet the requirements of terrain, enemy forces, and rules of engagement."[76] Combat operations that supported pacification were conducted in two sequential stages. The first was the clearing phase. Clearing operations were undertaken to eject enemy main-forces and eliminate the communists' ability to corrode government control by threatening or interfering with rural development and nation-building activities. During this phase, military forces conducted company- and battalion-sized operations to "clear" enemy forces from an area through sweeps, saturation patrolling, day and night operations, and search-and-destroy operations. Psychological operations (psyops) and short-term civic action programs were also an integral part of these missions. Theoretically, once a core area had been pacified, large friendly forces would hand off security duties to territorial forces and move outward to prevent NLF/NVA main-forces from reentering the area.

Securing operations constituted the next phase. The purpose of these missions was to "secure" an area that had already been cleared of enemy main-forces by eliminating any remaining vestige of communist presence, developing sufficient security forces and infrastructure to prevent a resurgence of communist influence, and establishing local government to implement social reform and foster community development. Much like clearing operations, securing operations focused on maintaining area security through "aggressive offensive operations." Moreover, they were to continue until the insurgents' capability was reduced to "sporadic" attacks and terrorist activities. During the securing phase, territorial forces replaced the large regular forces as the security situation improved and RVNAF and FWMAF were redeployed to peripheral areas to continue the attack against enemy forces. Securing operations involved mostly company- and platoon-sized operations and encompassed a variety of security functions, including establishing and maintaining security outposts, conducting day and night ambushes, and fulfilling other small-unit security missions. Psyops and civic action programs were also stepped up.[77]

Once NLF/NVA main-forces had been driven from an area through clearing and securing operations, the security burden then passed to Saigon's full-time territorial forces. The RF/PF—or "ruff puffs" as U.S. advisers often called them—conducted sustained security operations at the district and provincial level. Along with local police and part-time paramilitary units, the RF/PF maintained local security by protecting the population and preventing the reentry of communist guerrillas and political cadres. According to Thayer, territorial forces were an extremely important component within the allied war effort. A principal objective of the war, he argued, was to gain control of the rural population so that insurgent forces could not hide among them. "The territorial forces were the troops closest to that population and had the mission of protecting the people from communist attacks and terrorism," he wrote. "The permanent forces residing in the contested areas of South Vietnam were the territorial forces on each side: the communist guerrillas, local forces, and cadres versus the government's territorial forces and cadres. The outcome of the struggle to gain the support of the rural population depended heavily on the effectiveness of these opposing forces."[78] Allied policymakers agreed with Thayer's statement. Despite RF/PF's subordinate status in the overall chain of command and relatively limited areas of operation, these forces played an increasingly significant role in Saigon's efforts to extend government control over the countryside. Thus, their numbers grew from 320,000 in 1965 to 680,000 in 1971. By 1970, RF/PF units constituted nearly half of all RVNAF forces.[79]

The RF were organized into companies and placed under the command of province officials. The primary missions of RF units were to conduct combat operations against enemy local forces at the province level and serve as a mobile reaction force to relieve villages and hamlets that had come under sustained enemy attack. Unlike the Popular Forces, RF were not rooted to a particular village or hamlet and they often participated in joint operations with the ARVN and FWMAF outside populated areas. The number of RF companies varied from province to province depending on its size, population, level of security and presence of enemy forces, and the priority it was assigned by pacification officials. Although their primary mission was to support pacification anywhere within a given province through mobile offensive operations, RF companies were often sent to villages and hamlets to assist the PF in maintaining local security. Other RF duties included protecting lines of communication, political and economic infrastructure, and government installations.

The PF served at the district level and were assigned to specific villages to provide local security and defend the local population against infiltration and attack by local guerrillas. Organized into platoons and placed under the operational control of the village chief, PF were the principal security force at the hamlet and village level. Besides being tasked to provide around-the-clock security for the population and to prevent infiltration by local guerrillas operating within their assigned areas, PF units fulfilled additional security duties such as protecting lines of communication and fixed installations, maintaining outposts, gathering intelligence, establishing small bases for offensive operations, and conducting a variety of antiguerrilla operations, including patrols, ambushes, and sweeps.

Other village-level security forces operated at the low end of the combat spectrum, including PSDF, NPFF, and RD cadres. Also under the command of the village chief, these units conducted "developing" operations, which were initiated once clearing and holding operations had been concluded. Developing operations paired local security missions with hamlet development. Aside from providing security, village-level security forces were responsible for establishing government administration, solidifying civil authority, and building social, political, and economic institutions. In essence, this was pacification. An area was considered "developed" and pacification completed when all of its villages fulfilled eleven criteria assigned to New Life Hamlets (*Ap Doi Moi*).[80] These included identification and destruction of local VCI, removal of corrupt hamlet and village officials, rejuvenation of hamlet morale, organization of civic groups, organization of a village defense system, eradication of illiteracy, organization of a public health program, implementation of land reform, agricultural and handicraft development and improvement, development of a communications network, and meritorious treatment of combatants.[81] Once villages and hamlets had reached a high level of pacification and development as stipulated by New Life Hamlet criteria, RF/PF were replaced by the NPFF, whose main job was to root out remaining VCI and other clandestine operatives. RD cadres, meanwhile, trained and developed local militia and implemented long-range development programs such as education, public health, agriculture, land reform, and road construction.[82]

The PSDF were made up of unpaid citizens who served as local security forces within their own village. Grouped into company-sized units and given rudimentary training in community defense measures, they were local militias charged with providing small-scale defense and protecting village and hamlet officials. The PSDF program was created in in the wake of *Tet Mau Than* and the political fallout that followed the slaughter of thousands of South Vietnamese civilians at the hands of the NLF in the city of Hue. These atrocities spurred demands that the citizenry be allowed to arm for their own defense.[83] According to the 1969 Pacification and Development Plan, the PSDF consisted of self-organized "popular formations" intended to elicit the active participation of the population by mobilizing for local self-defense. Raised in relatively secure hamlets, the PSDF were trained and armed in a manner that would allow them to gradually assume security and public order functions from the territorial forces. PSDF duties included community defense, assisting local armed forces, and promoting community development and self-help activities in their respective villages and hamlets. To the Saigon regime, the PSDF were more than a tool for combating communist insurgency and subversion at the local level; they were an important mechanism for nation building. Officials believed that arming local populations and allowing them to defend their own villages helped build political community and apply what the Pacification and Development Plan called the "Principle of Community Spirit." It read: "To arm the people is to create a reciprocal confidence between the people and the government. This would make the people feel interested in national affairs, thus bringing about conditions favorable to the Pacification and Development Plan."[84] The PSDF were considered a vehicle for strengthening the bond between Saigon and the people of rural South Vietnam. "The weapon is more than a fighting instrument to be used against the enemy," stated the RVN minister of the interior. "It is a commitment by the man who holds it to support the Government—and it is much more than that. To all unarmed hamlet people organized into intelligence, first aid, firefighting and other groups, the armed self-defense force is a symbol of their determination to resist the enemy. To all hamlet people, the armed self-defense force is also a symbol of their Government's determination to provide them the physical and moral support they need to resist the enemy."[85]

The National Police included both uniformed and plainclothes officers. They were responsible, with help from the PSDF, for maintaining public order as security improved and RF/PF were progressively redeployed to less secure areas. Uniformed police performed civil functions such as investigating crimes and accidents, registering and classifying the local populace, and population control. Plainclothes police were responsible for establishing and directing local intelligence networks.[86] The National Police branch most closely involved with pacification was the NPFF, a highly trained cadre of "combat policemen" organized and deployed to eliminate VCI at the district, village, and hamlet level. The NPFF was assigned areas of responsibility after combat operations had been conducted by allied units. Trained in civil police functions as well as paramilitary operations, the NFPP worked in conjunction with other agencies as part of the Phung Hoang or "Phoenix" program—a campaign put in place by the RVN and the United States to destroy the VCI through intelligence gathering and targeting operations. Besides hunting communist agents and sympathizers, the NPFF was responsible for extending National Police activities into rural areas.[87]

Rounding out village-level security forces was the Revolutionary Development cadres. Trained and organized into thirty-man teams, RD cadres operated under the command of the Ministry of Revolutionary Development in Saigon.[88] Their overall objective was to motivate and organize local populations so that villagers could take charge of their own defense and improve their standard of living with help from the government. "The Revolutionary Development Cadre enact their political role by motivating the people to create for themselves a democratic life, and by instilling in the people an unshakable national conviction so that it is possible to insure the inevitable victory for all the people in the political struggle against the communists."[89] RD cadre teams were sent to villages to provide small-scale local defense and to stimulate social, political, and economic development once a certain modicum of security had been established by regular and territorial forces. Once security was restored, RD cadres helped build and organize village and hamlet administration and lay the foundation for local development programs.

RD cadre teams were responsible for upgrading security ratings in D- and E-rated hamlets. They fulfilled several specific functions.[90] First, they provided physical security and enforced population-control measures such as conducting censuses, issuing identity cards, and taking census-grievance—a process of interviewing the peasantry in order to learn their preferences for development projects and to hear their complaints and recommendations concerning village and hamlet administration. Next, RD cadres took steps to improve hamlet management, expand popular participation in the political process, and build popular organizations to increase the population's social and political education and activity. Lastly, they managed self-help community-development projects through a variety of separate functions, including education and literacy training, public health, land reform, agriculture, public works, and civic associations.[91] According to Rural Affairs officer Early Young, RD cadre teams were "the direct counterparts of the Viet Cong propaganda experts and political cadres."[92] As such, training strongly emphasized political motivation. Once recruited, RD cadres received 13 weeks of training and instruction in "political and psychological warfare techniques, local government administration, civic affairs, organization for self-defense, and development of self-help programs."[93] The black pajamas they wore were representative of Saigon's attempt to inculcate nationalist values. In the eyes of one contemporary, Revolutionary Development promoted "the mystique of patriotism and personal dedication to the humble, grass-roots level" of Vietnamese society.[94] Organized along paramilitary lines, the RD cadre was neither soldier nor official. "In theory," CORDS asserted, "he is a trained political organizer who works among the common people to win their support for a cause through his deeds as well as his words."[95] In 1968, it was estimated that there were 53,000 RD cadres working to expand government control. In 1969, as pacification increasingly took hold throughout the countryside and fewer villages were in need of the cadre treatment, these numbers began to drop. By the end of 1972, the number of RD cadres had been reduced to 23,000.[96]

Prior to 1968, the security framework MACV and the JGS laid out in their concept of operations resembled a series of concentric rings emanating outward from a geographical center. The use of different forces, distinct security missions, and separate (albeit often overlapping) areas of operations had, much like the French during the First Indochina War, produced a spatial configuration that resembled an expanding "oil spot" (*tache d'huile*). The area within each ring represented a specific security phase as well as an appropriate type of military force suited to that level of security. At the center was a secure or "pacified" area that included the PSDF, National Police, and a fully functioning local government. Here pacification and nation-building activities flourished with little or no disruption by communist guerrillas and political cadres. Moving outward into areas that were less firmly under government control, security diminished and the type of defense forces changed. PF units operated in the contiguous zone directly outside secure regions. Here they conducted saturation patrols, ambushes, and other antiguerrilla tactics. Within the next contiguous area, a region where government control and security levels were more tenuous, RF units used similar tactics and techniques to target main-force and local guerrilla units. The outermost and least secure ring contained allied main-forces. Here the ARVN and FWMAF conducted mobile strike operations against NLF/NVA main-force units to destroy enemy manpower and blunt the communists' ability to infiltrate or attack the more secure, densely populated interior areas.[97] This arrangement worked well theoretically. Its practical application, however, proved problematic. The security situation in South Vietnam was never so neat or symmetrical. As mentioned earlier, security levels and control varied widely throughout South Vietnam, producing a contested and ever shifting patchwork of secure areas directly adjacent to or intermingled with insecure areas.

In 1969, MACV and the JGS attempted to address this fragmented collage of influence and control by refining the RVN's territorial defense system and developing a new territory-based security model. The initiative came in response to the Nixon administration's concerns that security in rural South Vietnam was inadequate. According to the Office of the Assistant Secretary of Defense, there were a number of reasons for poor security in the countryside, including increased infiltration and incursion by NVA main-force units; allied preoccupation with the military threat; a failure to effectively counter communist political and military activities; and a lack of competence, interest, resolve, and motivation among RVN officials.[98] The

Area Security Concept was a new theoretical framework. It served as the foundation for allied strategy during the years of American withdrawal. The 1970 Combined Campaign Plan and the 1970 Pacification and Development Plan made area security the primary strategy for allied military efforts in South Vietnam.[99] The purpose of the new approach was to upgrade the security situation throughout the country. The overall objective was to consolidate existing security levels and raise all C hamlets to A or B status rather than expanding security to all remaining D, E, and V hamlets. The RVN had extended its control over the vast majority of the population and populated areas by the end of 1969. In January 1970, it was estimated that all but 7 percent of RVN hamlets were in the ABC category.[100] Thus, it seemed logical to work on consolidating existing government control rather than expanding it.

At its heart, the Area Security Concept was an attempt to apply the "one war" strategy and its principle of coordinating combat actions, pacification, and territorial security within an environmentally based concept of operations. The framework called for a method that not only paired the destruction of armed forces with population protection, but that synthesized several distinct yet integrated security tasks into a conceptual framework built on environmental factors. The concept did not introduce new approaches. Instead, it applied familiar methods toward the goal of consolidating and improving population security. To fulfill the Area Security Concept, allied forces were required to accomplish five primary missions: maintain a permanent security apparatus for the population already living in secure areas; extend security to the people living outside of secure areas; neutralize the enemy found among the people—the VCI, guerrillas, and local forces; eliminate enemy main-force units; and create an independent, viable area security system that did not rely on the continued presence of "outside" forces (i.e., American/FWMAF).[101]

Each of these tasks was to occur within a prescribed area. The Area Security Concept divided South Vietnam's rural areas into four zones, each with a relative level of security. The "Secure Area" encompassed populations living in A and B hamlets. The emphasis within these areas was on pacification and on population-control measures. Villages and hamlets within this zone had a functioning local government and were protected by the PF, PSDF, and National Police. Conceptually, a Secure Area was a pacified nucleus that served as a foundation for expanding and consolidating government control. According to MACV, an area was considered part of the "Secure Zone" when the "normal functions" of an effective local government were being conducted and the population had freedom of movement both day and night with the exception of administrative controls put in place by the RVN. There was also no imminent threat of enemy attack and only sporadic episodes of terrorism, indirect fire attacks, and other forms of insurgent violence.[102] The second area was the "Consolidation Zone," which included the population living in C hamlets. The goal here was to provide an outer belt of protection for the Secure Area and to raise the level of security within the zone from a C to an A or B. This was to be accomplished through offensive combat actions aimed at seeking out and destroying local enemy units, VCI, and weapons caches. The RF, PF, and NPFF were the primary security forces assigned to these areas, although ARVN units were called on to help with security threats when necessary. Outside of this area lay the "Clearing Zone" and the "Border Surveillance Zone." These regions tended to be sparsely populated and under varying degrees of communist control—D, E, and V hamlets. Thus, they were broken into areas of operation and put under the control of ARVN division commanders. Military operations within these zones consisted primarily of mobile offensive strike operations to destroy or disperse NLF/NVA main-force units and base areas. The purpose was to isolate communist forces from the population and inhibit their ability to infiltrate or attack populated areas.[103] Conceptually, the boundaries that divided the Area Security zones were to be treated as "phase lines." As explained in a memorandum on territorial security, the objective of expanding lines of advance was to "keep the entire boundary situation dynamic and forward moving" to maintain and expand security until it encompassed the "total population" of the RVN. Retention of Secure and Consolidation Zones was also a priority. "After they are brought under government protection," one American officer asserted, "the Vietnamese people are not, repeat not, to be given up to VC control."[104]

The Area Security Concept was panned by several American combatant commands and CORDS personnel. Some felt four territorial zones did not adequately reflect the volatile security situation that existed in many areas of South Vietnam. Others believed the system thwarted rather than supported the "one war" concept by undermining its unity of forces principle. Assigning different types of units to distinct territories, they argued, separated and compartmentalized territorial security and undermined large-scale operations. Despite such criticisms, however, General Abrams defended the concept and the conceptual changes it brought to the RVN's territorial defense system. Abrams believed the territorial division of labor would strengthen rather than diminish military operations and security ratings by holding each force type accountable for its own tactical area of operations. In response to Abrams's defense of the program, field commanders buried their opposition and incorporated the Area Security Concept into planning and operations.[105] Concerns about the plan, however, were not only well founded, but would be borne out in the years to come.

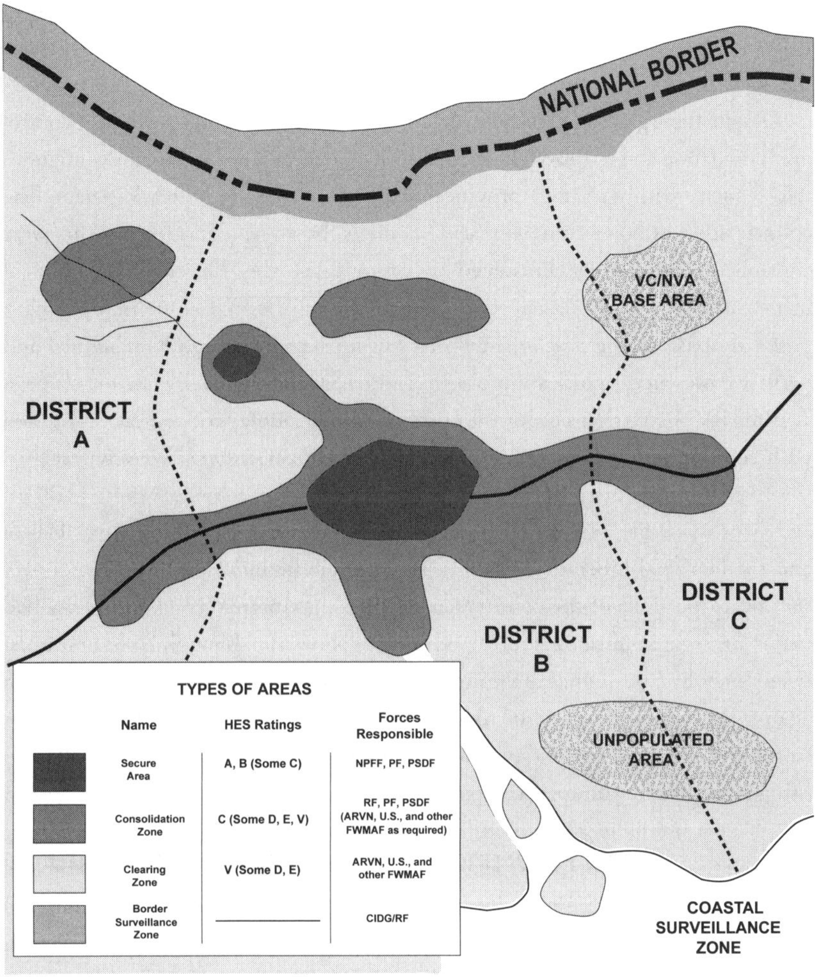

Area Security "One War" Concept.

Map by Erin Greb Cartography. Copyright © 2018 by the University of Oklahoma Press.

On the surface, the Area Security Concept seemed to work well. Data provided by HES reveal significant progress in pacification ratings after it was put into effect. For instance, between 1969 and 1971 the number of population living in A and B hamlets nationwide rose from 12.5 million (71 percent of the population) to 15.5 million (84 percent). The population living in C hamlets, meanwhile, dropped from 3.8 million (21 percent of the population) to 2.3 million (13 percent).[106] Gains in government control over rural areas also occurred at this time. The percentage of rural residents living in areas under RVN control jumped from 48 percent in 1969 to 76 percent in 1971. The percentage of people living in contested rural areas, meanwhile, dropped from 46 percent to 23 percent within the same time frame.[107]

Despite these gains, and in spite of a corresponding decline in communist control over population and territory, the Area Security Concept failed to provide contiguous A-B security within a single province or military region in South Vietnam. The countryside was honeycombed with C hamlets, base areas, and regions that were controlled, contested, or threatened by communist forces. Moreover, the political infrastructure, though damaged, remained intact. The VCI alone may not have posed an existential threat to the RVN, but it caused, through its sustained and pervasive presence, significant problems for the Saigon regime. By late 1971 (a peak year for RVN control), significant portions of the countryside remained infested with communist main-forces, guerrillas, and political cadres. "We have reached the maximum security we're going to have in this country and there's still an awful lot of enemy," an American adviser noted bitterly.[108] The ability of Hanoi and the National Liberation Front to retain sporadic influence and control over significant portions of the countryside despite "maximum" levels of security had major strategic implications in the years that followed. Although the communist insurgency had been greatly weakened from a half-decade-long war of attrition against the combined might of American and South Vietnamese forces, it was far from dead. The communist antistate had survived; and it worked tenaciously to bring down the Saigon regime.

THE CARROT: REVOLUTIONARY DEVELOPMENT IN SOUTH VIETNAM

Combat actions that destroyed communist military forces and political cadres were undoubtedly essential to the pacification process. On their own, however, they were inadequate to the task at hand. "To liberate the countryside from communist control, to destroy the VCI, and to maintain constant vigil against VC infiltration—these efforts are not enough," a 1969 State Department information memorandum declared. "Hand in hand must go vigorous and sustained action to involve the people in creating their own local government and, beyond that, generating economic and social development programs. The name given to this activity is Revolutionary Development (RD)."[109] Pacification was more than just armed struggle to cast out the communists and their legions from the body politic; it was a crusade to bury Vietnam's colonial past, create a national political community, and build a prosperous, modern, self-governing nation that enjoyed the consent and affection of those it governed. Local community development via social, political, and economic reform at the village and hamlet level was the means to achieve this end. Revolutionary Development was often called the "cutting edge" of pacification because it was intended to bring the population of South Vietnam under RVN control by providing them with a new and better life, one that transcended the social injustices, economic inequalities, and repressive political practices of the past. The purpose of RD activity was to develop local government, implement self-sustaining economic and political activity, and assist in the organization and implementation of civic-action and community-development projects. Together, these undertakings were meant to improve the welfare of the rural population by fulfilling their material needs and by nurturing their hopes and dreams of a more just, equitable, and democratic society.

Revolutionary Development had two primary components. The first was local self-government. Vietnam had a long tradition of village autonomy and local democracy. Democratization, decentralized decision making, and communal autonomy were a cornerstone of RD and a key element in every pacification program since Diem had introduced the Strategic Hamlet Program in the early 1960s.[110] "In a 'people's war,' military force is a necessary means," proclaimed the RVN's Central Pacification and Development Council (CPDC), "but the essential factor for a certain victory is the political struggle in which the masses play a key role. We are convinced that we will win over a large number of the population to the nationalist cause through the establishment of . . . a democratic system of government from its infrastructure to its superstructure [and] a strong and efficient administrative apparatus which concentrates its effort on serving he people's true interests."[111] In the fight against communist subversion, democratization was considered important at every strata of society. It was particularly critical, however, at the grassroots level. This was because the civilian environment was more than a place where the people and the landscape merged; it was a test site for social and political revolution and the medium through which both the RVN and the communists struggled to build a viable modern state. The villages and hamlets of South Vietnam were considered the foundation of the nation and essential for enacting nation-building processes. It was vital, therefore, that the government direct its political and administrative efforts toward these areas as it sought to defeat the enemy.

Local elections in secure areas were one aspect of rural self-government. Establishing a functioning governing and administrative body through universal suffrage was among the first steps taken after an area had been secured and cleared of hostile enemy forces. Once a village or hamlet had become pacified and security conditions permitted, local residents elected their hamlet chiefs, village councils, and other administrative bodies. Voting for local officeholders and selecting resident representatives were considered an exercise in self-sufficiency through self-government. Elections were more than a mechanism for building political community, however. They also served to tighten the government's grip on rural areas. Local elections were a means of solidifying government control over local communities and including the rural population in pacification and nation-building processes. Because the objective of the pacification program was to disintegrate communist control in the villages and hamlets where it existed, local elections were another way—besides military operations—to weaken the insurgent presence. Local elections, in short, allowed the RVN to supplant the communists' political and administrative substructure with one of its own. In 1968, the NLF began establishing "elected" administrative bodies in areas under its control. Deemed "liberation committees," the purpose of these organizations was to improve the NLF's bargaining position at the Paris peace talks. In light of this development, the RVN's push for rural democratization and local elections took on added urgency.[112]

Self-government in rural South Vietnam meant more than elections and establishing the rule of law. It also meant developing an effective political and administrative structure in thousands of local communities. To be successful, pacification needed strong leadership at the grass roots—provinces, districts, and villages—as well as at the national level. RVN officials believed pacification and nation building could not progress without an efficient administrative apparatus or popular support. The goal, therefore, was to improve local government and increase effectiveness through administrative training programs, enhanced organization and management techniques, and increased awareness of democratic values and the political process. Pacification plans drafted by the CPDC between 1969 and 1971 contained programs to strengthen village and hamlet government. The goal was to increase competence so that local administrators were more responsive to the people and able to manage the administrative machinery and security forces in areas under their jurisdiction. Village and hamlet officials were responsible for executing the policies of the national government and ensuring the population's needs and aspirations were met. Local officials were considered critical to the development of a national political community because they were the conduit that linked the citizenry to the central government. According to Saigon, elected village officials were more than mere managers; they were intermediaries who built bridges between the people and the Saigon regime. Transmitting the "true aspirations of the people" up to higher authorities was a critical component of local self-government. So too was the need to explain national programs and policies to the rural population and help bring them to fruition.[113] This explains why RVN officials at the village and hamlet level were prime targets in communist terror campaigns and frequent victims of insurgent violence, particularly assassination. Hanoi and the NLF reasoned that if the link connecting the citizen to the state could be severed, then Saigon's ability to expand government influence and control in rural areas would be fatally weakened. Aside from elections and strengthening local administration, other local self-government activities included the development of administrative, economic, and financial fields, and organizing local political groups, including hamlet cooperatives and associations such as farmers' and youth groups.[114]

The second component of RVN-led Revolutionary Development was local self-development. Pacification and nation building were dependent on the active involvement and support of the rural population. Ambassador Colby was correct; there could be no sense of nationhood in South Vietnam without first developing a sense of *local* political community—participation was a prerequisite for patriotism. Self-development, therefore, was critical to laying the foundation for *national* political community. "The exploitation of the natural wealth of our territory and the energy of our people can produce a happy, prosperous, and secure national community," the 1970 Pacification and Development Council declared.[115] Local self-development was meant to assist rural populations in meeting the requirements of rural life. It included a host of programs aimed at improving the lives and livelihoods of the peasantry through increased economic production, improved social services, and community development. Efforts to assist the citizenry in education, public health, and community life also played an important part.

Economic activities tied to local self-development included a land reform program that redistributed arable land to small farmers as well as an agricultural program that promoted crop diversification, modernization of equipment and technique, and self-sufficiency in rice production through the introduction of genetically modified high-yield variants. There was also an animal husbandry program that promoted new and improved techniques and a fishery program aimed at increasing production and improving living conditions for fishermen. Other local economic initiatives were undertaken to stabilize the cost of living, increase personal income, and improve local manpower through occupational development and specialized training. Social programs also focused on community social services, particularly health and education. Health initiatives included building and staffing local hospitals and dispensaries, providing health care to local communities, rehabilitating persons with disabilities, and improving maternal and child care. Preventive-medicine measures, including inoculation drives to eradicate communicable diseases, were also undertaken. Education programs provided free universal elementary education for all village and hamlet residents, improved and expanded opportunities for secondary and adult education, increased the number of secondary education teachers, and improved technical and vocational guidance programs. Other undertakings included programs to assist veterans and war victims. Public assistance to veterans, disabled servicemen, and war widows and their dependents was increased, while refugees and other persons displaced by the war were rehabilitated and resettled. Community development also included a wide range of self-help projects. Partially funded by the government with time and manpower coming from the communities themselves, these projects covered virtually every aspect of rural life. Construction projects built markets, schools, and other local infrastructure. Public works programs repaired roads, improved local water supplies, and increased energy supplies through electrification initiatives and the creation of rural energy cooperatives. Communication and postal programs were also implemented to construct and maintain railways and improve the mail system down to the village level. Irrigation projects such as dams, dikes, dredging, and canal construction were also prevalent in community-development efforts.[116]

Pacification was both a political and a military method for defeating the communist insurgency and projecting Saigon's power and authority into the countryside. It was also a means to forge a national political community built on local security, political freedom, and economic prosperity. Pacification was also a vision. Mirroring its antithesis—communist revolutionary warfare—the pacification program was more than a manifestation of the will to power. Both were crusades to build a brand-new society for the people of South Vietnam.

> Revolutionary war understood as a method of political struggle . . . can be considered a concrete example of nation building in process. Asian revolutionary wars are struggles between conflicting elite groups for control of territory and population. This kind of struggle, characteristic of our time, can turn on the mobilization and organization of rural populations.
>
> —David Wilson, *Nation Building and Revolutionary War*, 1962

> "Change life!" "Change society!" These precepts mean nothing without the production of an appropriate space.
>
> —Henri Lefebvre, *The Production of Space*, 1991

> Organized domination, which calls for continuous administration, requires that human conduct be conditioned to obedience towards those masters who claim to be the bearers of legitimate power.
>
> —Max Weber, *Politics as a Vocation*, 1919

3 COMPETING AND INCOMPATIBLE VISIONS

The Second Indochina War was in many ways a struggle to control space and the environment. The war in the villages and hamlets of rural South Vietnam was part of a political and military contest whose origins lay in the desire to construct a "modern" nation-state. As such, it was a product of what political scientist James Scott calls "high modernism" or "high modernist ideology"—an aspiration held by many nations over the past century and a half to pursue "the administrative ordering of nature and society" by way of "sweeping, rational engineering of all aspects of social life." As Scott indicates, the impulse that drove high modernism (one found in virtually all modern examples of large-scale social engineering) was a desire to reorder society in ways that would improve the human condition. High modernist ideology was initially a product of the centralized bureaucratic state. Over time, it was adopted by leftist revolutionaries. By the early twentieth century, it had become central to the utopian visions of socialist regimes throughout the world. Fueled by millennial expectations and the desire to radically restructure the existing social and political order, national revolutions such as those in China and Russia implemented high modernism and its blend of authoritarianism and centralized state planning.[1] These forces and impulses were also at work in Southeast Asia. The revolutionary civil war in Vietnam, aside from the will to power, was rooted in the desire to sweep away the detriments of the country's colonial past and create a more socially just and equitable society. To do so, the contestants relied on two primary mechanisms. One was the use of violence and coercion to seize political power. The other, which is the focus of this chapter, was social engineering and an administrative ordering of space and the environment. The goal was to remake society and create a happy and prosperous new life for the Vietnamese people.

THE STRUGGLE TO DEFINE SOUTH VIETNAMESE SOCIETY

During the Second Indochina War, the significance of space and the civilian environment transcended mere military matters. To both contestants, the countryside was a laboratory for social and political revolution, both communist and noncommunist. The war in South Vietnam was indeed, as Neil Jamieson had argued, a product of "competing and incompatible visions."[2] Vietnam scholar John McAlister reached a similar conclusion. "The revolutionary struggle in Vietnam," he wrote in 1970, "is a conflict between two modern cultures created by antagonists who still share an old culture in common."

> Both sides in the struggle want the same thing: a new Vietnam that is united, independent of foreign involvement, and modern in its economy and society. The differences between them over how to achieve this new Vietnam are not, however, merely political in the sense of being a disagreement over short-run strategies and priorities. These differences are much more profound. They are cultural, which is to say that they are based on sharply contrasting conceptions of the way a modern society ought to be organized. They clash over opposing ideas about social status and social cohesion, over what a man's place in society should be and what values should link him to other men.[3]

The immense human and material abundance of South Vietnam's rural areas placed them at the center of this struggle. The contested areas of the countryside, the villages and hamlets in which both sides fought for supremacy, were the primary site of spatial conflict. "The true battleground in the 'other war,'" Vietnam scholar Douglas Pike claimed, "is the so-called contested village, with its struggle for power between the communists and the government, between the two contending programs resting on two separate bases."[4] To the contestants, the South Vietnamese countryside was the key to attaining political power and maintaining it in the long run. The control war, as previously discussed, was a struggle for political domination—a contest to amass power by controlling manpower, combat support, supplies, and intelligence in rural areas while denying these assets to the enemy. Despite its material importance, however, the operational environment was more than a means of marshaling resources and a place where population and territory, physical and human geography, merged; it was the medium through which communist revolutionaries and the Saigon regime wrestled for ascendancy, each working to build a unified, independent, and modern state.

In rural areas, the initial task of the NLF and the RVN was to gain control through political and/or military means. Once in power, each side worked to establish "state" administrative functions to actualize its ideological blueprint for society. This meant putting into place a host of social, political, and economic programs and policies that reflected a national and political weltanschauung. Occupied areas were in essence testing grounds for the respective ideology of each contestant. As such, they were fashioned in ways that revealed divergent conceptions of how the communists and the RVN believed society should be organized. For Hanoi and the National Liberation Front, the endgame in Vietnam was a modern state founded on national unification and socialism. For the RVN and its allies, a modern Vietnam meant a prosperous, independent, noncommunist nation resting on democratic free-market principles and aligned with the West. One observer noted that in their attempts to win political legitimacy and a monopoly on power at the local level, both sides modified the social and political structure of South Vietnam "to conform to a new concept of political community—a change in the way who gets what, when, and how."[5] There is a great deal of truth in this statement. Occupation and administration of the civilian environment was more than a military expedient; it was an exercise in political power and a vehicle for putting ideology into practice.

Establishing sovereignty over the physical world was and is an indispensable precursor to the acquisition and consolidation of political power. Mobilizing the masses politically; procuring the necessary manpower and material resources for waging war; justifying the right to rule; demonstrating institutional or organizational viability; and establishing a particular social, political, and economic order all require space. Political movements and ideologies, like all spiritual endeavors, must create and maintain proprietary space if they are to be actualized and transcend the purely theoretical or abstract. According to Henri Lefebvre, any social or political movement that claimed or aspired to be "real" but failed to produce space of its own would "be a strange entity, a very peculiar kind of abstraction unable to escape from the ideological . . . realm. It would fall to the level of folklore and sooner or later disappear altogether."[6] Revolutions that do not produce their own space, he continued, fail because they do not change "life itself." Instead, they merely alter existing "ideological superstructures, institutions, or political apparatuses." In order for political and social movements to be truly revolutionary, Lefebvre argued, they must have a fundamental impact "on daily life, on language, and on space." Without the production of "an appropriate space," revolutionary causes and their social and political precepts for fundamentally transforming society mean nothing.[7]

In South Vietnam, communist revolutionaries and the RVN both needed to create and maintain tangible evidence of their power and authority if they were to gain recognition, win popular support, and advance their respective revolutions. Physical and ideational space, aside from the material and strategic benefits it offered, was indispensable to both sides because it helped sustain "the cause" on a physical and metaphysical level. For Hanoi and the NLF, "liberated" villages and hamlets and other areas proved to the people of South Vietnam, to their enemies, and to allies and adversaries around the world that the communist revolution was a genuine and, more importantly, viable social and political movement. The ability to produce and maintain a communist antistate within the national territory of the RVN concretely demonstrated the strength and vitality of the revolution and, conversely, the weakness and ineffectiveness of the RVN to control its territory and protect its people. For the RVN, the ability to maintain control over the operational environment was directly tied to issues of national sovereignty and political legitimacy. Saigon knew that failure to control the national territory or defend the country against enemy infiltration and attack undermined its credibility as well as its claim to power over the people and territory of South Vietnam.

The ability to establish de facto dominion over the civilian environment was an existential need for both the RVN and the communist insurgency. Claims to power were founded on the ability to control and administer space and the environment. "Political power is a fundamental matter of the revolution," one Vietnamese commentator wrote. "Under any regime political power is by nature a tool . . . to organize and build a new society."[8] Efforts to control territory and fashion the physical and abstract world in ways that would advance social, political, and ideological aims was a central feature of the Second Indochina War and the driving force behind both communist revolutionary warfare and the pacification program. Within the context of civil war and politico-military struggle at the village, hamlet, and district level, the National Liberation Front and the RVN fashioned and developed an appropriate space in accordance with their competing designs for a new Vietnam.

NLF GOVERNANCE AND ADMINISTRATION: THE "LIBERATED AREA"

In 1970, Le Duan penned an essay discussing the VWP's long-term vision for Vietnam. Among its many points, the first secretary argued that the fundamental problem facing the Vietnamese revolution was maintaining political supremacy once it had been achieved. "The successful seizure of power," he wrote,

> does not mean an end to the revolution, but only the beginning. Without toppling the ruling class's power and establishing proletarian dictatorship, the proletariat cannot advance the class struggle toward full victory, that is, toward successfully eliminating not only the exploiting classes, but also the root causes of exploitation; eliminating not only the state of mutual class opposition, but also every class difference; eliminating the causes of poverty and misery; and building the most plentiful and happiest life for the people. In short, the proletariat will advance toward successfully building a completely new society—communist, classless society in which everyone leads a materially plentiful and morally rich life.[9]

Le Duan's vision called for total societal transformation. Defeating the "puppet" regime in Saigon and seizing political power, he reasoned, would not be the revolution's culminating event. Rather, it would merely be the first step in a larger process of constructing a socialist utopia. This mixture of authoritarianism and the administrative ordering of society was an exposition of high modernist ideology. The ultimate objective, Le Duan and others asserted, was not simply to force the Americans out and crush the RVN, but to establish a new social and political order based on social justice and economic equality. "In this society," the People's Revolutionary Party trumpeted as it looked to the future, "class discrimination, differences between the rich and the poor, and exploitation of man by man will no longer exist. Everyone will be civilized and progressive. . . . [E]veryone will live in complete freedom, equality, love, and happiness."[10]

Hanoi and the NLF knew that to seize power and bring the revolution to fruition, they needed to do more than disintegrate Saigon's hold on rural areas through combat actions, terrorism, and other subversive activity. They also needed to implement a program of positive measures; a comprehensive system of governance and administration to control the rural population and win popular support. Thus, revolutionary warfare was fundamentally different from guerrilla war and other forms of armed rebellion.[11] In most cases, the latter pursued political goals exclusively through negative means—that is, violence and ruin. Revolution, however, paired negative measures (armed force and coercion) with positive political content and constructive methods to improve the daily lives of the people living under communist control. "A revolutionary war has a sharply defined, predetermined political goal, simultaneously constructive and destructive," noted one scholar. "Such a war seeks the utter destruction of the existing governmental system, administrative system, and above all, the social and economic structure of the society in which it is conducted, but it is not simply a negative phenomenon. It concurrently provides the citizenry with an alternative."[12] Le Duan agreed. "Socialist revolution," he proclaimed in 1970, "is the most thorough revolution in human society. It is not limited only to destroying the old order, but also is the enterprise of organizing and building a new society in every complicated aspect of life."[13] This, in essence, was the fundamental purpose of the communist insurgent antistate—to develop a socioeconomic structure that would abolish the existing society and build a new one. The goal was to develop an alternate governing/administrative apparatus in rural areas, one that would expand and eventually supplant the government's existing political infrastructure.

Liberated areas were, first and foremost, a means to support the battlefield. Their full significance, however, lay far beyond militarily expediency. "The liberated areas are not only firm guerrilla bases," Vo Nguyen Giap wrote in 1965, "but . . . shining models of a new life, of a new regime."[14] South Vietnam's liberated areas were vital because they strengthened the communists' military position and created territory where "increased construction in the political, economic, and cultural fields" could take place. This construction, many believed, was necessary for fulfilling the revolution's long-term goals.[15] Allied observation and interviews with communist defectors and prisoners of war corroborated communist intentions. "The Viet Cong treated villages under their complete or partial control primarily as sources of manpower, rice, and money with which to carry on the war," one intelligence analyst asserted. "In addition, they saw control of villages as an opportunity to create a new kind of society in South Vietnam."[16] Within liberated areas, the Vietnamese communists envisioned a new society and hoped to put into practice "a genuinely revolutionary regime . . . a regime of freedom, independence, and true democracy."[17]

As a mechanism for implementing high modernist ideology, liberated areas were considered places where significant social change could take place. "We are building socialism," declared an editorial in the VWP newspaper, *Nhan Dan*.[18] Giap supported this view. According to the North Vietnamese general, areas controlled by the communists allowed the people of South Vietnam to "build their lives along progressive lines."[19] Another *Nhan Dan* editorial declared that "in the liberated areas of South Vietnam, colonialist and feudal exploitation has . . . been done away with. Our people, with workers and collective farmers . . . have become the masters of the society and of their own destinies."[20] In the eyes of communist leaders and the rank and file, the practical application of socialism transformed liberated areas into proverbial cities on a hill. One editorial proclaimed, "In the liberated zone . . . the program of the National Liberation Front has been put into effect. There the dreams of every South Vietnamese patriot have come true."[21] Liberated areas were considered idyllic not only because they "built socialism" by putting principles into practice, but because many believed they dissolved, through land redistribution and the overthrow of the landlord class, the "colonial yoke" that had hung like a millstone around the neck of the Vietnamese people for centuries. Moreover, they were seen as a means to create an environment free from the corrosive social and cultural influences of the Saigon regime and its American ally. "The cadres, combatants, and compatriots in the newly liberated areas," declared another VWP editorial. "have also striven to comprehensively build a new life, have wiped out step by step the U.S. puppets' depraved and wicked cultural remains, have opened . . . general education schools and advanced cultural classes and tried every means to popularize the revolutionary and national culture . . . to overcome all social vices left by the enemy."[22]

For many, liberated areas offered a glimpse of what life would be like under communist rule. "The liberated areas," Giap boasted in 1965, "are precisely the first pictures of free and independent South Vietnam in the future; they will certainly be expanded and consolidated and will become the regions in which the future regime of the South will be applied after victory of the liberation war."[23] According to another VWP newspaper article, "the birth of a people's revolutionary power in the liberated regions foreshadows a future national union and democratic power for the whole of South Vietnam."[24] The DRV's state-run media, the Vietnam News Agency, made a similar proclamation in 1973. "A new life is blossoming," it declared, "and confidence in the future and the new regime is seen on everybody's face."[25] Liberated areas were also intended to serve as a beacon to those Vietnamese living under RVN control. It was hoped that "patriotic" elements living in government-controlled areas would be inspired to rise up against their neocolonial oppressors and seize power. "After building a sound and happy new life, the newly liberated areas will set a living example that urge people living in disputed areas and in areas still under the enemy's temporary control to extend their hearts to the revolution, to enthusiastically step up their resolute struggle against the U.S.-puppets and to arise to regain their mastership right."[26]

Specific plans and policies for constructing a socialist utopia and building "a sound and happy new life" for the people were laid out in numerous communist doctrines and manifestos. The NLF's 1960 and 1967 political programs, for example, were excellent examples of high modernist planning and centralized authoritarian social engineering. Both were comprehensive statements that contained specific tasks and objectives for remolding South Vietnamese society along socialist lines; they were formulas for achieving "national salvation." The 1960 program contained a ten-point "policy for action." In 1967 the policy was revised and expanded to thirteen points. Despite some superficial differences and a new tenor as a result of American intervention, the two programs were essentially the same. Together, they offered a comprehensive plan that detailed the NLF's social and political objectives as well as its ostensible plans for the future.[27] According to the 1967 program, "The people of South Vietnam are determined to defeat the U.S. aggressors and their lackeys, and to devote their might and main to build a political system that guarantees the independence and sovereignty of the nation and the freedom and happiness of the people, to heal the wounds of war, to liquidate the social evils left over by the U.S. puppet regime, to restore moral life and build an independent, democratic, peaceful, neutral, and prosperous South Vietnam."[28] The program contained a broad set of social and political policies meant to achieve a democratically elected "progressive" government; an independent and self-supporting economy; improved rural living conditions; land reform and national collectivization; a program for cultural and educational improvement; better public health; guaranteed rights for workers, laborers, and civil servants; social relief for war victims; and social equality for women and ethnic minorities. In tandem with military initiatives to strengthen the People's Liberation Armed Forces, liberate the people from government control, and defend the fatherland, these programs were intended to build an administrative apparatus at the local level, one that would displace Saigon's control, advance the revolution, and support the resistance war against the RVN and the United States.[29]

Within liberated villages and hamlets, the VWP implemented the plans and policies laid out in the 1960 and 1967 political programs. The party provided manpower and material support for the battlefield, facilitated agricultural production and community development, and put into place a system that promoted a new and democratic "people's" regime. "Our responsibility in the liberated area," read one NLF document, "is to gather the people and rule them; to strengthen them politically, militarily, economically, socially, and culturally; to establish order and security; to improve the people' morality and material living conditions; to intensify the struggle against the enemy . . . to mobilize manpower and resources so as to develop the revolutionary forces . . . and lay the foundations for a new political system."[30] The VWP promoted activities to establish order and security, promote the general welfare, and stimulate economic and social justice in the form of collectivism and socialist competition.[31] They implemented land reform, put in place political indoctrination and educational and cultural programs, and established farmers', women's, and youth organizations. The goal was to "enmesh" the peasantry in the NLF's administrative structure and grow the revolution's political base by closely tying the rural population to the communist cause.[32] These activities permitted the revolution to penetrate every aspect of rural life—social, political, economic, and cultural. As one scholar observed in 1968, liberated areas created an "artificial milieu" that allowed for the unfettered pursuit of collectivism and other socialist practices because it was "free of undesired external influences." Within these controlled and highly regulated environments, he argued, elements of the population were being "retrained and reoriented, ahead of time, for life in the totalitarian state that will function nationally upon successful consummation of the insurgency."[33] Ultimately, liberated areas were seen as a place not only where the revolution was protected and its military and political base strengthened, but where the theoretical tenets of Marxism-Leninism were reified.

Despite claims of happiness and a shiny new world, the reality of daily life in liberated areas was quite different. To begin with, there was no direct correlation between areas controlled by the NLF and popular support. Although backing for the Front and its system of progressive programs was robust in many of the villages and hamlets it controlled, such support was muted or nonexistent in others. The appeal of the NLF and its programs tended to be high in areas that had a long history of revolutionary activity or in areas that lacked unifying social or political institutions. Conversely, popular support for the Front and its efforts was often subdued in areas that were dominated by powerful self-identified social groups that had historically opposed the social and political intrusiveness of the communist revolution. Such areas were found in portions of the central highlands, home of the Montagnard people, and in some areas of the Mekong Delta, primarily those that contained large numbers of people from the Cao Dai and Hoa Hao religious sects. According to one report, the success or failure of the Front's appeal in rural areas was often dependent on the presence or absence of "grassroots" organizations at the local level, particularly religious groups. Communist insurgents and their social policies tended to enjoy higher levels of genuine popular support in areas that lacked popular organizations and strong identity groups outside of the family or the village administration. Such was the case in villages and hamlets that were predominantly Buddhist. On the other hand, the Front and its programs were not well received in areas with social organizations that generated strong communal identities, especially those dominated by Catholics, Cao Dai, or Hoa Hao. Unlike Buddhism, these identities served as a prophylaxis against socialist ideology.[34] "The appeal of [communist] revolutionaries," Samuel Huntington asserted in 1968, "depends not on economic deprivation but on political deprivation, that is, on the absence of an effective structure of authority. Where the latter exists, even though it be quite hierarchical and undemocratic, the Viet Cong make little progress."[35] Interviews with communist defectors and prisoners of war revealed a different impetus. They show class differences did in fact shape popular allegiance within liberated areas. The landless and poor tended to favor the NLF and its redistributive programs and policies. Conversely, "rich" landowners and farmers rejected collectivism.[36]

Life was hard for those living in liberated areas, regardless of the level of popular support for the revolution or receptivity to communist social engineering. To begin with, the NLF regulated every aspect of daily life once they gained control of a village or hamlet. They constructed fortifications and established "combat hamlets"; organized local guerrilla forces; drafted local men into PLAF units; arrested or executed government officials and other "collaborators"; strictly monitored movement to and from liberated areas; censored and controlled news media and other information sources; destroyed government papers and identification cards; appropriated agricultural and other labor for the revolution; and pressured individuals to join various Front organizations.[37] According to RAND analyst David Elliott, NLF programs were executed in a "blunt and uncompromising manner" because the population living under communist control were allegedly more "enlightened" than those living in government-controlled areas.[38] Testimony from former NLF members corroborated this claim. According to NLF defectors, the communists demanded absolute obedience. "The people had to put up with us," one declared. "They were told to help the Revolution. They had no choice. . . . [T]hey had to please us in order to avoid trouble."[39] According to another defector: "The people do not like heavy . . . taxes and having their sons drafted. . . . If the Front wants to win it has to stop oppressing the people."[40] In 1972, refugees who fled NLF-controlled areas in Kontum reported harsh treatment from the communist "government" in their village. "Property belonging to people who had fled was confiscated, their houses were destroyed or dismantled, and their oxen and buffalos were slaughtered," read an allied report on the incident. "The people who remained lost a substantial portion of their rice, money, and personal property. The VC distributed some food—one or two pieces of manioc per day and enough rice for one meal every two days—but the refugees reported being hungry most of the time. The VC also forced the villagers to increase their food production and to carry supplies form the villages into the mountains."[41]

People in liberated areas were called on to pay taxes, supply labor, and provide conscripts for the PLAF. Men of military age were drafted into armed service. Those unable to perform military duties—women, children, the elderly, and the infirm—were forced to contribute in other ways. Few individuals and families living in liberated areas escaped paying excises. Tax rates in communist-controlled areas were progressive. Moreover, individuals, families, and groups in poorer areas could provide labor or agricultural production in lieu of money or rice. Nonetheless, revenue collection was immensely burdensome and many peasants attempted to avoid paying. When tax collection became difficult, NLF cadres resorted to threats and intimidation. In response, many peasants fled or reduced economic production. In addition to taxes, people in liberated areas were expected to provide labor. "Everyone worked for the Front," one former resident remarked bitterly. Those who did not fight in local guerrilla or main-force units were required to perform some type of work. Noncombat obligations included providing support for local guerrillas; transporting food, ammunition, and other supplies to fighting units; removing dead and wounded from the battlefield; and other laborious tasks such as constructing fortifications, digging trenches, and building booby traps. Hamlet residents that were too old or weak to perform hard labor were assigned some type of duty. The elderly, women, and children were regularly recruited by the NLF to sharpen spikes and prepare local hamlet defenses.[42]

Another source of hardship for people living in communist-controlled areas was the Front's use of manipulation and coercion to keep local populations in line. Such methods were meant to modify thought and action. According to one report, the NLF sought "total control" over thinking and behavior. Thus, they implemented a program that involved three primary mechanisms: a vigorous propaganda campaign to motivate the people; an elaborate set of population and security control measures to prevent subversion and dissidence; and a wide range of punishments against those who "strayed from the prescribed path."[43] Naturally, NLF heavy-handedness alienated segments of the population. "They said the Front worked for the people, but in reality, the Front only worked for the Party, without ever thinking of the people," one peasant asserted. "They talked of freedom, but they deprived them of all freedom; those who didn't obey them were threatened, maltreated, and murdered."[44]

Political propaganda, including mass media and face-to-face dissemination, was meant to indoctrinate the population and secure support for the revolution.

Measures to stifle rumor, limit exposure to allied propaganda, and mitigate the inflow of outside information were also undertaken. In addition, the Front forbade large gatherings and meetings they did not control to prevent the transmittal of "nonapproved" messages from person to person. More importantly, they intentionally created an atmosphere of fear and mistrust to inhibit the expression of incriminating thoughts and ideas. Information gathered by the RAND Corporation indicates that communist efforts to control interpersonal communication were effective within areas they controlled. "No one dared make known his feelings," noted one villager. "We never trusted each other," stated another. Restricting free communication was advantageous because it blocked outside information and blunted internal dissidence. There was a price to be paid, however. The climate of fear and suspicion prevented the communists from knowing what the people under their control were really thinking or feeling. Although Hanoi and the NLF could (and did) boast of widespread support for the revolution in communist-controlled areas, the reality was that party members—much like RVN officials—never knew what truly lay in the hearts and minds of their constituency.[45]

Population control measures in Front-controlled areas included stringent travel restrictions and close surveillance of the peasantry. The former was meant to limit economic interaction and migration between liberated areas and regions controlled by the RVN. Much like efforts to control the spread of information, travel restrictions relied in part on physical restrictions but more so on terror. As noted by a number of former NLF cadres, peasants under communist domination eschewed traveling to government-controlled areas to visit relatives or conduct business because they were afraid they would be accused of espionage and arrested. Others avoided these activities because they feared family members would be imprisoned or have their property confiscated. Strenuous efforts were made to restrict local populations from fleeing liberated areas even when they were trying to escape allied bombing and shelling. In these instances, the NLF forbade peasants from leaving and urged them to, instead, build bunkers and other defensive fortifications to minimize risk during allied bombardment. However lamentable, the severe travel restrictions placed on peasants residing in Front-controlled areas was understandable. Every person lost to government-controlled areas tipped the balance of forces in the RVN's favor by draining the manpower and resources necessary to sustain revolutionary activities and fight the RVN and the United States.[46]

Surveillance was yet another means of population control in liberated areas. According to reports, villagers under NLF control believed they were being constantly watched and that secret agents were everywhere. Although the level of actual surveillance was likely less widespread and ubiquitous than many believed, there is evidence that the Front used secret intelligence networks and planted agents known as "backbone cadres" in the villages. These operatives remained permanently underground and were solely dedicated to monitoring local populations. According to one former NLF member, backbone cadres "mingled with the people and lived like one of them. Their job was to investigate other cadres. They watched for . . . ideological deviation or any pro-government thoughts. We didn't know who these security agents were or who was under investigation." Another technique for monitoring the population involved calling each villager to the security office and then giving him or her instructions on how to observe and report on neighbors. "The Front has urged families to watch one another," one villager noted. Known as "Each One Watches the Others," the program gathered information from villagers and then generated reports that were sent up the chain of command.[47]

A wide range of punitive sanctions were used to police the population and effectively govern villages and hamlets under communist domination. Communist "repression" was a means to consolidate communist control within liberated areas by regulating or eliminating "undesirable" or wayward citizens. Repression ranged from mild to severe and involved collective action (measures taken by the community) as well as individual action (measures taken by cadre or party members). Mild pressures and punishments included admonishment by party cadres, additional labor assignments, house arrest, restricted movement within the village, and confiscated property. Fairly severe actions included arrest, public criticism, forced enrollment in "reeducation" courses and "thought reform" camps, and compulsory participation in "reduction of prestige" sessions. In these public demonstrations, communist cadres forced the accused to denounce publicly his or her "crimes" against the revolution and beg for forgiveness in front of the entire village. Sentences were harsh, often involving execution.[48] "Death was, of course, the most severe form of punishment," noted RAND analyst W. P. Davison, "and the threat of death the most extreme kind of pressure."[49] A mixture of these techniques more than likely produced the best results. According to Davison, "the combination of apparent leniency and arbitrary severity probably made the Front's task of ensuring both acquiescence and active cooperation much simpler." Anyone living in liberated areas could be accused of crimes against the revolution. Thus, the best way to avoid suspicion was to outwardly lead the type of exemplary life the NLF expected. When RVN officials or allied military forces were in an area, villagers did their best not to do anything that would reveal the identity or location of Front members. In this way, they would avoid suspicion and punishment. As the war ground on, a great many people fled liberated areas for RVN-controlled villages and hamlets; many others remained. Those who did stay, however, did not necessarily do so because they believed in the revolution or supported the insurgency. Many people chose to remain in communist-controlled areas because the restrictions put in place by the NLF were too tight; because many were unwilling to leave their relatives, their property, and the graves of their ancestors; or because they feared the government more than they feared the Front. For those that stayed, the most common defense against possible incrimination and punishment was compliance and passivity.[50]

Villages and hamlets under communist control were also unhappy places because they were subject to fierce allied attack, particularly air and artillery bombardment. Those found in remote areas were also subject to "harassment and interdiction fire," a method of unobserved artillery attack that targeted known communist positions. Quantitative analysis shows that communist-controlled villages and hamlets were frequently exposed to intense allied bombardment during the war, far outstripping areas that were either contested or under government control.[51] Qualitative studies reveal the harsh nature of war on the civilian population living within NLF-controlled areas. "Bombing and artillery attacks only caused losses to the people," one defector told an American interrogator. "The victims were the villagers. We never lived in one place, but we moved around all the time. The villagers had to live in their houses, they didn't move around like we used to, therefore they were the ones to get hit."[52] Another defector claimed villagers living in liberated areas "feared sweep operations conducted by the American troops or ARVN forces. If there were any engagements in the hamlets, the people would suffer casualties and death. So, certainly they were sad rather than happy."[53] Allied bombardment of enemy villages belied the purported cheerful conditions in liberated areas advanced by the NLF. The intensity of the war in the countryside and the prevalent use of artillery shelling and airstrikes, needless to say, made life under Front domination much less "happy" than advertised.

Despite suffering many hardships, a portion of the South Vietnamese population living in Front-controlled areas remained committed to the revolution and its promise of a "brave new world." Although dissatisfaction with the NLF and exodus to RVN-controlled territory increased over time, particularly after the 1968 Tet Offensive when the intensity of the war in the countryside had increased and the labor and tax policies within liberated areas had become increasingly harsh, many peasants remained loyal to the Front and committed to the utopian vision advanced by the revolution. National pride and a firm belief in the possibilities of social justice, economic equality, land redistribution, and upward social mobility continued to exert a powerful hold on the imaginations and yearnings of the faithful.[54] Overall, public attitudes varied in Front-controlled areas. Some people favored the revolution while others did not. The opinions of the great majority, however, ranged somewhere in between these positions. Neutrality and a "wait and see" approach tended to dominate public attitudes, both in liberated areas and in those villages and hamlets that were still contested or under government control. Years of unfulfilled promises and sustained bloodletting had convinced most peasants, regardless of location, to withhold their political preferences and go along with whatever side was in control. "The Vietnamese peasant has always been quick to sense from which quarter the wind is blowing," noted one American journalist, "and they will make their own arrangements with the new rulers."[55] Still, given the insurgency's ability to hang on under the combined RVN-U.S. onslaught, it was obvious that a substantial number of people were actively and voluntarily supporting the Front. This seems to indicate that there were still many who firmly believed in the revolution and its assurances of a bright future. One RAND analyst reflected that while there was no "ideal harmony" between the NLF and the people of South Vietnam, there was enough goodwill to provide (à la Mao) the insurgent "fish" with the "water" he needed to sustain the struggle against the RVN and the United States. "The strains and stresses caused by the protraction of the war," he wrote in 1970, "were . . . caused by measures the [Front] had been forced to take that were, to some extent, against his principles; he had to tax more heavily than he wanted to in accordance with his desire to win the friendship of the people; he had to recruit to a greater extent; and his presence had drawn artillery and air bombardment. Nevertheless, the relationship between the fighting forces and the local population in the past was identified as one of the pillars in the cohesion the forces not only enjoyed, but appeared assured of for future operations."[56]

RVN GOVERNANCE AND ADMINISTRATION: THE "PACIFIED" AREA

The revolution that was espoused by the RVN and the United States was no less radical or transformative than that of its adversary. Pacification, much like communist revolutionary warfare, was meant to create a new and modern society in South Vietnam. In the eyes of allied officials, the war for the countryside transcended the fight against communist invasion, subversion, and terrorism; to them it was "nothing less than a social revolution to transform South Vietnam into a unified, democratic, viable nation."[57] The Revolutionary Development Council argued: "If the peace is to be won, a true social revolution must grow from the rice roots so the people themselves can keep their homes permanently secure . . . direct their own affairs through their own elected leaders . . . eat more and better food and generally live better lives . . . and educate their children as citizens of a developing democracy."[58] President Lyndon B. Johnson expressed similar sentiments. "They fight for dreams beyond the din of battle," he stated, "the dream of security in their village, a teacher for their children, food for their bodies, medicine for their sick, the right to worship in the way they choose. They fight for the essential rights of human existence, and only the callous or timid can ignore their cause."[59] In February 1966, high-ranking officials from the RVN and the United States, including Prime Minister Nguyen Cao Ky and President Johnson, attended the Honolulu Conference. The purpose of the meeting was to reemphasize the allies' commitment to pacification and nation building in South Vietnam at a time of intensified violence, increased American intervention, and heightened skepticism concerning the allied war effort. During and after this historic meeting, the two nations issued numerous declarations outlining their purpose in the current struggle as well as their visions for a new Vietnam. In addition to citing means and ends, these proclamations revealed the utopian aspirations and messianic impulses behind the pacification program.

Not unlike Le Duan's 1970 essay, many of the pronouncements made by Saigon officials vis-à-vis pacification and the Honolulu Conference were models of high modernist ideology. "We are a government—indeed a generation—of revolutionary transformation," an RVN pamphlet exclaimed. "We are determined to the eradication of social injustice among our people. We must bring about a true social revolution and construct a modern society in which every man can know that he has a future; that he has respect and dignity; that he has the opportunity for himself and for his children to live in an environment where . . . opportunities exist for the full expression of his talents and hopes."[60] A 1966 handbook for RD cadres stated: "We are determined to carry out a social revolution in the countryside." Such a revolution, it continued, would abolish "the present miserable life of our compatriots" and create a society that was new and improved. "The present life of our countrymen . . . is one where mourning, hatred, injustice prevails. . . . It is also a life where men live miserably in constant separation, hunger, illness, and ignorance. This life should be abolished! To substitute for it, a new life must be built! The new life is one of security, freedom, democracy and justice, and . . . happiness!"[61] Prime Minister Ky and others claimed that the fundamental purpose of social revolution in South Vietnam was to "build a true democracy for our land and our people" and to imbue in the people "a sense of national unity" and "a stronger commitment to civic responsibility." It was also intended to establish and maintain a stable, viable national economy and build a "better material life" for the people of South Vietnam.[62] Such calls for social revolution and political and economic transformation were central to the pacification program and RVN nation-building efforts in the countryside. "People in rural areas, stand up," the Revolutionary Development Council declared; "destroy the old life to build a new life!"[63] As these statements indicate, the RVN, much like its communist adversary, saw its struggle as a social revolution, one that would construct a new society from the ground up and ensure a more peaceful, prosperous, and equitable life for the Vietnamese people.

The United States was no less committed to sweeping societal change in South Vietnam. At Honolulu, American policymakers pledged to give their "full support to measures of social revolution . . . based upon the principle of building upward from the hopes and purposes of all the people of Vietnam."[64] President Johnson's welcome to the Saigon leaders at the conference reaffirmed America's commitment to the pacification program and the changes it would bring to the countryside. "We are making reality out of the hopes of the common people—hope for a better life," he declared. "We will talk of health and education, agriculture and economics and those other important aspects of a vital future for the people of Vietnam. In all of these endeavors we will give all the support possible to the energetic efforts of our Vietnamese allies."[65] Some were equating the war in Vietnam with modern progressivism, drawing parallels between Johnson's foreign and domestic political programs for eliminating economic deprivation and social injustice with the pacification program and ongoing social revolution in South Vietnam. "Just as the United States is pledged to play its part in the worldwide attack upon hunger, ignorance, and disease," read the Honolulu Declaration, "so in Vietnam it will give special support to the work of the people of that country to build even while they fight. We have helped and we will help them—to stabilize the economy, to increase the production of food, to spread the light of education, to stamp out disease."[66] In April 1966, two months after the Honolulu Conference convened, Vice President Hubert Humphrey informed the press: "I think there is a tremendous new opening here for realizing the dream of the Great Society in the great area of Asia, not just here at home."[67]

Allied officials viewed the war in South Vietnam as two distinct yet interdependent conflicts. The RVN and the United States, much like their adversaries, believed that they needed to do more than destroy the enemy and win military victories if they were to gain political power and maintain it in the long run. They would also need to implement a comprehensive system of governance and administration to win the support of the population. On the one hand, the struggle to control the countryside was a military war waged to defeat the enemy on the battlefield and eject communist political and military forces from the national territory. On the other hand, it was a political war undertaken to build political community and erect a new social and political order. "It is a military war and a war for the hearts of our people," Ky declared. "We cannot win one without winning the other. But the war for the hearts of the people is more than a military tactic. It is a moral principle For this we are trying to bring about a true social revolution, we are instituting a program for a better society."[68] President Johnson echoed this sentiment upon returning to the U.S. mainland after Honolulu. "They and their people understand—and we understand—that the war we are helping them fight must be won on two fronts. One front is military. The other front is the struggle against social injustice; against hunger, disease, and ignorance; against political apathy and indifference."[69] RVN and U.S. leaders knew that both campaigns were necessary if the peace was to be won in South Vietnam. According to an RVN pamphlet on Revolutionary Development, these two crusades—defeating communist aggression and revolutionizing Vietnamese society—shared "the single, large objective of transforming the Republic of Vietnam, changing it from an undeveloped parochial society to a modern nation at peace, a society of free Vietnamese conscious of their common heritage, their own role as nation-builders, and the cooperative nature of their partnership with government."[70]

The mechanism for fulfilling these utopian visions and enacting sweeping societal change was pacification. The process of social, political, and economic reform was the heart of the pacification program and revolutionary development. In order for democratization and economic development to take place, however, local security and government control had to be established in rural areas. The main thrust of the program, therefore, was to institute government control throughout the countryside and then transform what essentially had begun as a military occupation into a political process. The goal was to establish or reestablish local government and displace the communist administrative apparatus in the villages. Thus, the RVN would steadily reassert its authority over population and territory, disintegrate the insurgent antistate within its national borders, and institute sovereignty over the entire nation by transforming "liberated" areas into "pacified" regions under firm government control.

For the RVN and its American ally, the ideational significance of territory under their control was no less profound than for the communists; they too looked at these areas as means to achieve substantial and meaningful social change. South Vietnamese general Tran Dinh Tho believed pacified villages and hamlets were "centers of resistance" against armed insurrection and the corrosive effects of communist ideology. He argued that in addition to expanding national authority and controlling the nation's manpower and resources, government-controlled areas were a vehicle for advancing democracy, improving the welfare of the population, and meeting the aspirations of the peasantry through economic development and social betterment. According to Tho, those Vietnamese who opposed communism did so because they "adhered to a liberal, humanistic way of life sustained by cultural traditions and economic prosperity." Pacification, he believed, provided the RVN the best and most suitable opportunities to "preserve and develop this way of life" for the bulk of the South Vietnamese population, most of whom would have preferred such a system over "dehumanized, coercive collectivism" if given a choice that was free from violence and coercion.[71]

Pacified areas were a means to secure manpower and material resources for the battlefield and to diminish the insurgency's control and/or influence in the countryside. They were also an instrument for implementing high modernist ideology. Within government-controlled areas, pacification's battery of programs—particularly those geared toward developing village and hamlet security forces, instituting democratization and political reform, and spurring "rural development" through civic action and "self-help" community projects—were meant to improve the lives and livelihoods of the peasantry, and build an economically productive democratic society vis-à-vis the active participation of the population in government-sponsored nation-building activities. Ultimately, the purpose was "to transform South Vietnam into an anti-communist outpost of the free world, to heal the social ills inherited from colonialism and feudalism, and to make [the] country prosperous."[72]

After Ngo Dinh Diem's Strategic Hamlet Program had failed in the early 1960s, the primary instrument for the social engineering of South Vietnam's rural areas was the New Life Hamlets program—Ap Doi Moi. Seen as a mechanism for implementing socioeconomic change at the grass roots and building a new society from the ground up, Ap Doi Moi held both physical and ideational significance, as a practical exercise in high modernist ideology. In order to be designated Ap Doi Moi, hamlets had to fulfill eleven criteria within a two-stage process. The first four criteria were the foundation for hamlet development: (1) eliminate the communist political infrastructure, including its sympathizers, cadres, doctrine, and influence; (2) remove the "petty grievances and problems" among the local population that undermined community solidarity and reestablish understanding and cooperation among the people; (3) elect a hamlet government; and (4) organize a "people's" militia or local defense force. Once these had been met, the hamlet was designated a "Community of Responsibility." The next stage of development was achieved when hamlets fulfilled seven additional criteria: (1) improve the educational and cultural basis of the community, which included educating the population and fostering unity through community interaction; (2) implement a health and sanitation program; (3) enact land reform to increase production, land use, and the number of landowners; (4) improve agriculture, animal husbandry, fisheries, and handicrafts; (5) create hamlet cooperatives and associations; (6) institute public works, irrigation, and construction; and (7) recognize the contribution of the people. After these additional conditions had been met, the hamlet was designated a "Community of Prosperity." If and when a hamlet fulfilled all eleven criteria, it was designated Ap Doi Moi.[73]

The influence of high modernism on the pacification program can be seen in the annual plans disseminated by the CPDC. Much like the National Liberation Front's 1960 and 1967 political programs, the programs advanced by the CPDC were comprehensive plans for remolding South Vietnamese society. As such, they were model illustrations of state-sponsored social engineering and high modernist planning. All pacification plans called for shared responsibility and unity of effort between the government and the rural population. Beginning in 1969, RVN plans enshrined the idea of collective effort within the Community Spirit Principle, a doctrine that stressed solidarity and popular participation in pacification and development efforts. "Within the spirit of unity," proclaimed the *1970 Pacification and Development Plan*, "all the people of South Vietnam, together with their government, will succeed brilliantly in the struggle to protect the Fatherland and to restore peace and prosperity."[74] To forge a spirit of national cooperation, the 1969 and 1970 pacification plans contained eight objectives: expand territory and population under RVN control; neutralize the communist political infrastructure; improve local security by organizing, training, and equipping local militias; develop local administration and heighten its capacity for self-management in administrative, economic, and financial fields; increase defections among enemy forces; improve the lives of war victims through social welfare and public assistance; establish an effective information system to educate the people on their rights and duties as citizens and provide them with technical guidance to help improve their economic situations and raise their standard of living; and foster economic growth and development. "From this policy we will begin to enjoy a brighter life as a result of much work and much blood from our soldiers, citizens, cadre[s], and officials from every social class over many years," the 1970 plan proclaimed. "Given the wealth of national resources in this country, and the vast energy of its people, we can create a national community with happiness, prosperity, and security for all."[75]

In 1971, allied policymakers attempted to translate high modernism into an ideology that was authentically Vietnamese. The intent was to situate pacification and the struggle against communism within a social and cultural framework that could be easily understood and accepted by a majority of the population. Known as Tu Tuong—"thoughtful understanding"—its aim was to develop a "simple logic" that would not only explain the significance of the war in Vietnam but would encourage individuals to recognize their responsibilities to the nation and their community and decide how they could contribute their time and energy to the struggle against communism.

> The Vietnamese Tu Tuong is based on love for freedom and independence, and respect for the democratic process. This requires taking action against the communist inspired takeover that seeks to dominate the Vietnamese nation and to corrupt man's innate love of freedom. The people of Vietnam, especially the military, must be motivated by this Tu Tuong to build toward a unified will and a strong determination to achieve the immediate national goal of defeating communist aggression and subversion. Concurrently, Tu Tuong will guide us to the ultimate national goal of building freedom, building democracy, making economic advances, and maintaining peace—in the real sense of insuring justice for all citizens.[76]

According to the architects of Tu Tuong, most South Vietnamese resisted communism because it was a violent and foreign doctrine—"an alien ideology designed to downgrade their morality, eliminate national traditions, and disrupt the social order that for so long has been based upon respect for human dignity, familial affection, love for the native land, and loyalty to the fatherland."[77] Whether or not this claim is true is unknown. What is known, however, is that South Vietnamese society, even if the people were largely opposed to communism, was weakened by factionalism. The rural population remained politically fragmented and divided along religious, racial, regional, and socioeconomic lines. An extremely large number of political "cliques" in the cities and district capitals further undermined unity of action by splintering the population into a myriad of competing interest groups.[78] Tu Tuong was meant to remedy these centrifugal forces and unite the people in a common cause. What the RVN needed, the authors of Tu Tuong believed, was to develop a political philosophy or ideology that would build political consensus and unify the people in the struggle against communism.

On the whole, pacified areas were safer and more prosperous than liberated areas. In many places, the security provided by allied main-forces, Vietnamese territorial forces, and local militia allowed for a relatively safe environment removed from the ravages of large-scale combat and economic hardships experienced in communist-controlled areas. This was particularly true in the years following the 1968 Tet Offensive, when the tempo and intensity of the war decreased and areas designated "secure" and "relatively secure" by the Hamlet Evaluation System had increased to unprecedented levels. In December 1970, the Office of the Assistant Secretary of Defense reported the fighting in South Vietnam had fallen to a level far below previous years and that the RVN had entered a period of "relative peace."[79]

Areas under government control, despite their advantages, were beset by a host of problems that made life difficult for the rural peasantry. To begin with, the security situation, with some minor exceptions, was neither absolute nor permanent. "We have gone about as far as we can go in turning this country into an armed camp," Willard E. Chambers, the assistant deputy for CORDS, informed Ambassador Colby in 1970. "And yet there is no positive security. The VC still retains the capability to strike at times and places of their choosing, and there seems little possibility that this capability can be denied them in the immediate future."[80] Chambers's analysis was correct. NLF guerrillas and political cadres, despite suffering devastating losses in 1968 and 1969, continued to levy taxes, proselytize among the people, assassinate government officials, kidnap civilians, and launch attacks against refugee camps and populated areas. Within pacified areas themselves, the communist shadow government—the Viet Cong Infrastructure—was still capable of harassing and intimidating the population.[81] "The Viet Cong are far from defeated," one scholar stated in 1970. "Though, at present, they are unable to tax, recruit, or propagandize at anything near their pre-Tet rate, their infrastructure is still the most extensive and effective political network in the country."[82] A summary of the 1969 MACV Objectives Plan concurred:

> An assessment of the security situation in South Vietnam today shows that the population of the entire country is threatened by some form of VC/NVA action. There are many areas that enjoy a marked degree of security—but are still subject to indirect fire attacks and terrorism. In some GVN controlled areas, the population has only marginal security. This exists generally in proximity to the province or district

> headquarters, and is maintained only as long as local military superiority is provided. But in many places the GVN gives or provides only minimal "security" or—worse—the area is under VC control.[83]

The verity of these assessments can be seen in the high number of terror incidents throughout South Vietnam during the peak years of pacification, most of which were directed against local populations. After 1968, when the level of government control over the civilian environment had expanded to top levels, communist units began stepping up antipacification activities. Prominent among these were acts of terror in rural areas, including assassinations, bombings, abductions, small-scale assaults, and mortar and rocket attacks. According to MACV, there were an estimated 62,325 civilian victims of communist terror between 1969 and 1971, including 16,048 killed, 32,867 wounded, and 13,410 abducted.[84] Although some deaths were "collateral damage" from NLF actions against territorial forces and village officials, most were the result of deliberate targeting of the civilian population. The intent was to disrupt pacification programs in government-controlled areas, sow fear and create chaos among the population, and undermine the people's confidence in the government's ability to protect them.[85]

Combat operations also disrupted life in villages and hamlets under government control. Villagers often found themselves under RVN administration one day and communist rule the next. The NLF and RVN both made strenuous efforts to retake any area they lost. One of the most poignant examples of these seesaw shifts in control occurred during the 1972 Nguyen Hue Offensive, also known as the Easter Offensive because it commenced three days before the holiday. On March 30, three NVA divisions crossed into South Vietnam, initiating what was then the largest offensive launched by the communists against the RVN. Within three weeks, roughly ten NVA divisions were sparring with a similar number of ARVN divisions on three major fronts. Although the operation technically lasted until October, the NVA advance reached its zenith by May. Nonetheless, the communists had made significant gains in control over population and territory in a few weeks. According to allied analysts, population living in hamlets with HES designations of A and B dropped from 82.7 percent in February 1972 to 70.6 percent in April. The number of communist-controlled hamlets, meanwhile, jumped from 7 to 983 during the same period. By May, the number of hamlets under communist domination had peaked at 1,164. The Nguyen Hue Offensive brought nearly half a million people under communist control. February estimates showed approximately 6,000 South Vietnamese living in communist-controlled areas. By April, this number jumped to 564,000.[86] Saigon's largest loss of territory and population occurred in the northernmost province of Quang Tri. Just prior to the offensive, 76 percent of the population lived in A or B hamlets. By May, this number had plummeted to zero. An entire province had been consumed by the communist blitzkrieg.[87] An RVN counteroffensive to retake these areas and eject communist forces commenced immediately and lasted five months. RVN efforts to reclaim what was lost in the Nguyen Hue Offensive devastated the civilian environment in certain areas. In October, RVN military forces finally recaptured Quang Tri City. After months of bitter fighting and thousands of deaths, civilian as well as military, the RVN flag once again fluttered over the provincial capital, or at least what was left of it. "Quang Tri is no longer a city," the *Wall Street Journal* reported:

> It is a field of rubble, and circling over it in a helicopter recently, one searched the landscape less for any sign of life than for any sign that there had once been life. Quang Tri wasn't so much destroyed as obliterated. One can argue about what percentage of the city was obliterated by American bombers, what percentage by South Vietnamese artillery and what percentage by North Vietnamese cannons. But it seems almost academic. The war has progressed a stage beyond destroying cities to save them; the strategy now seems to be destroying the cities simply so the other side can't have them.[88]

Military offensives and counteroffensives made areas under government control uncomfortable and unhappy places. As we can see, pacified areas—contrary to their name and in sharp contrast to the visions hoped for by allied nation-builders—were often violent dystopian environments.

Brutal and insensitive behavior by allied pacifiers also made life difficult for people living in government-controlled areas. To begin with, ARVN soldiers frequently heaped abuses on inhabitants within their area of operation. Abusive behavior included arbitrary arrest by military police, armed robbery, harassment, assault, petty theft, extortion, and unauthorized/indiscriminate use of weapons, including firearms and hand grenades, inside of villages and hamlets. ARVN soldiers also plundered local markets, refused to pay for goods and services, and stole domestic animals and livestock. One villager reported that the ARVN unit in his area was known as the "chicken division" for its widespread theft of chicken and ducks.[89] American soldiers were often no better; they too indulged in brutal conduct within pacified areas. Many, despite their mission to defend and care for the people of South Vietnam, were hostile and abusive toward those under their charge. Although untold numbers of GIs clearly demonstrated a commitment to help and protect the civilian population, and cared deeply for the plight of the South Vietnamese people, many demonstrated a tangible disrespect if not outright hostility. Sadly, troop misconduct and abuse of civilians were common in pacified areas, such as drunken and disorderly behavior; reckless driving, including running Vietnamese motorists, cyclists, and pedestrians off roads; throwing sticks, cans, and rocks at pedestrians from moving vehicles; physical abuse of local inhabitants; indiscriminate and unjustified firing of weapons in populated areas; killing water buffalo, pigs, and cattle; and petty theft. More serious offenses, such as rape and murder, were also reported.[90]

Overall, American intervention had a debilitating effect on South Vietnamese society. To begin with, daily operations, even those that did not involve combat, were highly destructive. In November 1969, farmers in Binh Duong Province welcomed the news that U.S. troops were withdrawing. Local combat operations, they argued, hurt rather than improved their lives by destroying homes and damaging crops.[91] The largesse and lifestyle of American GIs produced economic inflation and encouraged corruption. It also exacerbated social ills such as prostitution, drug abuse, and street crime. The American presence also had a disruptive effect on local economies and the labor market, drawing rural residents away from the countryside and agricultural production toward the cities, where a host of service industries were established to meet the needs of U.S. military personnel. Together, these influences wreaked havoc on the traditional and conservative social fabric of South Vietnam. In 1969, Vietnam expert Gerald Hickey bitterly noted that the infusion of large numbers of Americans into Southeast Asia had produced "a very disrupted society." Although he admitted that U.S. forces had saved the RVN from military defeat in 1965, Hickey maintained that the presence of an estimated 700,000 foreigners had greatly exacerbated the social disorder already generated by the war. "Given the revolutionary nature of the conflict, this is extremely important to consider," he wrote, "for one of the aims of the revolutionaries is to destroy the existing social order as part of the process of bringing about the social revolution and building a new society. We are aiding them in achieving this objective."[92]

Even without such problems, life in pacified areas was often far removed from the idyllic scenarios depicted by pacification proponents. By 1970, most province studies had shown that the RVN was beginning to implement much-needed social and economic programs, including strengthening local government, initiating land reform, launching community-development projects, and improving agriculture. Many of these, however, were implemented intermittently or undermined by a number of factors such as poorly motivated or corrupt officials, a lack of popular enthusiasm, an inefficient administrative system, and the absence of an inspiring or unifying ideology.[93] "The GVN has made promises of reform programs in the past," MACV noted in 1969, "but it has done little to implement these programs."[94] A hamlet inspection report generated by the Pacification Studies Group for Thua Thien Province was even more damning. It read: "Projects are unfinished and tainted by corruption; government representatives are incapable, ineffective, incompetent, or dishonest; a strong enemy infrastructure is in full bloom; and the initially receptive population has been alienated to the point that voluntary cooperation with the GVN is non-existent. The records which show Duong Mong as a finished Ap Doi Moi are but a mirage."[95]

Government-controlled areas also fell short of the utopian visions of their architects because pacification failed to create a truly national political community. Despite its attempts to garner widespread support, Saigon failed to foster a strong sense of nationalism or patriotism throughout South Vietnam. Increased democratization and social and economic reform had stimulated only superficial identification with the national government at the grass roots. Very little progress was achieved toward building bridges between the villages and higher levels of government. "Village-focused political loyalty does not expand readily to broader dimensions," one expert wrote in 1970. "Rather, it tends to remain a stubborn, conservative force resisting the wider demands and pressures of nation building. As the emperor's writ traditionally did not traverse the village hedge, more modern forms of central authority also encountered great difficulty in dealing with the old spirit of parochialism."[96] Although the RVN under the leadership of President Nguyen Van Thieu had achieved some minor success in building political institutions at the village level, it had failed to achieve the political affiliation it needed at the national level. According to one postwar study, pacification failed, in part, because South Vietnam consisted of "an undigested lump of peoples" that was largely excluded from the nonmilitary struggle against the communists. Part of the reason behind this was a continuation of top-down governance by the Saigon regime, a model very similar to the centuries of "mandarin" rule. In essence, the RVN had become a *state* before it had become a *nation* and as a result, large segments of the population felt "no higher loyalty than allegiance to their village, their priest, their family, or their livelihood."[97]

During the post-Tet period, the people of South Vietnam had increasingly turned away from the National Liberation Front, the promises it had made, and the utopian vision it claimed was just on the horizon. Much of this loss in popular support stemmed from unfulfilled promises, increasingly harsh treatment at the hands of communist cadres, and sustained levels of death and destruction in the countryside. Most of it, however, was the product of war-weariness and a desire for peace and security. Although far from perfect, RVN-controlled areas offered an environment that was much safer and that enjoyed a much higher standard of living than communist-controlled areas. This exodus from liberated to pacified areas, however, did not signal a shift in popular support or political loyalties; it reflected instead a cold pragmatism that stemmed from a practical decision to increase chances for survival and to hedge one's bets until the Mandate of Heaven exposed the legitimate power in Vietnam. A Confucian concept first practiced by the Chinese, the Mandate of Heaven tied the right to rule to political legitimacy in the eyes of the people. It was believed that those who governed wisely and justly earned the mandate, were recognized as the legitimate authority, and won the contest for political control. Those who were corrupt or oppressive lost the mandate and were overthrown.

As noted by an American analyst in 1970, this political world, one driven primarily by self-interest, had created a political community that was in constant flux. "Today's arrangements will change as people strive to better their position," he wrote.

> The man who helped you today may betray you tomorrow. Therefore, gloomy VC prospects at present do not indicate a secure following for the GVN. The VC's loss has not been the GVN's permanent gain. People will follow the GVN now because it is strong and offers a chance to make a living. Should it be unable to maintain both security and a reasonable livelihood for individuals, it will be deserted as was Macbeth by his thanes. In no conversation during these last few days was there mention of a loyalty transcending self-interest or participation in an organization out of belief in some cause or principle. . . . Villagers all readily acknowledged that if the VC came back in force, recalculation of behavior would be in order.[98]

The fact that so many civilians living in South Vietnam's rural areas withheld their commitment to either revolution—communist and noncommunist—reveals the limitations of high modernist ideology as it was applied in South Vietnam. Despite tremendous effort and an immense expenditure in blood and toil, neither side was capable of building widespread popular support or a broad national political community. At most, the RVN and the NLF each held the cooperation (although not necessarily the loyalties or affections) of the rural population under its control.

The hearts and minds of the rural population, however, remained up for grabs. Although communist revolutionary warfare and pacification both succeeded to some degree in the administrative ordering of society, neither could convert the physical transformation of the civilian environment into a crusade that invested the peasantry in their respective revolutions on an emotional or spiritual level.

The significance of politico-military spaces produced by communist revolutionary warfare and government-sponsored pacification transcended mere military purposes and the procurement of manpower and material support for the battlefield. Larger issues, particularly the creation of a new society that would close the book on Vietnam's colonial past and usher in a shining new world of peace, prosperity, and modernity, also played a significant role in efforts to control and shape the civilian environment. Despite such efforts, however, and in spite of lofty dreams for creating a more perfect society, the competing utopias advanced by the NLF and the RVN within areas under their control were often wrecked by the hard hand of war. As noted by an American journalist in 1969, many of the villages and hamlets of rural South Vietnam had repeatedly been "invaded, fought over, reinvaded, taxed, and cajoled" throughout the conflict. This, he observed, produced a reticent and stubborn lack of political commitment among the rural population that largely defied determined efforts to win hearts and minds. Popular support was widely touted as the essential component in fulfilling these revolutions, and yet it was largely missing. "Both sides have promised revolutions," he wrote, "but both revolutions have died in the exhaustion of war. Today, as before, the rice grows and the cock crows, the children multiply and the young men go to the army on one side or another. Food is stored away against famine and allegiance is withheld as a hedge against change." To a recurrent visitor, he concluded, any changes in the social fabric of these rural microcosms would seem like "mere scratches on the surface" of a way of life and thought that—despite decades of determined efforts at armed social engineering—had gone on largely unchanged for two thousand years.[99]

In 1969, an analyst from the U.S. Agency for International Development underscored the reality that a majority of the population remained indifferent to communist and noncommunist assurances of a better world. This, he argued, was the case despite years of strenuous effort to force allegiance and garner support:

> Most Vietnamese are seeking a salvation which is more concrete and realistic than the promise of a foreign and abstract ideology. They are seeking personal security and a degree of predictability in a chaotic world. They believe in results not words. Above all they want peace and

> if possible a better life. Today the majority of the people are uncommitted whether they are under Viet Cong or government control. The individual is under heavy pressure from both sides and his mind and heart are troubled and his life threatened because harmony does not exist. In the final analysis he will truly commit himself only when he is certain which side will win or what the outcome will be.[100]

The question of which social and political vision would triumph in South Vietnam was answered in April 1975, when communist victory and the collapse of the RVN birthed the Socialist Republic of Vietnam. Ironically, however, it was not the design or implementation of rural revolution that resolved the conflict or brought a more egalitarian society. It was, instead, the desolation wrought by war. As historian David Elliott has shown, the widespread devastation of the countryside and mass displacement of the rural population were the mechanisms that destroyed landlordism, ended the old feudal structure of rural society, and opened up new avenues for social, political, and economic advancement. "By clearing away the detritus of the old society," he wrote, "the war laid the groundwork for a more participatory and egalitarian society sometime in the future."[101] In the end, it was the ravages of combat and the corrosive effects of war on the social structure of rural South Vietnam rather than the tenets and methods of utopian revolution that swept away the vestiges of Vietnam's colonial past and ushered in a new social and political order in Southeast Asia.

The revolutionary road for the South is the road of revolutionary violence.
—Resolution of the Twenty-first Conference of the Party Central Committee, 1973

When you have them by the balls, their hearts and minds will follow.
—Charles Colson, special counsel to President Richard Nixon

From the war's intensity [has] come . . . an emphasis by both sides in the struggle on coercion in winning political allegiance. Terror and brutality have mounted as techniques for obtaining popular bases of power as the war has escalated. It is impossible and undoubtedly useless to attempt to assess which side has been the most brutal. Neither side has clean hands. Their conflict has made pawns out of fellow Vietnamese; the score is now kept by the number of their countrymen each side claim to "control."
—John T. McAlister, *The Vietnamese and Their Revolution*, 1970

4 VIOLENT PACIFICATION AND REVOLUTIONARY REPRESSION

Nation building in a time of war and civil conflict presented a fundamental and intractable set of problems for the Vietnamese communists and the RVN. Both sides wanted the same thing: a peaceful, unified, independent, and modern state free from foreign involvement and founded on the political allegiance and willful participation of the people. Yet despite visions of a happy, prosperous, and cohesive society built on popular support and national political community, each side advanced its respective revolution largely through force. The political means of acquiring political legitimacy through economic and social action (often identified as the key to "victory" in Vietnam) were subordinated to a method that sought to expand political control through violence and raw coercive power. Although, at its heart, the conflict was a contest between competing political and social systems, and support of the population was of longstanding political and strategic importance, this was war. As such, positive actions that addressed the long-term hopes and aspirations of the population were obscured by the more immediate objectives of eliminating the enemy's military forces and political administrative apparatus in rural areas and securing physical and/or psychological dominance over the civilian environment. Rather than seeking to win the enthusiastic backing and affection of the populace largely through nonviolent competition and political, social, and economic policy, both sides instead relied on military force and other forms of violence and coercion to expand their influence and control over territory and population. Political scientist Stathis Kalyvas has shown there is a "logic" to the use of violence in civil wars. During such conflicts, he argues, war-related methods, most notably violence, "tend to replace the provision of material and nonmaterial benefits, inducing individuals, for whom survival is important, to collaborate less with the political actor they prefer and more with the political actor they fear; in other words, the provision of benefits loses out gradually to the effective use of violence."[1] This logic was a driving factor in the struggle for South Vietnam.

REVOLUTIONARY VIOLENCE: MILITARY FORCE AND REPRESSION

Violence and force had been a central tenet of socialist revolution since its inception. Vladimir Lenin once stated, "The philosophy of revolutionary violence is the basis of the whole Marx-Engels doctrine. The bourgeois state will never yield to the proletarian state and proletarian dictatorship through self-destruction. . . . [T]hese can be achieved only through a violent revolution."[2] The Vietnamese communists, like all of Lenin's disciples, faithfully adhered to these instructions. "Marxism-Leninism has taught the Vietnamese people a lesson which has been confirmed by their own experiences," North Vietnamese general Hoang Van Thai declared. "To free themselves from the yoke of the imperialists and their agents, they have no other way than to use revolutionary violence to smash the rulers' administrative machinery and build up for themselves a people's dictatorship."[3] General Vo Nguyen Giap agreed. "The experiences gained in the revolutionary struggle put up by . . . our southern compatriots over the past ten years," he wrote in 1965, "have proved that revolutionary violence and revolutionary war is the correct path followed by the peoples who want to rise up and smash the domination of the colonialist imperialists and their lackeys; in definite conditions they are the real possibilities for these peoples to liberate themselves."[4] In 1967, Le Duan informed the VWP that violence was synonymous with political power. "To speak of power is fundamentally to speak of violence," he wrote in loose facsimile of Mao Zedong's dictum that political power grows out of the barrel of a gun. "Power means violence, without which our Party, no matter what it wields, has no authority and strength."[5] The Central Office of South Vietnam, the organ that directed the southern insurgency, concurred.

"The fundamental objective of the Revolution is political power," proclaimed one COSVN directive. "In order to oust the puppet government and seize power for ourselves, we should resort to revolutionary violence."[6]

Violence and coercion were an effective mechanism for extending influence and control over the civilian environment. According to its architects, revolutionary violence consisted of two distinct yet interdependent approaches—one military and the other political. "Revolutionary violence," the VWP stated in 1963, "can be either political struggle or armed struggle, or political struggle combined with armed struggle."[7] Within the context of the control war, revolutionary violence assumed two primary approaches. The first involved combat operations and armed force to seize control of populated regions via military means. The second approach included low-level violence and intimidation to gradually erode the government presence in rural areas. Communist revolutionary warfare, as previously discussed, involved a sophisticated blend of conventional military operations, guerrilla tactics, and other violent actions to increase control over the civilian environment. The means was a blend of military and quasi-military activities.

Military operations were a primary means to damage pacification and grow the insurgent state. The communist main-force war experienced a marked decline in the years following *Tet Mau Than*, as NLF/NVA main-forces were destroyed, dispersed, or forced to withdraw from populated areas due to high casualties and incessant allied pressure. Nonetheless, the VWP continued to use military force when possible, including periodic large-scale offensives, to attack pacification and grow the insurgency. Hanoi and the NLF's reversion back to a protracted war strategy in 1969 did not preclude the use of combat actions to attack population centers and overrun villages and hamlets. In fact, the 1972 and 1975 general offensives along with a host of smaller operations were intended to rapidly seize territory and population and to transform government-controlled and contested regions into liberated areas. Despite a marked reduction in the communist main-force war beginning in 1969, conventional and irregular combat actions endured. Revolutionary forces were ordered to sustain military attacks against the RVNAF/FWMAF when possible and destroy allied military and political forces as a means to defeat pacification, "paralyze" the RVN administrative apparatus in rural areas, and grow the insurgent state. According to COSVN Resolution 9, military offensives to destroy enemy forces and attack the pacification program remained a fundamental part of communist strategy despite an overall reduction in battalion-sized combat operations between 1969 and 1972. The objective, the resolution maintained, was to "annihilate the enemy and expand the liberated areas, thereby broadening our human and material resources while drying up the puppets' replenishment resources."[8]

Small-scale violence and intimidation at the local level—referred to as "repression" in communist parlance—was the second primary means to attack pacification. Military force, or "armed struggle" (*dau tranh*) as it was often called by communist leadership, constituted only half of the communists' hybrid prescription for seizing political power. Political violence in South Vietnam's villages and hamlets was also used to expand the NLF's presence and seize control of the civilian environment. "Violence does not base itself solely on the armed forces and does not consist of a single form, namely armed struggle," Le Duan wrote in 1970. "It necessarily needs the political forces and political struggle."[9] Repression was designed to weaken and delegitimize the RVN, particularly in rural areas. Beginning in the early 1960s, NLF guerrillas and political cadres initiated an imposing campaign of harassment and terror in the countryside along with stepped up attacks against province-, district-, and village-level security forces. According to a RAND Corporation report, communist repression was meant to influence and control the peasantry through the "systematic elimination, intimidation, and subversion of those supporting the defense against communism."[10] Initially directing their attacks toward government officials and local self-defense forces, the communists increasingly targeted ordinary civilians over time as their hold on the civilian environment eroded under the withering impact of main-force warfare and increased emphasis of the U.S./RVN on pacification operations.

The violence program implemented by communist operatives at the local level used a sophisticated blend of individualistic violence (terrorism, assassination, kidnapping, sabotage, and other forms of personal intimidation) and undiscriminating acts (bombings, assaults on villages and hamlets, and indirect fire and mortar and rocket attacks on refugee camps, government installations, and populated areas). Repression was not "a casual circumstance of war," as one pundit put it, but rather a "highly-developed, highly-refined political weapon" used to seize political power in the villages and expand rural revolution.[11] In short, communist repression was a strategy—a method of violence and coercion to eliminate the competition, build an alternate administrative structure at the local level, and establish political suzerainty throughout the countryside. "We must snatch each citizen, each inch of land from the enemy," the People's Revolutionary Party declared.[12]

Repression was used for political rather than military purposes. Although it sometimes took on the appearance of military or guerrilla operations, repression was a tool used to wreck pacification, wrest control of the civilian environment, and seize political power in rural areas. Violence (or the threat of it) was central to communist repression; however, it was only one element among many. Other forms of coercion, including private and public warning, house arrest, and reform camps were also used.[13] Repression and revolutionary violence were applied differently within different contexts. In "rear" areas that were under the predominant or total control of the NLF, the overall level of violence tended to be minimal; warnings, surveillance, incarceration, public denunciation, and indoctrination sessions were most commonly utilized in liberated villages and hamlets. In "front" areas controlled or dominated by the RVN, levels of repression and violence were more acute; assassinations and abductions most frequently occurred in villages where government control was robust and where the NLF's presence was weakest. Finally, in contested areas, where both the NLF and RVN vied for dominance, forms of repression and levels of violence varied depending on the situation and on the relative power of each side.[14]

Repression was also a product of circumstance. Communist revolutionary warfare, as previously discussed, incorporated guerrilla warfare and used economy-of-force tactics to offset the numerical and material superiority of the enemy. "Select the tactic of seeming to come from the east and attacking from the west," Mao once declared, "avoid the solid, attack the hollow; attack; withdraw; deliver a lightening blow. . . . [T]he enemy's . . . vulnerable spots are his vital points and there he must be harassed, attacked, dispersed, exhausted, and annihilated."[15] Unexpected assassinations, abductions, surreptitious bombings, and sudden rocket or mortar attacks fit hand in glove with such an approach. Repression had been a fundamental part of the communist revolution in Vietnam since the Vietminh used it against the French. It assumed particular emphasis, however, after *Tet Mau Than*, a period when the attenuating impact of sustained combat, the devastating setbacks suffered in the failed 1968 and 1969 general offensives, and subsequent expansion and intensification of U.S./RVN efforts in the countryside forced Hanoi and the NLF to return to political struggle and low-level guerrilla war aimed at inflicting casualties and disrupting the pacification program.[16] Pacification, after years of neglect, mismanagement, and ineffectiveness, had become a threat to the NLF by late 1968. The severe attenuation of communist guerrillas and political cadres as a result of *Tet Mau Than* and an intensified pacification campaign posed serious difficulties by eroding the communists' hold on the operational environment. Captured documents reveal COSVN's concern that the accelerated pacification program brought "great danger" to insurgent forces and that efforts to destroy the NLF were steadily increasing.[17] In light of this growing threat, Hanoi and the NLF made antipacification operations a top priority. "Their vicious measures and schemes can create a number of difficulties for us," the PRP declared. "If we do not exert our effort to attack the enemy and break their scheme, we will encounter many difficulties. Therefore, we should rapidly and in detail work out specific plans to attack and frustrate their pacification program. Our present requirement is that all activities are aimed at serving the destruction of the enemy's pacification plan."[18]

By 1969, antipacification operations and a reversion from main-force to protracted warfare (a development that was in consonance with the mutating nature of the General Offensive and General Uprising) had become paramount. COSVN Resolutions 9 and 14 laid out the new approach. "The enemy . . . still controls all cities and heavily populated areas. Particularly, he is concentrating his force on his pacification program to expand his control over the rural areas and the population, in competition with us," Resolution 14 stated. "Therefore, our military operations should rely upon the development of guerrilla warfare so as to be able to protect and consolidate our rear areas to create confusion in the enemy rear area and disintegrate his units therein. This action would bring the masses in rural and urban areas into a high tide of uprising to wrest back mastership, to liberate the rural areas and to basically defeat the enemy's pacification program."[19] To achieve these ends, the NLF relied on small-scale harassment, indirect-fire and sapper attacks against U.S./RVN forces, and terrorism aimed at government officials and select segments of the rural population. As Resolution 14 revealed, the purpose was to wrest control of villages and hamlets away from the government and to consolidate and strengthen the communists' position in rural areas.

Empirical evidence on the ground verified COSVN edicts. "There are some indications the enemy is shifting back to protracted war strategy," MACV noted in December 1969. "Consequently, a significant rising trend in terrorism . . . with continuing low levels of conventional activity may indicate a reversion to guerrilla warfare tactics aimed at controlling the rural population."[20] CORDS monthly province and pacification overview reports reaffirm the trend toward lower levels of conflict. The following passage was typical of such reports: "Throughout the [Military Region], the enemy is continuing to avoid major contact while resupplying and refitting. The enemy is expected to continue to utilize economy of force tactics (terrorist, sapper, and indirect fire attacks) against isolated RF/PF outposts, US airfields, storage areas, and [forward support bases]. Activity also will come in the form of harassment and propaganda operations directed against the GVN pacification program."[21] Although there were instances of battalion-sized attacks and attempts to wrest control of select areas through conventional combat operations, communist efforts were predominantly marked by the techniques of protracted guerrilla warfare.

The purpose of revolutionary violence in the years following *Tet Mau Than* was not to annihilate U.S./RVN military forces (an impossible task) but to speed American withdrawal and, especially, to undermine the pacification program. The goal of antipacification efforts was to expand insurgent influence and control over the operational environment by inhibiting RVN growth in rural areas and corroding the government presence where it already existed.[22] When paired with the use of military force, repression was a primary mechanism for achieving these objectives. "A return to terrorism may be a sign, in classical warfare of weakness," one report declared in 1970. "But at this time and in this place, it makes excellent sense and will be most difficult to thwart. It can also undo much of the recently achieved progress in pacification at a much reduced cost to the enemy."[23] According to American intelligence estimates, violence and coercion were the means the NLF traditionally used to force entry or reentry into populated areas. Provinces with large populations or close to communist base areas experienced high levels of terrorism and other repressive activity. "It is in these areas that the VC feel they can rekindle the fires of latent sympathy for their cause, negate the effect of GVN presence, and make inroads into GVN control," declared a report issued by the Systems Analysis Division of the Department of Defense. The "liberal use of terror," the report continued, was an integral part of the struggle to control the civilian environment. Within rural areas, repression weakened and destroyed local government, strengthened the insurgent apparatus, and eroded RVN control and influence.[24]

Communist guerrillas and political cadres used a judicious combination of selective and general violence to attack pacification and seize control at the local level. Captured documents and COSVN resolutions reveal that communist repression in government-controlled and contested areas was intended to corrode the RVN's hold on rural areas. According to one NLF directive, communist "security" detachments were ordered to "continuously intensify the destruction of tyrants, punishment of puppet administrative personnel, spies, and other enemy oppressive control personnel according to our policy. In addition, we should also destroy whole groups of local puppet administrative personnel and tyrants, to confuse them and disrupt the enemy organization in hamlets and villages."[25] One seasoned analyst concluded, "The primary objective of repression is the disintegration of the GVN. Through the physical elimination, demoralization, and subversion of personnel at all levels of the government's structure, the Viet Cong seeks progressively to erode and eventually to paralyze the GVN's capability to govern and secure those areas now under its control, as well as to blunt any attempts by the GVN to extend its control or influence into the contested or communist-held areas."[26]

Revolutionary violence and coercion served several key functions. First, it physically eliminated key adversaries. According to communist strategists, eliminating "tyrants" and repressing "antirevolutionaries" through incarceration, assassination, or execution after capture were the best ways to dissolve the government presence in rural areas and create conditions that would allow the communists to fill the vacuum and expand their own control. "The village and hamlet guerrillas, secret agents, security agents, penetration agents, and local troops should conduct attacks . . . to kill ringleaders and tyrants, annihilate cruel units, smash the Puppet's oppressive control machinery to gain power for the people so they can rise up to liberate their hamlets," read a directive issued by the Quang Ngai Province Party Committee.[27] Another party cadre was just as blunt: "Elimination of traitors and tyrants to disintegrate the enemy ranks is an important policy of the Party in weak areas. This mission aims at breaking the enemy's control and weakening their prestige and, in addition, raising our revolutionary prestige. It also aims at encouraging the people's movement to break the enemy ruling machine and gain the administration power in villages and hamlets."[28] COSVN Resolution 9, issued in July 1969, urged special operation teams to conduct repression in areas that were either contested or "under enemy oppression." Trained in armed propaganda, assassination and terror techniques, and political organizing, these teams were charged with liquidating government officials, propagandizing and motivating the masses, and initiating struggle against the RVN's local administrative apparatus to win political control at the local level, create and/or expand liberated areas, and transform government villages and hamlets into "fiercely contested" areas. "At present, the enemy is still maintaining his control grip on a number of densely populated rural areas," it read. "It is necessary to . . . transform these areas into areas of struggle against enemy oppression, for the purpose of winning political control, liberating the people from enemy schemes . . . destroying the enemy sources of manpower and material resources and turning these enemy sources into ours, broadening the liberated areas and areas where we have political control, and creating a new position of encirclement and pressure against the enemy."[29]

As part of an effort to disintegrate RVN influence and control, communist repression targeted groups and individuals closely tied to the pacification program, including Saigon officials, district and hamlet chiefs, civil servants, village militia and local paramilitary forces, police personnel, refugees, former NLF guerrillas and political cadres who had defected, and ordinary citizens who either aligned with or were sympathetic to the government. In 1969 MACV observed that individuals "associated with the pacification effort, as well as pro-GVN local officials and informants continue to be the primary targets for VC selective terrorism."[30] Captured Front documents corroborated this claim. "Local troops must concentrate their efforts on eliminating a number of ringleaders in hamlets, villages, the district, and enemy police sub-stations," read one NLF directive. "The killing of . . . pacification personnel and tyrannical Regional and Popular Force elements in the district is of vital importance."[31] Most often referred to as "tyrants," "reactionaries," or "traitors" within communist edicts, individuals involved in the pacification program (along with anyone holding thoughts that were not "in consonance" with party ideology) were considered a threat to the revolution and subsequently singled out and targeted for punishment by NLF guerrillas and political cadres.[32] "We must wipe out all hamlet and village puppet administrative personnel, spies, tyrants and reactionaries who operate openly or secretly against the revolution," a captured document declared.[33]

Statistics show ordinary citizens bore the brunt of insurgent terrorism.[34] They also confirm, however, that significant numbers of people tied to the pacification program were victims of revolutionary violence.[35] Within the context of a war to establish political control at the local level, pacification personnel were not only a symbol of government authority; they were the embodiment of anticommunist revolution and agents for antisocialist nation building. "It is the village/hamlet official, not a faraway district or province chief, who personifies the government of Vietnam to the rural citizen, to whom he may look for assistance and advice," a handbook for American military advisers claimed in 1970. "For this reason," it continued, "the village and hamlet governments were a principal target of the enemy insurgency in its early stages and remain an important target today. By neutralizing the village governments, the enemy was able to cut off nearly all communication between the central government and the rural population. Where he was successful in this effort, he could then harness the rural population to his own war machine."[36] Between 1968 and 1971, more than 3,300 government officials were assassinated. In 1970 alone, 1,376 were killed by revolutionary violence.[37] Virtually everyone connected to the pacification program was a target of communist repression.[38] Village and hamlet chiefs, however, were particularly important because they were the link that bound local villages and hamlets to the national government. "Kill the hamlet chief," read one study, "and you destroy a great part of the control that the government maintains in a virtually isolated area."[39] Territorial forces, local militia, and police personnel were also targets for attack, not only because they inhibited NLF recruitment and tax collection and were charged with protecting the population, but because they personified pacification and efforts to unite the people and the government through nation-building activities.[40]

A second principal function of repression was intimidation. Assassination, kidnapping, warnings, and other forms of coercion were designed not only to discourage individuals from actively supporting the RVN or participating in activities that gave the government legitimacy (elections, community development, and membership in local militia and security forces), but were intended to terrorize the population into submission and prevent the people from resisting NLF demands for taxes and manpower. Violence also deterred civilians from assisting the government and providing intelligence on insurgent members and their activities. This explains why the NLF went to great lengths to punish and repress individuals publicly. Executions and denunciations in full view of families, friends, and neighbors was an effective means to intimidate and modify behavior through fear. Beheading a village chief, stabbing a school teacher to death, or shooting a police officer before an assemblage of villagers left an indelible impression on the local community. "Such savagery, fully documented in one's midst," noted one journalist, "leaves an unmistakable scar of terror."[41] The communists themselves knew the power of repression. "It is necessary to control the villagers," proclaimed one former cadre. "Without terror we could control our own followers but not the whole population. We sought not killing for killing's sake, but to break completely the people's will to resist and to think for themselves."[42] Terrorism and fear were fundamental to communist revolutionary warfare as a means to control the population and seize power. "The elimination of traitors," recalled another former cadre, "will frighten them and make them stop their activities and reduce their sabotage against the Revolution. We will push them to secretly give in to our side."[43] Interviews with communist prisoners and defectors lend credence to the use of repression to intimidate. One former party member claimed acts of repression and terror were used to demonstrate the NLF's power and exact obedience from the peasantry. "As long as the people dread terrorist actions," he told his interrogator in 1967, "the Viet Cong can tell them what to do."[44] Some scholars believe "political-coercive" techniques such as terrorism, kidnapping, and harassment allowed the insurgency to maintain psychological influence in rural areas. Repression allowed the NLF to exercise "veto control" and keep people in line by highlighting the dangers of supporting the government. This veto control, some argue, successfully undermined RVN efforts to build political community in many areas.[45]

A third function of repression was to destroy the RVN's credibility and undermine public confidence by demonstrating the government's inability to protect the population from insurgent violence. "The enemy's terrorist activities," stated one report, "play a large role in his attempt to discredit the ability of the government to provide security for the people."[46] In many instances, the purpose of NLF activities was simply to dissolve the people's faith in the RVN. "The recent rocket attack on Phu Cuong city," one senior American adviser reported, "is an indication by the enemy to prove that no area is really 100 percent secure." Terrorist attacks of this sort, he continued, were not meant to achieve military objectives but to present a psychological threat that would "play on the fears of the people."[47] The ability of the NLF to strike fear into the population was concerning to allied officials. NLF terrorism was not only effective; it was a threat that had to be addressed. "The message the communists want to get across," one RVN official declared, "is, 'don't cooperate with the government—it won't help you and it can't protect you.' Their strategy is very, very effective. We have to . . . win the confidence of the people again."[48] Within the context of agrarian civil war, terror possessed the ability to disorient and psychologically isolate the individual. It also made the government appear powerless either to stop revolutionary violence or to provide security to the masses. A single assassination in a hamlet or village, one analyst argued, "not only frightens villagers but destroys part of the structure which previously was a guarantee of security for the people." Terror, he continued, "removes the underpinnings of the orderly system, creates and sows confusion as much as fear. A civilian expects safety and order in his society and when these are destroyed he becomes disoriented, for he can no longer draw strength from social support. Instead he becomes fragmented completely within while desperately trying to fix his own personal security."[49]

Violence aimed at choice individuals and applied throughout the countryside was a highly effective force multiplier that undermined government authority by inciting fear and making the power and reach of the NLF appear omnipresent. "A villager risks death when he openly cooperates with the GVN because in most hamlets there are Viet Cong agents or sympathizers—popularly called infrastructure—who will betray him," declared Bing West, a marine officer who spent considerable time during the war developing local Vietnamese security forces. "The fear of death hangs over the hamlets and lingers on, long after the VC have gone."[50] Even when RVN forces were "nominally present," West argued, the NLF infrastructure could control the people using the same techniques the Mafia used to rule Sicily. "By its invisible and ubiquitous presence," he wrote, "the Big Brother of the VC infrastructure influences the public conduct of the people."[51] Stuart Herrington, a former U.S. intelligence adviser, also testified to the NLF's ostensible omnipresence. "The Viet Cong weren't literally everywhere as their propaganda would have liked people to believe," he wrote, "but their organization was sufficiently developed so that one could never be certain whether or not there was a communist agent in a given group. This organizational achievement enabled the Viet Cong to accomplish many feats that would have been otherwise impossible—the most significant of which was . . . control of the country's rural population by a relatively small elite."[52] Combined with propaganda, subversion, and political activities, assassination, kidnapping, and harassment were meant to sow confusion, distrust, discontentment, and disenchantment in the government while reinforcing the political influence of the NLF among the population. In many cases, the strategy worked. "What I thought we had accomplished—security, confidence in the government, a better life—seems to be lost," lamented a former village chief whose village had been recently devastated by NLF attack. "I was mistaken before about our success. I will not make that mistake again. I have seen what a dozen terrorists can do in a month or two without any effort." He added, "My people have also seen what the terrorists can do."[53]

Closely related to repression were communist efforts to fill depleted guerrilla units through impressment. Village and hamlet youths were routinely forced to serve as hamlet, village, and district guerrillas. Many were kidnapped and placed in local and main-force units. "The cadres conducted propaganda in my village," noted one NLF defector. "They said that if I did not volunteer for the Front forces I would be drafted."[54] Some were brought into the communist fold with threats or mild coercion. Others, however, were kidnapped or arrested and taken to NLF-controlled areas for indoctrination. Although persuasion was the Front's first choice in securing recruits, violence or the threat of violence was used when appeals to patriotism, community, and other forms of positive persuasion didn't work.[55] As the war wound on and circumstances became more desperate, the NLF increasingly resorted to impressment on a large scale to keep units up to strength. One report estimated that forced abduction and coercion (threats against family members to compel an individual to join a guerrilla unit were common) accounted for nearly three-fourths of all NLF recruitments during the first six months of 1969.[56] Statistics compiled by MACV and the RVN show that nearly 30,000 civilians had been abducted in South Vietnam between 1968 and 1972. Although records indicate that some of these individuals were only temporarily detained to attend indoctrination programs, testimony from defectors and prisoners of war indicate that many were pressed into military service or forced to perform manual labor for guerrilla units.[57] Most communist directives deliberately avoided the term "conscription" for propaganda purposes. The objective was to place NLF activities in a positive light. Still, orders to "step up recruitment" littered the directives and edicts issued to forward-deployed units.

Another form of revolutionary violence intended to damage pacification was forced relocation. NLF/NVA units regularly targeted refugee camps and other settlements to herd people into communist-controlled areas. Closely linked to terrorizing the populace and demonstrating the government's inability to protect its citizenry, mortar and rocket attacks on temporary resettlement camps and permanent villages and hamlets were an integral part of Hanoi and the NLF's protracted war strategy after 1968. Indirect-fire attacks on refugee camps were intended to demonstrate the government's impotency and the NLF's ubiquity, potency, and ability to disrupt pacification. Most importantly, they were designed to force displaced persons out of RVN resettlement areas into areas that were under the total or predominant control of revolutionary forces. "The VC terrorist attack on a Danang refugee camp has reinforced enemy attempts to discredit GVN security claims," read one report. "Enemy attempts to resettle the homeless population in VC 'liberated' areas have been reinforced."[58] Although they appeared indiscriminate in nature, insurgent attacks on refugee camps were a methodical and integral part of the communists' strategic effort to disrupt the pacification program.[59] According to one captured document, the NLF considered the destruction of refugee camps and RVN-controlled areas urgent. "In the face of the enemy and friendly situation in resettlement areas and with our present advantages and disadvantages," it read, "we should . . . destroy resettlement areas and return the people to their native villages."[60] A 1969 study by MACV expressed the belief that the NLF were aware of a decrease in popular support among the rural population, and that pacification had to be "thoroughly disrupted and destroyed" if this trend was to be reversed. "The VC plan," it stated, "is to force the people back to VC-controlled areas and out of the GVN resettlement areas." Aside from "promises of land, security, and better government," the report continued, the best way for the NLF to achieve this result was through "selected terrorism" aimed at the population.[61]

Terrorism and revolutionary violence shed a great deal of blood in South Vietnam during these years. MACV estimated nearly 113,000 people were killed, wounded, or abducted by the NLF between January 1968 and April 1972. Of these, 24,114 were killed, including 1,646 village and hamlet officials and 18,360 ordinary citizens. Over the same period, nearly 55,000 people were wounded. Here too the preponderance of victims came from the general populace rather than from the ranks of government officials or civil security forces.[62] In 1969 alone, officials recorded 12,491 victims of NLF terrorism, including 6,202 killed and 6,289 abducted.[63] An additional 15,000 were wounded. According to testimony before a U.S. Senate committee, of the more than 6,000 people killed by insurgent terror in 1969, more than 1,200 occurred in "selective assassinations." Also, among the dead were "some 90 village chiefs and officials, 240 hamlet chiefs and officials, 229 refugees, and 4,350 of the general populace."[64] Other reports reveal similar trends and provide anecdotal evidence. For example, one study that tracked NLF terrorist incidents in early 1972 revealed 77 hamlet officials and police personnel had been killed between January and April along with 240 local militia members and more than 1,000 noncombatants. In January that year, the attempted assassination of a province chief at a youth rally resulted in 17 civilians killed and 100 injured. In March, insurgent forces entered the home of a hamlet chief in Long An Province, assassinating him and his entire family. That same month, an NLF mine killed two schoolteachers and two civilians, and wounded twenty-two children on their way to school. During a ten-day period in April, insurgent forces attacked a number of Return-to-Village and resettlement (refugee) sites. In Quang Nam Province, one site was nearly 50 percent destroyed, creating 6,000 refugees. Meanwhile in Quang Ngai Province, eleven resettlement sites were completely destroyed and another sixteen seriously damaged. According to the report, these attacks created approximately 25,000 refugees.[65] Such examples are but a small sampling of the consequences of the Front's widespread use of terrorism.

The use of violence, force, and coercion yielded some positive results for the NLF in the control war and the politico-military struggle for rural South Vietnam. Judging from information provided in a number of sources, repression appears to have been an effective and resourceful method for attacking pacification. Terrorism and harassment tied up government resources, kept government officials and local security forces on the defensive, and agitated an already difficult and highly problematic refugee situation. Most importantly, revolutionary violence influenced the attitudes and behaviors of the rural population and prevented numerous individuals from cooperating with or supporting the RVN. "The people are for the most part afraid to cooperate with the government," stated one report submitted by a senior district adviser. "The VC," he continued, "partially control Kanh Loc by fear."[66] According to some analysts, harassment and terrorism allowed the NLF to leverage power and extend their influence throughout the countryside to a degree that was vastly disproportionate to their actual numbers.[67] "The willingness of the Viet Cong to resort to terror tactics permits them to 'influence' the population to a far greater degree than their numbers or political power base would otherwise permit," stated a 1969 White House national security study.[68] The Department of Defense Systems Analysis Division offered the following analysis: "The unwillingness of GVN officials to travel in the countryside or spend the night in villages indicates the VC campaign is successful despite the low individual risk. Besides eliminating the best (and worst) government officials, the VC/NVA use assassinations and abduction as a psychological tool to influence the behavior of other employees and the civilian population. Selective terror and harassment can be very effective in demoralizing the general populace and destroying faith in the GVN."[69] MACV claimed communist terrorism was highly effective in that it hurt "both the effectiveness of the government's pacification campaign and the prestige of the government."[70] Interviews with NLF defectors and South Vietnamese civilians conducted in 1970 and 1971 confirm this, illustrating that threats, abductions, assassinations, and other forms of repression instilled fear in the rural population and undermined the RVN's credibility in protecting the population from revolutionary violence.[71]

Former ARVN officers Nguyen Duy Hinh and Tran Dinh Tho claimed that communist repression had a detrimental impact on pacification because it successfully intimidated village and hamlet security personnel and pushed large numbers of people who may have otherwise supported the government into political passivity. Hinh and Tho stated that although local militia and territorial forces despised the communists for their campaigns of terror and intimidation, many swallowed their hatred and maintained a "fence-sitting" stance to survive. "The situation was such that, over a period of time, many village cadres and, according to intelligence reports, even some RF and PF outposts, assumed a neutral position and merely gave a perfunctory performance."[72] The CPDC repeated this claim in 1970: "In some areas the people are reluctant to associate with the GVN for fear of retaliation by the enemy. Civil officials often become the target of enemy terrorism and assassination and thus are reluctant to carry on the business of government. Some police hesitate to conduct operations against the VCI because they fear retaliation, and local security forces, under the threat of terrorism, often accommodate rather than resist the enemy."[73] Although it is impossible to precisely gauge the impact of insurgent violence on pacification, reports generated between 1968 and 1972 as well as postwar regional and provincial studies showed that terrorism and harassment sowed fear among segments of the rural population and undermined the peasantry's confidence in the RVN by demonstrating the government's inability to protect its citizenry or to defeat what appeared to be a persistent, indefatigable, and ubiquitous insurgent presence. "The people fear the VC . . . particularly the local armed guerrilla units.

These units continue to disregard the law, terrorize, and create insecurity among the people," noted one RVN cadre.[74]

Despite its benefits—eliminating enemies, disrupting allied efforts, and inspiring the faithful—revolutionary violence also had drawbacks: repression produced a backlash among those segments of the peasantry subject to its ravages. This adverse reaction undermined NLF efforts to win support and expand its power and influence throughout the countryside. Beginning in 1968, the proliferation of terrorism and harassment (combined with increased taxation, forced conscription, and labor impressment) disenchanted a significant portion of the rural population and prompted a rejection of the NLF and its cause in many areas. Perhaps the most egregious example of revolutionary repression to elicit negative fallout occurred in February 1968. During the preliminary attacks of *Tet Mau Than*, communist guerrillas and political cadres executed thousands of Vietnamese civilians in the city of Hue.[75] Some quickly recognized the folly of such a massacre. "Many people here don't particularly like the Saigon government . . . but since Tet 1968 all of them fear and hate the VC."[76] According to an August 1968 study, peasants in one Mekong Delta village were beginning to express "anti-VC sentiment" in response to the high taxes and labor demands made by local insurgents. "These harsh and coercive tactics," it read, "are not winning any hearts and minds for the VC. For the first time villagers who had traditionally opposed the presence of GVN and Allied forces as a nuisance are now openly expressing their hopes for increased friendly troop strength to provide a shield against unwarranted VC incursion."[77] As David Elliott points out, many South Vietnamese peasants who might otherwise have supported the revolution "on patriotic grounds" turned away from the insurgency in the years after *Tet Mau Than* in response to communist brutality.[78] As the populace became increasingly disillusioned with the revolution and its repressive modus operandi, growing numbers of civilians rejected the NLF. Subsequently, many began helping the government root out and eliminate communist guerrillas and political cadres—this despite a lack of enthusiasm or love for the RVN. "The tide of popular favor appears to have turned against the little men in black pajamas who organized and led the rebellion," wrote one veteran war correspondent. "The Viet Cong's claim to be the 'authentic representative' of the South Vietnamese people is a mockery of the truth to most of these people. The people still fear the Viet Cong from the destruction and cruelty they are capable of wreaking, but fewer support them than formerly, even under coercion."[79]

In 1970, American analysts concluded that terrorism was "generally counterproductive" when applied in areas that were traditionally unsympathetic to the NLF. Rather than exact compliance, repression instead stiffened resistance to the insurgency in these areas.[80] This observation was corroborated by events on the ground. According to one report submitted by a senior deputy for CORDS, the surge in revolutionary violence in his military region was "self-defeating" because it embittered the populace and prompted growing numbers of rural residents to provide intelligence on insurgent activities. "Already there is considerable evidence that a backlash effect is being generated among the people," he wrote. "Incidents are being reported more frequently and promptly; taxes and materials are not being contributed so frequently to the enemy; more and more people, by both word and deed, are committing themselves to the support of GVN; and fewer people are by choice remaining in enemy controlled areas."[81] Although care must be taken to not overstate the frequency of such events, it can be safely assumed that such occurrences were not rare, and that revolutionary violence against civilians garnered a negative backlash against the NLF.

Repression was an integral part of the political struggle for South Vietnam and efforts to control the operational environment. In the eyes of communist leaders and the rank and file, localized violence was not only a legitimate but a necessary means to advance the revolution and expand the Front's hold on the countryside. To the NLF and its masters in Hanoi, repression was a cost-effective means of winning and maintaining political suzerainty at the local level at a time of diminishing power and resources. When applied gratuitously, however, the Front's extensive use of violence, force, and coercion yielded mixed results. On the one hand, repression advanced revolutionary aims by disrupting pacification programs, cowing the rural population, discrediting the Saigon government in the eyes of its constituency, eroding RVN control, and weakening the government's administrative structure in the countryside. On the other hand, revolutionary violence was counterproductive; its sheer brutality often worked at cross-purposes to the goal of growing popular support and strengthening the NLF's political and administrative apparatus in rural areas. Although assassinations, kidnappings, and indiscriminate mortar and rocket attacks on populated areas damaged pacification and allied nation-building efforts, they also hurt the insurgency by alienating the rural population and repulsing a political constituency the Front desperately needed to attract. In fact, many of the difficulties the communists faced after *Tet Mau Than* stemmed from increasing disillusionment with the NLF. Blowback and unintended consequences from repression, however, were not the most significant factor in the decline of NLF fortunes during the post-Tet period. As will be discussed next, the violence program enacted by the RVN and the United States played a larger part in the communists' loss of control over rural areas.

Violence and coercion were an effective mechanism for extending communist influence and control over the civilian environment. The two primary methods of revolutionary violence—military force and repression—did not receive equal weight, however, after 1968. The asymmetrical nature of the war and shift in communist operations toward lower-level conflict forced Hanoi and the NLF to adopt an approach that largely avoided large-scale combat operations and favored repression and guerrilla activity at the local level. Although Hanoi and the NLF periodically resurrected the General Offensive and General Uprising and used main-force combat to overrun select areas and grow the insurgent state, these instances where eclipsed by an approach that was dominated by repression and small-scale violence and coercion at the local level—a method that allowed the insurgency to conserve manpower, rebuild its shattered forces, and maintain pressure on the RVN and its allies as they battled to expand control over the countryside and prepared for future offensives against the Saigon government.

VIOLENT PACIFICATION

Over the years, a number of historians, participants, and analysts of American intervention in Vietnam have argued that 1968–1972 constituted a "better war," where the conduct, character, and trajectory of the conflict took a positive and constructive turn for the RVN and its American ally, thanks to new leadership and a new strategy that broke sharply with the counterproductive policies of the past. The revival and reemphasis of pacification and the civil/political side of the war in the wake of *Tet Mau Than*, these observers believe, along with a purported shift in strategy from William Westmoreland's war of "big battalions" to Creighton Abrams's "one war" concept signaled a critical turn of events. It was here, they contend, that the United States and the RVN rejected a strategy of attrition and brute force for a more nuanced military policy, one that supplanted body count and the profligate use of firepower with population security, pacification, and a deft combination of military, civil, and economic policy. This argument is largely inaccurate.

Under Abrams, military support for pacification was just as reliant on heavy firepower and main-force operations as it was under Westmoreland—there was no fundamental change in strategy, as some revisionist historians have argued, once Abrams replaced Westmoreland as head of MACV.[82] The purported shift from "attrition" and the profligate use of force toward a nuanced, less heavy-handed approach simply did not happen. Abrams inherited a highly favorable strategic position inside of South Vietnam. The 1968 and 1969 general offensives devastated the ranks of the southern insurgency and left a political and military vacuum in the countryside. Together, these developments provided Abrams with an unprecedented opportunity to undertake more pacification operations and focus on population security at the local level. Moreover, the generally improved situation in South Vietnam was followed by a sharp decline in communist main-force activity. This allowed MACV to dramatically cut the number of large-scale operations (battalion-size and larger) beginning in 1970.[83] Otherwise, the methods used under Abrams to advance pacification were no different than those used during Westmoreland's tenure. Like his predecessor, Abrams utilized sweeps, clearing operations, and search-and-destroy missions to annihilate or eject communist forces as a means to pacify the countryside and expand RVN control over territory and population. Rather than initiate fundamental change at the operational and strategic levels, Abrams instead retained the status quo. The "counterforce mission," as historian Andrew Birtle puts it, remained a "central feature" of allied strategy after Tet.[84]

Pacification, despite its name, was no less violent or coercive than communist efforts to extend politico-military control over the civilian environment. Contrary to euphemistic shibboleths such as "winning hearts and minds" and "building democracy," the program was not a passive popularity contest. Military support of pacification relied as much on the "stick" of combat operations as it did on the "carrot" of land reform, civic action, community development, and national and local elections. Contrary to what some believe, pacification was never a gentler, kinder form of warfare. Clear-and-hold operations conducted in the villages and hamlets of South Vietnam under the rubric of pacification were just as deadly and destructive as search-and-destroy missions in the main-force war. Moreover, they had an equally devastating, if not greater, impact on the civilian population. The frequent use of overwhelming firepower in and around populated areas, particularly airstrikes and artillery fire, caused high numbers of civilian casualties[85] and generated thousands of refugees,[86] as peasants fled their homes for the safety of government-controlled areas.

Much like revolutionary violence and repression, the use of armed force and coercion to advance pacification and expand politico-military control was the product of circumstance and design. To begin with, the communist insurgency remained an entrenched, dangerous, and disruptive presence that could only be eliminated through force. Although the NLF suffered tremendous losses in 1968, and its power and influence atrophied in the years following *Tet Mau Than*, main-force units and local guerrillas—along with increasing numbers of NVA fillers from the DRV to supplement depleted guerrilla ranks—continued to operate in varying degrees throughout South Vietnam's forty-four provinces. According to a 1969 Special National Intelligence Estimate, anywhere from 50,000 to 65,000 communist combat forces continued to operate in South Vietnam and adjacent areas. These were supplemented by 40,000 to 50,000 administration service personnel and anywhere from 50,000 to 100,000 active guerrillas.[87] Thus, the communists retained "a substantial capability to sustain military operations. The Viet Cong infrastructure, which plays a vital role in supporting the war effort, continued to function effectively despite some attrition and reduced access to the population and resources of the South."[88] Faced with a tenacious and substantive armed threat embedded within the social fabric of rural South Vietnam, allied military forces had no choice but to root out and destroy insurgent elements from contested and communist-controlled villages and hamlets using the hard hand of war.

The use of conventional combat operations to advance pacification and extend government control over the operational environment was rooted in existing military doctrine and practice.[89] The tactics and techniques used to support the pacification program were identical to those used in other operations; they included an array of offensive "strike" missions supplemented by air and artillery fire, including sweeps, ambushes, raids, saturation patrolling, and both search-and-destroy and cordon-and-search operations. Contemporary military doctrine wedded military force and violence to pacification through the concept of area security. "Clearing operations" were meant to cleanse an area of hostile elements so that civil, economic, and social measures could be implemented. In the eyes of allied military planners, area security was "key" to allied success in rural areas and an essential first step in pacifying the countryside through civic-action and community-development projects, as well as other programs meant to build political community and foster cooperation between Saigon and the rural population. Pacification and nation building could only take root and flourish within a secure environment. Military action and "aggressive offensive operations" that destroyed or drove off enemy forces created that environment. "Pacification is not possible unless there is security," a U.S. 9th Infantry Division report asserted. "Strike operations by US and RVNAF . . . provide this necessary ingredient."[90]

The laws of war governing U.S. forces in the field solidified the use of traditional combat tactics and techniques during pacification operations. Military commanders repeatedly dictated to subordinates that civilian casualties and collateral damage were to be avoided when possible. Moreover, rules of engagement (ROE) placed stringent restrictions on indiscriminate force in populated areas and required that allied troops issue warnings to civilians when an attack on an area was forthcoming. Nonetheless, MACV allowed these restrictions, along with limitations on artillery and airstrikes, to be overridden when dictated by military necessity. For example, Directive 525-18 stipulated that fire missions directed against known or suspected enemy targets in populated areas could be waived and that villages and hamlets could be attacked without warning if deemed necessary or relevant to the success of an operation.[91] Combatant commands that supported pacification agreed. On the one hand, commanders openly recognized the wisdom of limiting artillery and airstrikes in populated areas. On the other hand, they argued that overwhelming firepower was not only necessary but permissible under certain circumstances. According to one American officer, U.S. and RVN forces had no choice at times but to "lower the boom" and use massive force against guerrilla forces *wherever* they were located. Battalion commanders charged with supporting pacification, he argued, were authorized to use heavy firepower against guerrillas in populated areas when small arms were ineffective and the situation met ROE requirements.[92] Studies conducted by MACV show that even in situations where the ROE forbade the use of indiscriminate fire, American and RVN units frequently violated artillery and air bombardment restrictions in populated areas either because they did not understand the rules or because they chose to disregard established protocol when retaliating against NLF attacks. "Recent tactical operations," declared one American officer, "have resulted in death and injuries to a considerable number of civilians. The frequency and number of such incidents clearly dictates that responsible personnel are either unaware of existing MACV and RVNAF rules of engagement or fail to adhere to these principles."[93] Instances where Vietnamese civilians were killed or wounded as a result of ROE violations, it continued, were not only detrimental to the pacification effort; they were counterproductive to U.S. goals in Southeast Asia.

The use of conventional combat operations to advance pacification was also sustained by a persistent and deep-seated belief in the utility of force. "The payoff," General Abrams declared shortly after taking command of MACV, "is getting a hold of this fellow [NLF/NVA] and killing as many of them as you can." Lieutenant General Fred C. Weyand, commander of II Field Force, agreed. "I couldn't believe more in the truth of what you said about killing people," he responded. "This is right now what's got to be done—people got to be killed. It isn't so much his supplies and his food, it's just this son of a bitch running around with an AK-47. We've got to find him and kill him—or capture him."[94] When queried about the continued usefulness of search-and-destroy operations by a fellow officer in October 1968, General Andrew Goodpaster responded that the purpose of such missions was "to win the war in South Vietnam, and you can do it by killing troops and you can do it by going after him out in the bush."[95] Despite a strategic realignment toward Vietnamization and reemphasis on pacification and the civil/political side of the war after *Tet Mau Than*, American and South Vietnamese commanders remained reluctant to abandon their reliance on firepower and mobility. According to Vietnam scholar Guenter Lewy, allied commanders "had a penchant for calling in artillery and air support at the slightest sign of enemy resistance, using air strikes in particular as a substitute for rather than in support of infantry forces, and much of the time they seemed quite oblivious of the destruction and suffering these weapons inflicted on the civilian population."[96] Many general officers expressed the long-held conviction that annihilating the enemy was the raison d'être of military forces. Although a growing number of officers recognized the importance of civic action and political and economic reform, and many freely acknowledged the primacy of civil considerations and the need to subordinate military to political action, commanders continued to place a premium on force to limit allied casualties and eliminate the enemy. In 1969, Robert Thompson lamented this fact, noting that the "Americans seem to have been influenced only by the very worst interpretation of Clausewitz's doctrines and by the Prussian example of Moltke, so that the sole aim of most American orthodox military commanders has always been 'the destruction of the enemy's main-forces on the battlefield.'"[97]

During the 1968–72 period, no American officer personified Thompson's point better than Lieutenant General J. Julian Ewell, commander of the 9th Infantry Division. Pacification, Ewell argued, was meaningless unless the enemy was eliminated. "I had two rules," he said. "One is that you would try to get a very close meshing of pacification . . . and military operations. The other rule is the military operations would be given first priority in every case. That doesn't mean you wouldn't do pacification, but this gets at what you might call winning the hearts and minds of the people. I'm all for that. It's a nice concept, but in fighting the Viet Cong and the NVA, if you don't break their military machine you might as well forget winning the hearts and minds of the people."[98] In his final report as commander of the 9th Division, Ewell stated that the "hearts and minds approach" was "overdone" in Vietnam. "Some units do a tremendous job with pacification support," he wrote, "but put so much effort into it that their military operations suffer unduly. I prefer the opposite approach. Put maximum military pressure on the enemy. This helps pacification more than anything else." The only way to overcome guerrilla control and terror in the countryside, he argued, "is by brute force applied against the VC."[99]

Although it would be unfair to portray all general officers in the same light as Ewell—an infamously aggressive and hard-driving commander who earned the sobriquet "Butcher of the Delta" during his tenure with the 9th—similar attitudes toward lavish firepower and the utility of force were representative of most American officers who commanded troops in Vietnam. In fact, postwar surveys revealed that a large majority of those who served in Southeast Asia believed the levels of close air support and artillery support used to assist ground forces were appropriate. Only a substantial minority believed the amount of firepower used in combat was excessive given the irregular nature of the war.[100]

Efforts to rapidly expand pacification and RVN control of the civilian environment through "brute force" were revived and intensified in the spring of 1968 following the first phase of the 1968 General Offensive and General Uprising. Less than a month after *Tet Mau Than* had begun, General Earle G. Wheeler, the chairman of the Joint Chiefs of Staff, urged RVN and U.S. forces to "restore security in the heavily populated areas of the countryside" and to "regain the initiative through offensive operations."[101] Two months later, Ambassador William Colby argued 1968 was a critical year for pacification and that the RVN should rapidly expand its control in the countryside to undermine communist claims to territory and population.[102] On 1 November, U.S. and RVN military forces launched the Accelerated Pacification Campaign, a three-month nationwide blitz meant to rapidly expand government control in the countryside and improve the security ratings of nearly 1,000 hamlets that were either contested or controlled by the NLF.[103] The campaign was followed over the next several years by a number of pacification programs that applied the same military principles.[104] The APC and subsequent campaigns meshed pacification and military operations as part of the "one war" strategy. Moreover, they assigned U.S./RVN military forces a prominent role in assisting pacification.

The use of overwhelming firepower and large-scale operations to advance pacification was best illustrated during Operation Speedy Express. Launched in November 1968 as part of the APC, the mission focused on pacifying four key provinces of the Mekong Delta. Military support was assigned to elements of the U.S. 9th Infantry Division and three ARVN divisions.[105] According to the 9th Division's after-action report, the purpose of Speedy Express was to eliminate guerrilla forces and increase government control over more than four hundred designated hamlets that were either "contested" or fully occupied by the NLF.[106] Although the aim was to support pacification efforts in portions of the Delta, its "fundamental purpose" was "the destruction of the enemy" in Military Region 4.[107] To accomplish this mission, U.S. and ARVN forces conducted a six-month series of intensive offensive operations, lasting from December 1968 through May 1969. Speedy Express employed a variety of standard combat tactics and techniques designed to harass and destroy NLF forces, including patrols, assaults, sweeps, ambushes, sniper team activities, and search-and-destroy operations. Ground operations were lavishly reinforced by airstrikes, artillery, and naval riverine operations.[108] Over the course of the operation, artillery support fired hundreds of thousands of rounds in direct support of combat units and to interdict enemy infiltration routes.[109] Meanwhile, tactical air support flew anywhere from 95 to 120 combat sorties a day, incorporating helicopter gunships, fixed-wing aircraft, B-52 strikes, and a variety of air cavalry and/or airmobile missions sent to find, fix, and destroy NLF guerrillas and secure areas for pacification.[110]

Judged by the primary objective of expanding government control over rural areas and the civilian environment, Speedy Express achieved notable results. To begin with, hamlet security ratings in the four provinces within the 9th Division's tactical area of operations showed remarkable improvement. According to HES figures, the number of hamlets under communist control in Dinh Tuong, Go Cong, Kien Hoa, and Long An Provinces dropped from 1,061 in January 1968 to 398 in July 1969. The numbers of people living in "liberated" areas also dropped significantly, from approximately 662,000 to 239,000 within the same time frame.[111] After-action reports of the 9th Division show combat results were equally if not more imposing. As of 3 June 1969, the division reported 10,899 enemy killed and another 2,579 captured. U.S. and RVNAF forces, meanwhile, suffered only 242 killed over the course of the seven-month engagement; an impressive kill ratio of 45 to 1.[112]

Despite inflicting a spectacularly high number of casualties during Speedy Express, U.S./RVN forces recovered only 748 enemy weapons. The discrepancy in body count and captured weapons, along with the acknowledged aggressiveness of 9th Division leadership, raised concern that many of the "enemy" forces killed during the operation were in fact civilians rather than NLF guerrillas. "Units of the 9th were intensely active, possessed awesome firepower, and operated over a large area of operations," stated one report on the division's activities in MR 4. "It would be naïve to assume that all activity would be precisely and exclusively directed at enemy forces. According to . . . advisors in the area, U.S. commanders may have placed undue emphasis on body count, and there may have been too great a tendency to view anyone near the VC controlled areas as hostile."[113] Postcombat analysis of the 9th Division's activities in Kien Hoa and Dinh Tuong Provinces during 1968–69 reached similar conclusions. "It was felt by advisory personnel," stated an October 1969 report generated by the Pacification Studies Group, "that the high body counts achieved by the 9th were not composed exclusively of active VC. The normal ratio of three or four enemy KIA for every weapon captured was raised at times to fifty to one; this leads to the suspicion that many VC supporters, willing or unwilling, and innocent bystanders were also eliminated."[114] Although it would be difficult if not impossible to determine the exact number of noncombatant deaths inflicted by allied actions during Speedy Express, one American official estimated that of the roughly 11,000 people killed by allied forces during the operation, at least 5,000 were civilians. "Noncombatants," he explained, "were always mixed with the Viet Cong." According to a villager from Kien Hoa Province, U.S. forces actually killed more civilians than NLF guerrillas during the operation.[115]

Speedy Express was hardly unique in its use of conventional combat operations. Other pacification operations conducted between 1968 and 1972 were just as reliant on intensive infantry operations, artillery bombardments, and airstrikes to eliminate guerrilla forces, establish territorial security, and "pacify" rural areas—particularly villages and hamlets that were contested or under the predominant or total control of communist insurgents.[116] "In order to expel the VC rapidly with minimum casualties to friendly forces," read a 1968 report on pacification efforts in Thua Thien Province, "it was necessary in many instances to inflict heavy damages on the occupied hamlets."[117] Monthly pacification reports generated by CORDS, including province and military region field overviews, further indicate that combat operations in support of pacification severely damaged villages and hamlets, inflicted civilian casualties, and generated thousands of refugees in the process. For instance, when an estimated 200–300 NVA troops occupied three pacified hamlets in Long An Province "for the stated intent of disrupting the pacification program and destroying the district town" of Thu Thua in May 1969, subsequent efforts to dislodge them by U.S./RVN forces partially destroyed the hamlets. In addition to killing 19 civilians and wounding another 50, airstrikes and artillery damaged or destroyed three hundred homes and rendered nearly 3,000 residents homeless.[118] Similar efforts to expel insurgents and expand security in Kien Hoa Province in May and June 1969 generated 2,200 new refugees in the densely populated district of Ham Long. During the same period, allied operations to eject NLF/NVA forces from two occupied hamlets in Chau Doc Province destroyed or damaged over nine hundred homes in Tan Chau District. According to the CORDS Refugee Division, allied military operations in Kien Hoa and Chau Doc Provinces destroyed more than a thousand homes and generated over six thousand war victims during May and June.[119]

Vietnamization and the withdrawal of American combat forces did little to lessen the use of firepower and conventional combat operations. Although the overall size, scope, and intensity of major combat operations diminished between 1969 and 1972 (with the exception of the 1972 Nguyen Hue Offensive), the drawdown of American units and increase in RVNAF main-force responsibilities did little to curb the use of overwhelming counterforce to eject communist forces from occupied villages and hamlets. Because the RVNAF had been organized, trained, and equipped to fight like the Americans (reliant on technology, firepower, and mobility), military support of pacification during the Vietnamization phase looked very much like it did when American combat forces were at their peak. As pointed out in one postwar study on firepower and limited war, the RVNAF followed the example of their ally, who "relied on helicopters to get to battle and on firepower to destroy the enemy."[120] One former South Vietnamese officer, Hoang Ngoc Lung, agrees. He believes that beginning in 1969, after nearly two decades of advice and support from the U.S. military, it would have been impossible for the RVNAF to abandon the high-tech, high-firepower approach of its ally. Lung argued that Vietnamization really meant the "Americanization" of RVNAF combat tactics and techniques. "It was going to have to be fought the American way," he wrote. "This meant that firepower was the dominant element of the strategy."[121]

Operations undertaken northwest of Saigon in the fall of 1972 illustrate well the RVNAF's reliance on overwhelming counterforce to support pacification efforts during the Vietnamization phase. In early October, six main-force NVA regiments supplemented by local guerrillas launched an "all-out offensive" in Binh Duong and Bien Hoa Provinces as part of an effort to "attack the GVN pacification program" and "control population centers" throughout the two provinces. After successfully contesting and occupying more than a hundred hamlets, communist forces were driven out by the RVNAF after two months of bitter fighting and tremendous material cost. "In the contested hamlets," read the October province report for Binh Duong, "a great deal of destruction has occurred." According to CORDS advisers, 8,000 homes had been destroyed and another 4,000 damaged in Binh Duong and Bien Hoa by the end of November. The former province bore the brunt of the battle, with approximately 5,000 houses destroyed and an additional 2,000 damaged. The human toll was also high. Although civilian casualties were limited (a majority of the population had fled their villages as communist forces entered), 50,000 new refugees were created in the process of ejecting communist forces from these provinces, and 1,100 communist forces were killed—more than half by tactical airstrikes—in Binh Duong alone. By the time November came to a close, the RVNAF had successfully driven NLF/NVA forces from all of the hamlets and villages that they had occupied since the first week in October.[122]

The operations discussed here illustrate that overwhelming firepower and conventional force clearly advanced the short-term goal of pacification—control—by eliminating communist forces and expanding the RVN presence in rural areas. "Where we are in Vietnam," Abrams informed the White House National Security Council in September 1969, "is due to the application of raw power. That is why the enemy is where he is, why pacification has moved. . . . When you turn off the power you have got an entirely new ball game."[123] Abrams's predecessor agreed. "At the time we entered, the government held only a few enclaves," Westmoreland informed the House of Representatives Committee on Appropriations in February 1970. "It took a great deal of effort, sweat and blood to expand the enclaves to the point that the countryside opened up. In the process of that, we have had to grind down the guerrilla force, which has been done. Pacification has been extended bringing more people and area under government control. The indigenous enemy in the south has been reduced. The enemy now has been required to take refuge in remote areas of South Vietnam, in Cambodia, in Laos, and in the DMZ area."[124] Such assessments were accurate; the American combat experience in Vietnam demonstrated that large conventional combat operations such as Speedy Express had a place in modern counterinsurgency warfare. These missions assisted pacification by decimating communist main-force units and destroying NLF/NVA base areas, sanctuaries, and safe havens. Search-and-destroy and other "big-unit" combat actions not only saved the RVN from imminent collapse in 1965; they helped pave the way for pacification progress and greater government control over the countryside in the years that followed. By the end of 1969, search-and-destroy missions and other main-force combat operations helped bring Saigon's control of population and territory to peak levels. According to a study produced by the Vietnam Special Studies Group in 1970, conventional military operations were the foundation for the unprecedented expansion of RVN control in the wake of *Tet Mau Than*. "Throughout late 1968 and 1969, enemy main-forces were destroyed, dispersed, and forced to withdraw from populated areas of the countryside," the report stated. "In most provinces studied, this change in the main-force environment was a prerequisite to the RVN control gains of late 1968 and 1969."[125]

Overwhelming firepower and conventional combat operations, despite their short-term benefits, frequently undermined the long-term goals of pacification and nation building. This was particularly true of efforts to construct a national political consciousness among the peasantry. Although conventional military force possessed an obvious utility in expanding RVN control over population and territory, high levels of physical destruction and collateral damage along with war-weariness incurred by a seemingly endless conflict damaged allied efforts to win popular support. Military force had utility in South Vietnam; combat operations clearly advanced pacification by eliminating insurgents, improving security, and expanding government control over territory and population. Nonetheless, the use of force corroded the very essence of pacification, which was to build a nation founded on unity, community, comity, and cooperation between the government and the rural population.

Besides causing physical destruction, allied military operations damaged the social and cultural fabric of rural South Vietnam. According to one report, the disruption of the everyday life of the peasant was "almost total" in many areas. Frequent ground sweeps, air and artillery bombardments, the destruction of crops through chemical defoliation or "Rome plow" ground clearing, and other intrusive counterguerrilla operations made it virtually impossible for the rural population to perform the daily routines and normal functions of their lives. "In many areas of Vietnam schools have been shut down since they appear to be prime targets—groups of persons tend to draw attacks," the report concluded. "Farmers till their fields at night if at all. What little capital goods the Vietnamese farmers had—pigs, buffalo, ox carts, etc.—have been either killed or destroyed. The traditional family life is placed under severe strain. The consequences of these changes are hard to evaluate; but, they appear to be largely dysfunctional with regard to the avowed . . . interest in building a more stable, prosperous South Vietnam."[126]

The sheer destruction caused by allied military operations was highly disconcerting. In 1970, Congressman Pete McCloskey delivered a statement before the Senate Committee on Foreign Relations that decried the desolation he witnessed during a recent trip to South Vietnam. "In waging counterinsurgency war in this area over the past several years," he stated, "I was advised that we had destroyed 307 of the original 555 hamlets of Quang Nam Province. I flew over miles and miles of area south and southwest of Danang where nearly every hamlet had been destroyed that was not within a mile of the main highway or larger settlements. Tree-lines, hedgerows, and gravesites are all that remain. The purpose of this destruction was to deny rice, cover, and support to the Viet Cong, as well as to the North Vietnamese units operating in the adjacent mountainous areas."[127] Journalist Tom Buckley reported similar scenes of devastation in the Mekong Delta in November 1969. Buckley reported that a young American he had interviewed had informed him that U.S./RVN forces were "blasting" Front-occupied villages and hamlets "right out from under them" during pacification operations. Reflecting on this encounter, Buckley wrote: "To fly over the Delta or the central lowlands is to see the truth of that statement—bomb craters beyond counting, the dead gray and black fields, forests that have been defoliated and scorched by napalm, land that has been plowed flat to destroy Viet Cong hiding places. And everywhere can be seen the piles of ashes, forming the outlines of huts and houses, to show where hamlets once stood."[128] Three years later, John Paul Vann voiced similar concerns about the devastating impact of allied firepower on rural areas. In an April 1972 memorandum to U.S. advisers in Military Region 3, Vann stated that he personally witnessed hundreds of villages and hamlets destroyed through combat over the past decade. Such widespread destruction, he argued, was predominantly the result of "heavy friendly fires."[129]

According to critics, the use of overwhelming counterforce in populated areas not only damaged physical infrastructure; it hurt pacification by inciting the rural population. "One of the major things you must keep in mind in this conflict in Vietnam," one commander warned, "is the confidence that the people have in the government and the progress in pacification. Often, we defeat our own efforts when we use more firepower than the enemy situation warrants."[130] Although the NLF usually received an equal share of animosity when villages and hamlets were damaged and civilians killed, the allies were routinely denounced for the death and destruction they caused. Such hostility only served to undermine pacification. "Blame and resentment are emotional reactions which are inimical to US/RVN goals," noted one RAND analyst.[131] In February 1968, as the allied counteroffensive against *Tet Mau Than* was heating up, Vann warned General Weyand that collateral damage caused by "friendly firepower" was driving the local population into the arms of the NLF. "I have never detected an air of greater hostility by any group of Vietnamese, even those in hamlets under long-term VC control," he wrote of an inspection made shortly after the village was bombed. Vann knew peasants made homeless by artillery bombardments and airstrikes became embittered toward those responsible. This, he argued, made them amenable to insurgent persuasion and likely to abandon the government in favor of the NLF. "The logistical and administrative facts of life," he wrote, "are such that the family is without their own home for a number of months and possibly for an indefinite period. This causes a resentment that grows each day they remain in this status, fanned naturally by enemy agents and enemy propaganda activities. They then become highly susceptible to switching sides (if they were GVN supporters) or siding with the NLF (if they were previously uncommitted)."[132]

In May 1969, Vann expressed similar sentiments in a policy statement that denounced the use of overwhelming counterforce in populated areas. Writing in response to an incident where helicopter gunships destroyed a Buddhist pagoda and killed or wounded a number of civilians as they responded to insurgent sniper fire,

Vann argued that while CORDS advisers were expected "to influence an aggressive, hard-hitting military offensive against the enemy at all times," they were at the same time "charged with the conscience of the U.S. effort in Vietnam and of insuring to the extent possible the security and welfare of the civilian population."[133] Although Vann admitted that there were occasions when "security requirements" justified attacks on populated areas, he maintained such instances were rare and—most importantly—counterproductive to pacification and the allied mission in South Vietnam. "The point I wish to emphasize," he stated, "is that the net result of this action was to make more difficult the task of eventually restoring GVN influence in this area. . . . It is a well known tactic of the Viet Cong to deliberately stimulate fire attacks upon the population and upon built-up areas as a means of creating dislike for the government. Regardless of the logic which considers the Viet Cong presence as the basic causative factor, my experience has indicated that the Viet Cong end up gaining and we end up losing when we fall for their trap."[134]

Three years later, Vann again lamented the unresolved habit of responding to guerrilla attacks from populated areas—a tactic designed to deliberately provoke an overwhelming allied response—with artillery bombardment and airstrikes. "I have witnessed the enemy's employment of this tactic for the past 10 years," he wrote in April 1972. "His specific objective is to get our friendly forces to engage in suicidal destruction of hard-won pacification gains. Invariably, he is successful since in the heat of battle rational thinking and long term effects usually play second fiddle to short term objectives."[135] Although Vann once again insisted there were "rare instances" where military operations dictated the use of artillery bombardment and airstrikes in populated areas, the "destruction of a hamlet through friendly firepower," he concluded, "is an event that will always be remembered and practically never forgiven" by those whose lives and property were destroyed.[136]

Given the fact that support for the NLF declined and the rural population had become less—not more—sympathetic to the communist revolution, in the years after *Tet Mau Than*, Vann clearly overstated the case that allied actions were helping fuel the insurgency and filling its ranks by generating widespread antipathy for the RVN. Although this may seem like a reasonable assumption given the resentment most people feel when their property and their means of living are destroyed, there is little proof to support Vann's argument. Nonetheless, the death and destruction caused by allied military operations did little to build bridges between the peasantry and the government. Nor did these operations advance pacification's ultimate goal of developing a national political community. Instead, the persistent use of force helped perpetuate the war-weariness and apathy that gripped the population and sapped its enthusiasm for the Saigon regime. The reality was that despite a decline in popular support for the NLF, there was no commensurate increase in support for the RVN. The rural population, for the most part, remained largely skeptical of both sides and unwilling to commit to either until the Mandate of Heaven was revealed and the legitimate power in South Vietnam became evident.

The use of overwhelming counterforce to eliminate insurgents and advance allied efforts in the villages and hamlets of South Vietnam continued despite the physical and psychological damage it caused and criticism it generated. The destructiveness of conventional military force and the condemnation it provoked had little impact on the conduct and character of the "other war" in the years following *Tet Mau Than*. Because overwhelming firepower decimated communist units and paved the way for greater government control over the countryside, it remained a fundamental element within the military's operational and tactical repertoire during the post-Tet period. Although the tempo and intensity of the war tapered after 1968, as American forces gradually withdrew and the communists resumed low-level protracted warfare in the countryside, the "other war" in Vietnam remained a violent and deadly affair throughout the period of Vietnamization and America's last years in Vietnam.

The communists and the allies both freely applied violence and force as needed and in appropriate measure to advance their position in the control war and expand their influence and/or control over the civilian environment. Circumstances and a disparity in the balance of forces in the field after *Tet Mau Than*, however, led each side to stress a particular approach—repression and political violence for the communists; and conventional military force and combat operations for the allies. This difference in emphasis was in large part a product of the asymmetrical nature of the conflict and the overwhelming material superiority held by allied military forces. It was also a result of timing and the different force requirements and methods for maintaining and expanding politico-military control that were inherent within the insurgent-counterinsurgent struggle for South Vietnam. To begin with, the NLF had metastasized and established a presence in nearly every village and hamlet long before the RVN and the Americans had launched an accelerated pacification program in the autumn of 1968. This posed immense problems for Saigon and Washington once Vietnamization and an American withdrawal had become reality. With the clock running and military assistance from U.S. forces diminishing, the allies had no choice but to pursue a hard-and-fast pacification strategy—an approach that demanded heavy combat actions to rapidly remove insurgent forces from rural areas. The communists, meanwhile, had no such time restrictions; protracted war by nature allowed Hanoi and the NLF to take the indirect approach and apply a far more patient method of warfare.

Next, the strategy of communist revolutionary warfare and its use of base areas, guerrilla tactics, and blend of conventional and irregular war had enabled Hanoi and the NLF to set the tempo, the scope, and the intensity of the war. The ability to pick and choose when, where, and how the control war was fought not only allowed the communists to husband their resources, strengthen their forces for future offensives, and stretch out and exhaust a much larger enemy at a much lower cost and with relatively fewer forces; it afforded them the opportunity to alternate between repression and conventional combat as needed. The communists also enjoyed strategic mobility and an ability to concentrate their forces on those areas where the government was most vulnerable. The RVN and the Americans, in contrast, had to literally defend everywhere and were required to offer an armed riposte anywhere insurgent forces were located. The political obligation to remove the insurgency root and branch from the national soil and to annihilate the enemy wherever he was found was a burden the communists simply did not have to bear. Together, these disparities helped shape the different emphasis on force and violence during the post–*Tet Mau Than* period.

Notwithstanding what some have argued, the events that transpired between 1968 and 1972 did not constitute a "better war," particularly for the population living in thousands of villages and hamlets throughout South Vietnam. In fact, beginning with *Tet Mau Than*, the conflict became an increasingly deadly and destructive affair for those living in rural areas. Despite the overdue precedence given to pacification and the civil/political side of the war in the years following the Tet Offensive, traditional combat tactics and overwhelming firepower continued to ravage the countryside. Military force helped pacification progress by bringing increasing numbers of villages and hamlets into the government's fold; yet it sustained in many places a cycle of death, destruction, and social disruption brought about by nearly two decades of war. Revolutionary violence and repression, likewise, exacted a deadly price on inhabitants of the countryside. Although less intense than allied efforts (with the exception of the 1972 Easter Offensive), the NLF's violence program was equally brutal.

The net effect of the control war was a protracted conflict that brought devastation, death, and displacement to South Vietnam's densely populated villages and hamlets. As noted by two former ARVN officers, the civilian environment was the crucible of war and bitter revolutionary struggle. "The countryside of

South Vietnam was the major arena of contest and, as such, suffered unspeakable destruction," they recollected.

> The rural areas were the warehouse of manpower and material resources. . . . Our opponents regarded the rural areas as a strategic objective to be captured, which was the reason why the RVN and its allies defended them at all costs. The participation of US and FWMAF forces, with their tremendous firepower, added to the destruction. Many villages were completely obliterated from the surface of the earth. Throughout the war, it could be said that no corner of the countryside had been spared the destructiveness of bombs and shells; many areas had in fact changed hands many times, and each time the destruction was worse. The end result was that houses were reduced to rubble, innocent people were killed, untold numbers became displaced, riceland was abandoned, and as much as one-half of the population of the countryside had fled to the security of cities, province capitals, and district towns at some time during the war. A small minority of rural people with resources adapted themselves quickly to city life, but most languished in abject poverty in refugee camps or as outcasts in city streets. When the situation seemed to have improved, the people returned to their home villages, only to flee again at the next battle. This process was repeated again and again over the war years.[137]

To the poor suffering peasants caught in the crossfire of the control war, the increasingly harsh contest to establish and expand politico-military domination over the operational environment had made the post-Tet period a worse—not better—war. Within the villages and hamlets of the countryside, the battle between communist revolutionary warfare and pacification confirmed the old proverb that "when elephants fight, it is the grass that suffers." Or, as one individual grimly observed in 1968, "the terror of Vietnam is being in the middle."[138]

PART II
The Control War
in South Vietnam,
1968–1975

5 PEACE POLITICS AND THE TRANSFORMATION OF THE SECOND INDOCHINA WAR

THE IMPACT OF THE 1968 TET OFFENSIVE

The final seven years of the Second Indochina War and the control war cannot be understood without first examining the political and diplomatic context that framed the battle for hegemony in South Vietnam. After the "Americanization" of the war and several years of main-force combat dominated by a grinding war of attrition, the politico-military struggle for the countryside had assumed unprecedented emphasis vis-à-vis the advent of peace talks in Paris, new leadership in Washington, and the U.S. decision to reverse what had been up to that point an open-ended commitment to preserving the Saigon government. This chapter explores the larger currents that shaped the war in Southeast Asia and explains why efforts to expand control over territory and population had become imperative in the years following *Tet Mau Than*.

NEW INITIATIVES AND NEW TRAJECTORIES IN THE WAKE OF TET MAU THAN

On 30 January 1968, NLF main and local forces aided by NVA regulars launched *Tet Mau Than*, the largest and most closely coordinated offensive of the war. The attacks targeted thousands of villages and hamlets, thirty-four provincial capitals, sixty-four district towns, and five of the RVN's largest cities, including Saigon and the ancient imperial city of Hue. The Tet Offensive, so named because it was launched during the Vietnamese New Year, was part of an overall plan to push the South Vietnamese government to collapse, force an American withdrawal, and impel the formation of a coalition government dominated by the NLF. As with other past communist offensives, the attacks were meant as a vehicle to topple the Saigon regime and reunite the two Vietnams under communist rule.[1] The means to achieve these ends was the mythic General Offensive and General Uprising, a coordinated politico-military attack that meshed an international propaganda campaign with offensive military operations and popular uprisings launched simultaneously against urban and rural environments throughout South Vietnam. "The coming General Offensive and General Uprising will be the first stage in a process involving a very fierce and complex strategic offensive which will combine military and political attacks to be carried out by all the three forces in all three strategic zones in combination with the diplomatic offensive," a Party Central Committee resolution stated in the fall of 1967. "Military attacks by the armed forces on the main fronts and the uprising of the people in large urban areas," the resolution continued, "are the two main thrusts."[2]

The communists launched *Tet Mau Than* in hopes of achieving three primary objectives: first, to disintegrate a large portion of the South Vietnamese armed forces; next, to overthrow the RVN administration at all levels; and finally, to provoke a political crisis in the United States that would crush Washington's political will and culminate in American withdrawal from Southeast Asia.[3] Despite careful planning and a rapid, unexpected, and violent execution during the initial phase, the series of countrywide attacks failed to achieve their immediate goals. The RVNAF were not crippled and the Saigon government did not crumble. Moreover, the much hoped for general uprising by the "oppressed" and disgruntled masses never materialized. Within days most of the attacks on urban areas were blunted or reversed. Subsequent communist offensives, one in May and the other in August, were also rebuffed by strong allied countermeasures. Perhaps even more damning, the population did not initiate a popular uprising. Rather than rise up in rebellion against the neocolonial "puppet" government in Saigon, much of the South Vietnamese population instead turned away from the NLF as a result of the failed attacks and the destruction and suffering they unleashed in populated areas. "It is increasingly evident that the people continue to harbor resentment against the VC/NVA for the Tet attacks, and for the disruption to normal living caused by them," noted an American adviser.[4] In short, *Tet Mau Than* was a major military defeat, both for the NLF and for Hanoi. Besides failing to achieve short-term objectives, the series of communist offensives and allied counteroffensives decimated the Front's political infrastructure and its fighting forces (NVA forces were held in abeyance during the first wave of attacks and therefore suffered few casualties). According to some estimates, the NLF lost approximately 80 percent of its fighting force and suffered as many as 50,000 casualties.[5] These losses would have a profoundly negative impact on communist efforts in the years following Tet.

Despite its failures, *Tet Mau Than* turned out to be a long-term political victory for Hanoi if not necessarily for the NLF. The attacks were the catalyst that undermined Washington's commitment and spurred American withdrawal. Nearly twenty years after the war in Vietnam had ended, North Vietnamese general Tran Van Tra identified the central paradox of Tet: "The General Offensive and Uprising of Tet Mau Than . . . matched the conditions at the time," he wrote. "Despite the fact that the General Uprising did not take place and power was not turned over to the people, Tet Mau Than was still a great victory, creating the most important strategic turning point of the war, eventually leading us to total victory."[6] Tra was correct; Tet set in motion a series of events that helped Hanoi overwhelm and eventually defeat the RVN and its armed forces. The attacks, through their unexpected suddenness and sheer audacity, fundamentally transformed the conflict in Vietnam by shifting its emphasis from the politics of war to the politics of peace.

Tet Mau Than profoundly changed the dynamics of the war in Southeast Asia. Although the 1968 campaign failed to break the politico-military stalemate in South Vietnam or present a clear-cut advantage for either side in the conflict (despite claims to the contrary by both camps), it was, nonetheless, a watershed in the Second Indochina War. On the operational level, Tet was indecisive; it produced no winners. Yet it was a heuristic event, a point of no return that shaped and determined, in the words of one scholar, "the succeeding stages of peace politics for the losers."[7] For Hanoi and the NLF, Tet laid bare the reality that when combined the allied armies were just too large and strong to defeat militarily. Moreover, it revealed that the revolution could not sustain the extremely high casualties and material losses suffered over the past three years and survive. Nor could the communists count on the masses to rise up and initiate a popular insurrection in support of the General

Offensive and General Uprising with the Americans still around. As long as U.S. forces lingered and the Saigon regime remained strong, the NLF and Hanoi were largely on their own; they would get little help from the majority of the population in South Vietnam, most of whom remained apprehensive and uncommitted to either the NLF or the RVN despite concerted efforts by both sides to win them over. Together, these circumstances produced temporary retrenchment and a shift in strategy from a predominantly military approach to a method that fused low-level protracted war with diplomatic struggle.

Tet Mau Than was equally jarring to the allies. For the United States the attacks validated the argument that the war in South Vietnam could not be won on the battlefield and that current policy had to be reconsidered and revised in light of growing war-weariness among the American people and an increasingly problematic domestic political situation. Such recognition prompted Washington to significantly alter American policy at both the strategic and grand strategic levels. After Tet, the United States not only refocused and accelerated the pacification program and efforts to expand government control over rural areas; it capped its overall commitment to the RVN and began extracting American combat forces from Southeast Asia. For Saigon, Tet demonstrated that the government and people of South Vietnam would have to assume primary responsibility for prosecuting the war; they simply had no choice but to accept the increased demands that would be placed on them in light of new developments and shifting circumstances both inside and outside the country.[8] Once it became evident that a U.S. withdrawal was imminent, the RVN had to increasingly undertake the burden of its own defense and prepare for the day when U.S. combat forces would no longer be present to assist them.

In sum, Tet proved to each adversary that the war could not be settled decisively through military force, at least not at that particular moment. Issues related to political power and who would rule Vietnam, they realized, could only be mediated through new and unprecedented initiatives that included diplomacy and a reemphasis on politico-military struggle at the local level. "Politically as well as militarily, stalemate increasingly appears as the unavoidable outcome of the Vietnam struggle," the *New York Times* declared on 8 February 1968. "Neither side is entitled any longer to illusions about military victory, nor is there evidence that either side is achieving political ascendancy. A negotiated settlement seeking a political accommodation . . . remains the alternative to a prolonged war of attrition, a war that neither side can win."[9] CBS anchorman Walter Cronkite made a similar assessment that week. "To say we are closer to victory today," Cronkite said, "is to believe . . . optimists who have been wrong in the past. To suggest we are on the edge of defeat is to yield to unreasonable pessimists. To say we are mired in stalemate seems the only realistic yet unsatisfactory conclusion. . . . It is increasingly clear . . . that the only rational way out would be to negotiate."[10] Coming from Cronkite—a journalist so highly esteemed it earned him the sobriquet "the most trusted man in America"—the grim assessment had a profound impact on the Johnson administration and the American public.

The effects of *Tet Mau Than* on larger political and military currents were felt almost immediately. Mere weeks after the first leg commenced and before the shock of the initial attacks had worn off, significant changes were under way. On 22 March, President Johnson formally rejected General Westmoreland's request for an additional 206,000 troops for Vietnam despite grave warnings from the chairman of the Joint Chiefs of Staff that a denial might result either in military defeat or an indefinite continuation of the war.[11] That same day, the Johnson administration announced that Westmoreland, the beleaguered head of MACV, was being recalled from Southeast Asia to serve as army chief of staff.[12] Given that the general was essentially the architect of American ground strategy in South Vietnam from 1965 to 1968 and the face of the American war in Vietnam, this development was significant. It was also in March that the Johnson administration had informed Saigon's top leadership, President Nguyen Van Thieu and Vice President Nguyen Cao Ky, that the United States—although it would continue to provide assistance—expected the RVN to assume greater responsibility for its own defense. According to historian George C. Herring, this decision represented a significant shift in American policy. It was, he argued, "a return, at least in part, to the principle that had governed [U.S.] involvement before 1965 and adoption, at least in rudimentary fashion, of the concept of Vietnamization which would be introduced with much fanfare by the Nixon administration a year later."[13] Finally, on 31 March, Johnson announced publicly that the United States was halting the bombing campaign against the DRV (with the exception of the area immediately above the Demilitarized Zone separating the two Vietnams) and initiating formal peace talks with Hanoi. "We are prepared to move immediately toward peace through negotiations," the president informed his audience. "Tonight, in the hope that this action will lead to early talks, I am taking the first step to deescalate the conflict. We are reducing—substantially reducing—the present level of hostilities. And we are doing so unilaterally and at once."[14] Also significant in Johnson's 31 March speech was his announcement of another key change in leadership—his own. The immense problems facing the nation both at home and abroad and the "awesome duties" of the presidency, he argued, precluded any "personal partisan causes" (i.e., running for reelection). He would, therefore, neither seek nor accept the nomination of his party for another term as president. Johnson's speech was profoundly important because it essentially ended America's open-ended commitment to the defense of South Vietnam and set the stage for a number of subsequent developments that would exert tremendous pressures on the control war and the struggle to establish dominion over the operational environment. These included the commencement of peace talks in Paris, a change of leadership in Washington, the possibility of a negotiated settlement to end the war, Vietnamization, and ultimately, American withdrawal.

THE CRUX OF THE PARIS PEACE TALKS: WITHDRAWAL, COALITION, AND REGIME CHANGE

The peace talks that opened in Paris on 7 May 1968 were a direct result of *Tet Mau Than* and Johnson's subsequent call for negotiations. More importantly, the talks reflected the practical realization that despite an immense expenditure of blood and treasure, the war had achieved nothing more than military and political stalemate. Both camps agreed to formal negotiations because each realized, as did the *New York Times* on 8 February, that military victory was unattainable under the current circumstances and that neither the communists nor Saigon had been able to achieve political ascendancy in South Vietnam after nearly a decade of war. The ultimate decision to enter into negotiations, however, lay largely with Hanoi and Washington. Although Saigon and the NLF were touted as equal parties at the negotiating table and on the battlefield, the reality was that the DRV and the United States were the senior partner in each coalition and the fountainheads from which the lifeblood for the contending southern factions flowed. As such, Hanoi and Washington assumed the leading role in negotiations, with Saigon and the newly formed Provisional Revolutionary Government (PRG) playing second fiddle.

The United States entered into formal negotiations in Paris because Tet underscored the reality that American military power had its limitations and that brute force and conventional combat operations had not achieved policy objectives in Vietnam. A number of American officials considered the general offensive's failure a tactical victory for the allies, one that opened up new possibilities for pressing the attack (particularly in pacification);[15] others believed it was an indicator of the bankruptcy of American strategy. "The Tet offensive and its aftermath starkly revealed the strategic vacuum in which United States policy had been operating since 1966," Westmoreland's onetime intelligence chief stated.[16] The Johnson administration's decision to enter into peace talks in Paris stemmed largely from a series of national security meetings in which a majority of the president's top advisers on Vietnam, collectively known as the "Wise Men,"[17] had determined that the war was unwinnable under current conditions and that the best policy for the White House was to move immediately toward a negotiated settlement.[18] The man held most responsible for the Wise Men's verdict on Vietnam and Johnson's subsequent decision to deescalate was Clark Clifford, the newly appointed secretary of defense. Clifford, a staunch critic of American policy in Vietnam, was eager for a way out of the war in Southeast Asia. "I was convinced that the military course we were pursuing was not only endless, but hopeless," he wrote in July 1969. "A further substantial increase in American forces could only increase the devastation and the Americanization of the war, and thus leave us even further from our goal of a peace that would permit the people of South Vietnam to fashion their own political and economic institutions. Henceforth, I was also convinced, our primary goal should be to level off our involvement, and to work toward gradual disengagement."[19] Once begun, the peace talks in Paris went nowhere. Negotiations deadlocked almost immediately, and absolutely no progress was made during the final months of the Johnson administration. The burden for pursuing and (hopefully) reaching a negotiated settlement to end the war in Vietnam, therefore, fell on the incoming president, Richard Nixon, and his top adviser, Henry Kissinger.

The objective of peace talks for the United States and the RVN was simple: to bring the war in Southeast Asia to a conclusion and preserve the national sovereignty of the RVN. Another goal, of course, was to create conditions that would allow the United States to "honorably" extricate itself from an exceedingly unpopular and divisive war. Ultimately, the overall aim was to end armed hostilities between Saigon and the NLF and create a postwar environment that was free from outside interference and allowed for the peaceful, democratic, and volitional mediation of power between the contending Vietnamese factions. According to President Nixon, the basic terms of the settlement that Washington and Saigon sought to negotiate in Paris were modest: the mutual withdrawal of NVA and FWMAF from South Vietnam and the ability of the Vietnamese people to freely choose their own government. "I believe that the long-term interests of peace require that we insist on no less and that the realities of the situation require that we seek no more," the president declared on 14 May 1969.[20] In order to achieve these objectives, the United States and the RVN set out to win two primary concessions from Hanoi and the PRG: a bilateral withdrawal of all American and North Vietnamese forces from South Vietnam; and the preservation of Thieu and Ky as heads of state.[21] Both of these conditions, the allies believed, were critical if a brokered and, more importantly, lasting peace was to be achieved in Southeast Asia. The former demand was vital for the military survival of the RVN. Should only U.S. and other allied military forces be removed, the communists would likely launch a major military offensive and conquer the RVN by force. It was vital, therefore, that all "outside" military forces, including those from the DRV, withdraw. The latter condition was necessary if the RVN was to survive politically. Should Thieu and Ky leave office and be replaced by a coalition government dominated by the NLF (both were key communist demands), the RVN would be politically outmaneuvered and eventually overrun, albeit through "democratic" rather than military means. Thus, the mutual withdrawal of American and North Vietnamese armed forces and the retention of the Thieu-Ky government were considered mandatory if peace was to be preserved and the RVN was to have a fighting chance at survival after U.S. forces had left.

The DRV had agreed to formal peace talks in the spring of 1968 because it too recognized the bankruptcy of its military policies under current conditions. Although Hanoi's leaders proved much more reluctant to relinquish their military strategy—as evinced in the May and August phases of *Tet Mau Than* and again during the spring and summer attacks of 1969—they eventually acknowledged that the General Offensive and General Uprising would likely fail so long as American combat forces remained in South Vietnam. It was only after the communists had exhausted their forces in the series of failed attacks that Le Duan and the Politburo decided to enter into serious diplomatic talks.[22] Hanoi's willingness to negotiate was also influenced by communist setbacks in the wake of *Tet Mau Than.* The allied "one war" strategy, which combined a revitalized pacification program with unrelenting allied military offensives, caused serious damage to the southern insurgency. Robert Thompson believed Hanoi had begun to negotiate seriously because it knew the revolution was losing ground in the countryside. This vital strategic area, Thompson argued, would be extremely difficult to recapture given the weakened state and diminished number of communist guerrillas and political cadres in rural areas. "It was one of the reasons they had to negotiate," he wrote. "They were about to lose territory which they had gained, because they could not possibly have held it with what they had left."[23] Historian Lien-Hang Nguyen confirms this assertion. "With reports reaching Hanoi of the deteriorating military picture in South Vietnam," she argues, "the Party leadership had no choice but to advance the diplomatic struggle in Paris in order to provide much-needed support for the weakened military and political struggle in South Vietnam."[24]

For Hanoi, the overall objective of peace talks was not to achieve a compromise solution to the war, but to win it. Despite their willingness to enter into talks, the communists refused to renounce the fundamental purpose of the revolution, which was to liberate the South, defend the North, and unify the country under communist rule.[25] Winning the war, however, first meant negotiating a favorable set of circumstances that would improve the communists' chances of toppling the RVN. The most important among these was the removal of all U.S. combat forces. American withdrawal had been a possibility as early as 1968. By mid-1969, however, it had become reality. The process, however, was gradual and there were no guarantees that Washington would not reverse course if necessary or try to leave a residual force in South Vietnam. Hanoi's negotiating policy, therefore, was aimed at prolonging the deliberations so that the military and political struggle could develop and pressure could be exerted on the United States to make vital concessions, including a precipitous, permanent, and—most importantly—unilateral withdrawal. Although it appeared on the surface that Hanoi wanted to seriously explore the possibilities of a negotiated settlement, the Nixon administration feared the communists would never accept a deal that did not, at a minimum, guarantee the complete extrication of all U.S. forces and a fairly clear shot at political control in South Vietnam once the Americans had departed.[26] According to Lien-Hang Nguyen, recently declassified documents reveal Hanoi's envoys in Paris were ordered to "appear cooperative at the talks but not overly eager in working toward a settlement." The Politburo, Nguyen asserts, "advised the negotiators to consider the Paris talks as another battleground where territory should not be given up without a fight. Nonetheless, the military struggle to topple the Thieu regime would remain the Party's most important objective."[27] Historian William Duiker also asserts Hanoi's primary objective in Paris was "to string out the negotiating process by imposing conditions until the United States had been driven by military defeat and domestic unrest to make fundamental concessions at the conference table."[28]

The communists' central negotiating plank during the Paris talks called for three primary conditions: the complete, unilateral withdrawal of all non-Vietnamese military forces; the removal of the current leadership in Saigon; and the formation of a "peace cabinet" or coalition government in South Vietnam.[29] The purpose of these demands, however, was not to achieve a peaceful settlement but to provide the communists with an advantage they could later use to overwhelm the RVN and achieve victory. "We should fully understand," the PRG declared, "that the purpose of the new diplomatic offensive is to coordinate efforts and provide conditions to change the balance of forces between us and the enemy, but not as a step toward a peaceful solution."[30]

The deliberately intractable nature of communist demands along with the Nixon administration's stubborn refusal to agree to any conditions that would undermine its quest for a "just and honorable peace" delayed an agreement for more than three years. According to Kissinger, the administration considered Hanoi and the

National Liberation Front's negotiating position to be nothing less than a demand for unconditional surrender and political desertion. "With an immutable deadline set for our withdrawal, with all known non-communist leaders excluded from political participation, and with the existing government overthrown," Kissinger recalled bitterly after the war, "there was no possible outcome except a communist takeover in which we were asked to collude. It was not just a matter of our pulling out and letting the chips fall where they may, nor of pulling out abruptly without any quid pro quo whatever. *Our unilateral exit was not enough*; we had to engineer a political turnover before we left, or else the war could not end. . . . Our dilemma was that Hanoi maintained this position until October 1972. As long as it did so, no negotiated settlement was available."[31]

A complete, unilateral, and unconditional American withdrawal had been the central aim of communist efforts since 1965. It was only natural, therefore, that Hanoi and the NLF pursued this objective at the bargaining table once the "diplomatic front" had officially opened in May 1968. Even before *Tet Mau Than*, the communists insisted that the removal of all U.S. combat forces was a central requirement for any negotiated settlement. "If the enemy wants to negotiate, he must . . . cease his war of destruction against the North, withdraw his troops from the South, and dismantle his military bases," North Vietnamese general Nguyen Van Vinh declared in 1967.[32]

The United States initially rejected a unilateral withdrawal for two reasons. First, it feared the communists would exploit the situation and attempt to conquer the RVN by force once U.S. combat forces had departed. Second, the Nixon administration believed American combat power could still win the day in South Vietnam. Despite these fears and hopes, and in spite of its haggling and tough talk in Paris, the Nixon administration capitulated to the communist demand for a unilateral American withdrawal by quietly dropping its call for the mutual extrication of all non-ARVN forces. In May 1971, Nixon authorized Kissinger to concede this point at a secret meeting with the North Vietnamese delegation and retract the demand for mutual withdrawal. The concession would allow North Vietnamese forces to remain in South Vietnam after a peace agreement had been signed.[33] On 31 May, the seven-point "final" proposal for a settlement presented by the American delegation in Paris made no mention of the removal of communist forces. The long-standing American demand for mutual withdrawal, a cornerstone of America's diplomatic strategy for ending the war in Vietnam, had been dropped.[34]

This concession rested, in part, on the administration's conclusion that the only way to break the logjam in Paris and get the communists to approve a negotiated settlement was to agree to either a coalition government or an in-place ceasefire followed by territorial accommodation—an arrangement that would allow communist forces to remain inside South Vietnam. If the United States stuck to its current negotiating stance, however, there would likely be no breakthrough in the Paris negotiations. "The other side might not really insist on both its conditions of unilateral U.S. withdrawal and coalition government," Kissinger informed Nixon in July 1970, "but it will not budge without concessions on at least one of them."[35] The decision to abandon mutual withdrawal also stemmed from the administration's belief that its bargaining power would wane over time given Vietnamization and the ongoing extraction of U.S. forces. "At some point our present policy will turn into either negotiations, with our withdrawal schedule part of the bargaining, or into a unilateral withdrawal, with the pace non-negotiable. We will have to choose either to seek actively a settlement while our remaining forces can be used as leverage, or to leave the political settlement strictly to the South Vietnamese, whether by negotiation or force of arms, while we withdraw more or less unilaterally." Agreeing to a "unilateral disengagement," Kissinger argued, "could mean either a slow winding down of the war along present lines or disengagement on a fixed timetable that would give the GVN a fair chance."[36]

Kissinger later recalled that by late 1970 it was obvious that the U.S. Congress would never have allowed American forces to remain in Vietnam, even if Hanoi had refused to extricate its troops following an armistice. Given congressional resistance to maintaining a permanent American force in Southeast Asia, the administration simply had no choice but to proffer a deal that provided the communists discretion in removing their military forces from southern soil. "The decision to propose a standstill ceasefire in 1970 thus implied the solution of 1972," Kissinger recalled. "That North Vietnamese forces would remain in the South," he continued, "was implicit in the standstill proposal; no negotiation would be able to remove them if we had not been able to expel them by force of arms."[37] Self-serving and self-absolving as this statement was, Kissinger was correct; the chances the United States could wait out the communists on the matter of mutual withdrawal or that either a residual U.S. force could be left in South Vietnam or, more importantly, the American drawdown could be reversed by reintroducing American combat forces should Hanoi refuse to extract its forces had become highly unlikely by 1970 given the American public's continued dissatisfaction and disenchantment with the war. As it turned out, the "solution of 1972" and the decision to allow NVA units to remain within RVN borders after the Paris Peace Accords had gone into effect in January 1973 played a significant role in the control war and bore bitter fruit in the years that followed.

The formation of a coalition government was the second key communist condition for a negotiated settlement. According to the NLF, such an arrangement was absolutely necessary if a peaceful resolution to the conflict was to be achieved. The function of a coalition government, it argued, would be to implement an agreement once all outside forces had been removed, to unite the southern population regardless of class or political affiliation, and to hold elections so that the people of Vietnam could exercise self-determination and decide democratically what type of government they wanted. It would also be responsible for adjudicating the reunification of Vietnam's northern and southern halves through peaceful means.

Despite paeans to democracy and self-determination, the communists pursued coalition government solely as a means to seize political power outright; they had no intention of allowing true democratic processes to mediate the struggle in South Vietnam. The objective was not to pursue political supremacy through consensus, but to manipulate the system at the local level and gradually move into a position of dominance. In 1970, MACV declared that "it is noteworthy that the term 'coalition government' is a misnomer, in that the [GVN] is specifically excluded from [the] coalition. The communists view coalition government as a transitory vehicle for change from the GVN to eventual full communist control in the South."[38] According to Vietnam expert Douglas Pike, coalition government as understood and pursued by the communists in Vietnam did not mean power sharing. Unlike its practice under parliamentary systems, coalition government in the eyes of Hanoi and the NLF was not an arrangement for joint authority. Rather, it was a means—a "technique"—for achieving absolute political power for the party. Once inside and allowed to work within the provisional coalition government, the communists would then use infiltration and superior organizational work at the grassroots level to gradually take over the reins of power.[39] Captured communist documents verify Pike's argument and reveal that while the communists intended to preserve a "broad union of all strata" at the top of any coalition government (more than likely for propaganda purposes), the lower echelons—villages, districts, and provinces—would consist of governing bodies that were not only handpicked by Hanoi and the NLF but dominated by prorevolutionary factions.[40] The communists were not meant merely to participate in a coalition government, but to dominate it.[41] According to COSVN, "The national, democratic coalition regime, which is our immediate fighting objective, is . . . conditioned on . . . the NLF playing the key role."[42]

The United States rejected calls for a coalition government because they understood that Hanoi and the NLF intended to use it as a political mechanism for overthrowing the RVN. They also knew the Saigon regime would more than likely be unable to stop the steady erosion of its power should a governing coalition be implemented. Kissinger later recalled that while communist demands for such an arrangement seemed reasonable on the surface, a more in-depth exploration revealed conditions for a coalition government were much more insidious. In addition to demanding that all American forces be withdrawn and the current regime in Saigon be deposed, the communists reserved for themselves the right to determine which political groups would be allowed to serve in the new government. Naturally, this was unacceptable to the Nixon administration. "The operational content," Kissinger wrote, "was that after we had decapitated the government of South Vietnam and demoralized the population by total and unconditional withdrawal, we would then collude with the structure containing the NLF and whatever groups the communists alone would define as acceptable. And that new coalition government was to be negotiated between it and the NLF, backed by Hanoi's army. Such was the communist definition of a 'just' political settlement."[43] In September 1970, Kissinger informed Nixon that the communists were also demanding that any coalition government be arranged in a three-tiered, or "troika," framework. This, he stated, was additional evidence of their intent to game the system by dividing the Saigon government into separate strata. According to Kissinger, the division was intended to "emasculate" the anticommunist forces in South Vietnam.[44] Allied fears concerning a coalition government were also rooted in empirical evidence; recent history in the region had given Washington and Saigon reason to be skeptical. As pointed out by Kissinger in 1969, the United States dreaded a coalition government because, as evinced in Laos during the early 1960s, such an arrangement would more than likely create two competing regimes rather than a true governing coalition. Each would have its own territory, administrative apparatus, and military forces, and each would carry on its own business—thus perpetuating rather than settling what was at its heart a civil war. This prediction proved prescient. The Nixon administration also feared that a coalition government not only would help destroy the existing political structure of South Vietnam and facilitate an eventual communist takeover, but would remove Washington's ability to influence events by ending American involvement. "Negotiations seeking to impose a coalition from the outside are likely to change markedly and irreversibly the political processes in South Vietnam—as Vietnamese who believe that a coalition government cannot work quickly choose sides," Kissinger wrote. "We would, in effect, be settling the war on an issue least amenable to outside influence . . . and the long-term implications of which are most problematical."[45]

The third—and perhaps most irreconcilable—negotiating demand made by the communists called for the removal of the current regime in Saigon headed by Thieu, Ky, and Prime Minister Tran Thien Khiem. These men, Hanoi and the PRG insisted, had to be replaced before any settlement could be reached. "The U.S. should stop supporting Thieu-Ky-Khiem so that there may be set up in Saigon a new Administration standing for peace, independence, neutrality, and democracy. The Provisional Revolutionary Government of the Republic of South Vietnam will enter into talks with that administration to settle the internal affairs of South Vietnam and to achieve national concord."[46] According to the communist delegation in Paris, the Thieu-Ky-Khiem "clique" was a "bellicose" and "oppressive" administration that suppressed those parties in South Vietnam that were genuinely committed to peace and willing to work with the communists to reach an agreement. More importantly, it was an illegitimate regime whose very existence depended on U.S. aid and the presence of American combat forces. Thieu and Ky, they argued, were American "puppets" and pretenders to the throne who obstructed a peaceful solution; they had to be removed if progress was to be made toward a settlement.[47] Lurking beneath these public pronouncements, however, lay a shrewder and more pragmatic reason why the communists found the current regime unacceptable. Hanoi and the NLF knew they could not game the system nor advance the revolution inside a coalition that included anticommunist stalwarts such as Thieu and Ky.

The United States refused to capitulate to the communists' demand for the forced deposition of the Thieu-Ky regime for obvious reasons. In June 1969, Kissinger informed Soviet ambassador Anatoly Dobrynin that the Nixon administration dismissed Hanoi and the NLF's demand that the current RVN regime be dismantled because such a concession would virtually guarantee a RVN defeat/NLF victory. "We would never accept that," Kissinger told the Soviet diplomat. "We would agree to a fair political contest—not to what [Nixon] had called a disguised defeat."[48] Kissinger later confessed that while the United States found all three of the communists' primary conditions unacceptable, he and Nixon considered the ultimatum that Thieu and Ky be removed to be the most objectionable. It was this demand, Kissinger later recalled, that stymied allied willingness to agree to a compromise settlement the longest. "We were not prepared to do for the communists what they could not do for themselves," Kissinger stated after the war. "Our refusal to overthrow an allied government remained the single and crucial issue that deadlocked all negotiations until 8 October 1972 when Hanoi withdrew the demand."[49]

The stubborn refusal of both sides to acquiesce to conditions that would deny them what they wanted most stalled the peace talks in Paris for more than four years. For Hanoi, the objective was to win concessions that would swing the balance of forces in its favor and improve communists' chances for total victory once the Americans had left Southeast Asia. For the United States, the goal was more modest; it was looking for an "honorable" exit from Vietnam and a deal that would give the RVN at least a fighting chance at national self-determination and survival. Washington, unlike Hanoi, was no longer pursuing "victory" as the concept is traditionally understood. Although, as Nixon later admitted, the initial goal of his administration was to "go for broke" and attempt to "win the war one way or the other—either by negotiated agreement or by an increased use of force," the overall objective became much more circumscribed after sustained military pressure and hardnosed negotiating failed to break Hanoi's political will.[50] "It is not our objective to 'win the war' in the conventional sense," Ambassador Ellsworth Bunker informed the Council on Foreign Relations. "When we talk of winning the war, what we aim at is a negotiated settlement." The endgame in Vietnam, Bunker argued, was no longer to guarantee a noncommunist South Vietnam but to provide the Vietnamese people "the opportunity . . . to choose freely (on the basis of one man, one vote) the form of government under which they want to live."[51] Nixon himself underscored this shift in objectives early in his administration. "We seek the opportunity for the South Vietnamese people to determine their own political future without outside interference," the president announced in May 1969. "We are prepared to accept any government in South Vietnam that results from the free choice of the South Vietnamese people themselves."[52] Despite the Nixon administration's muted goals and strategic shift from securing a noncommunist South Vietnam to ensuring self-determination for the South Vietnamese people, the White House remained doggedly committed to winning diplomatic concessions that would, at a minimum, preserve the RVN as an independent nation (and America's international prestige along with it) for a period of time after U.S. troops had left Vietnam. This decision helped produce stalemate in Paris, one that remained unbroken until October 1972, when the first signs of movement appeared at the negotiating table.

IN-PLACE CEASEFIRE AND TERRITORIAL ACCOMMODATION

Any negotiated settlement to end the war in Vietnam would have included a ceasefire agreement. Throughout history ceasefires have been looked on as a welcome development, particularly in conventional wars. In Vietnam, however, the cessation of hostilities presented a highly problematic scenario laden with enormous implications. The problem lay with the war's "mosaic" nature and the patchwork state of control that existed throughout South Vietnam. "When two armies oppose each other on either side of a clearly defined front line, a ceasefire is much easier to arrange," noted one journalist. "Each side continues to hold its ground while the shooting dies away. In Vietnam, however, . . . there is no front line, no clearly demarcated area belonging to each of the combatants. Rather than a well-defined zone across the country, a depiction of government-held areas and those controlled by the Viet Cong would, on a map, appear to be a series of intermixed blotches."[53] This was a common concern after 1968. According to another observer, "Areas of relative peace for both the communists and the nationalists are likely to be unevenly formed zones of local influence. No clear, single demarcation line will emerge from the checkerboard of both Viet Cong and GVN 'liberated' areas."[54] In 1969, the chairman of the Joint Chiefs of Staff expressed concern that an in-place ceasefire would be dangerous given the highly fragmented nature of South Vietnam's conflict environment. "If we had a frontal war we could do so," General Earle Wheeler stated in a national security meeting with the president. "In this war where the enemy is pock-marked in the countryside, unless you have verified withdrawal plus other factors, you are giving the enemy the ultimate advantage."[55] The possibility of a ceasefire in Vietnam clearly raised a number of practical concerns regarding territorial apportionment and the designation of respective zones of control south of the seventeenth parallel. It was obvious that should a ceasefire be declared, both the NLF and the RVN would claim authority over the territory and hamlets they controlled. Moreover, they would both seek to rapidly expand their influence and control prior to a ceasefire and make every effort to "fly the flag" over the areas they occupied and in as many contested villages and hamlets as possible to claim larger jurisdiction after a truce had gone into effect. A ceasefire, most observers understood, would more than likely perpetuate hostilities, albeit in a different form, rather than reduce them. "A ceasefire in South Vietnam," noted one American official, "does not imply a shift from war to peace. It means instead a shift from military to political conflict."[56]

Because the war in South Vietnam was a civil war in which a partially homegrown insurgency had metastasized and established entrenched enclaves throughout the host nation, the likelihood of a ceasefire with regroupment like the one after Geneva in 1954 was highly unlikely. Thus, the only alternative was an in-place, or "standstill," ceasefire accompanied by territorial accommodation between the RVN and NLF. In such a situation each side would be expected to remain in place once hostilities had ended and govern the territory it occupied at the time of the ceasefire. Territorial accommodation, which was first introduced as a possible solution by Harvard professor Samuel P. Huntington in 1968 and later embraced by former U.S. ambassador Cyrus Vance and members of the Nixon administration, proposed partitioning the country into respective zones of control. Under such a deal communist forces and political cadres would be permitted to administer the areas they currently occupied in exchange for NLF recognition that Saigon was the sole, legitimate national authority. The purpose was to integrate the NLF into the Saigon regime's political structure and allow for an interim distribution of power at the local level.[57] The plan's major emphasis, therefore, was on political competition and accommodation at the village, hamlet, and district level; questions concerning communist political participation at the national level would be left until a later date and settled through elections. Territorial accommodation would only take place once a ceasefire was declared, all U.S. and North Vietnamese forces were removed, and the Vietnamese people could determine their own future without outside influence. In a practical sense, the arrangement was intended to recognize the realities of territorial control in South Vietnam and implement what Nixon called "a fair political solution" that reflected "the existing relationship of forces" inside the country—a status of mixed control where NLF enclaves and government territory were interspersed.[58] According to U.S. officials, territorial accommodation would help facilitate negotiations by legitimizing areas of communist control and granting the NLF a role in the political process. Moreover, it would assimilate the NLF into the body politic while preserving the overall structure and integrity of RVN sovereignty. "The advantages of this alternative," one official informed Kissinger, "are that the GVN would retain its national authority and constitutional legitimacy, and that it might serve as a flexible basis for negotiations, since the Viet Cong would be offered a large measure of local control of at least part of the country." Territorial accommodation risked a de facto partition of the country, something that had the potential to undermine Saigon's national authority. Nonetheless, U.S. policymakers considered this option because they believed the risks were minimal as the plan guaranteed the political legitimacy of the RVN and granted the NLF a limited, carefully proscribed role in the overall political structure.[59]

The Nixon administration had considered a standstill ceasefire as early as 1969. In September, Kissinger urged the president to consider an alternative strategy for settling the war aside from elections or a coalition government, one that would freeze opposing forces in place and apportion territory to the side that exercised effective control over the zone it occupied. Although Kissinger believed the current negotiating strategy and its emphasis on a political settlement through "free, fair and supervised elections" to be sound in many respects, he felt, nonetheless, that allied diplomacy was seriously undermined by one major weakness: time. The communists, he alleged, would never accept elections held under the auspices and authority of the RVN. On the other hand, the Thieu regime would never accept any deal that included the possibility of a coalition government. Meanwhile, the patience of the American people would be increasingly strained. Unless some "great outside pressure" forced both sides to move toward a settlement, he argued, negotiations would stagnate because Thieu and the communists would both refuse to negotiate any agreement that did not guarantee their preferred mechanism for political competition. "If there is not rapid movement in Paris," Kissinger wrote, "we just will not have the time with American opinion fully to play out this strategy. . . . [T]here will be building pressures in this country for further compromises . . . or accelerated troop withdrawals or a ceasefire." Kissinger's conviction that the current diplomatic strategy provided little incentive for either side to negotiate seriously or reach an early settlement prompted him to propose a ceasefire and accommodation.[60]

Kissinger and others assumed that a negotiating policy that pursued an in-place ceasefire followed by territorial accommodation was the best way to avoid the temporal dangers inherent within the current diplomatic strategy (protraction and gridlock). Such an arrangement, they argued, would expedite an agreement by assuaging the communists' concerns about elections and convincing Hanoi and the NLF that Saigon was willing to share power in South Vietnam. "Territorial accommodation should hold many attractions for them, both in terms of short range consolidation of local power and a longer term shot at national control," Kissinger informed the president. The willingness to cede control of territory and share political power with the communists appeared extreme. Kissinger, therefore, assured Nixon that territorial accommodation would do nothing more than place a legal imprimatur on a de facto situation that had existed for some time. "The situation which has existed for many years in South Vietnam would be given a measure of legal status. . . . Territorial accommodation would invoke implicit acceptance of the status quo and would seek to rule out efforts to change it by force. The NLF and the GVN would retain control over the territory and population in South Vietnam they now dominate. Power would be shared in contested areas." Kissinger argued that the surest way to achieve accommodation was through a negotiated ceasefire; and this, he believed, was the best way to buy time for settling the war on favorable terms. "We would be buying time with the US public by being forthcoming in Paris negotiations and moderating our military tactics, as well as continuing a careful phasing of troop withdrawals," he wrote. "If, indeed, a general ceasefire in place were put into effect, the psychological effect on American opinion would probably give our policy a major new lease on life."[61] As Kissinger would find out, mollifying the American public by adopting accommodation was one thing; convincing Saigon and the American military of its merits, however, was something quite different.

Kissinger's optimistic assessment of an in-place ceasefire and territorial accommodation was not shared by all. In fact, Kissinger himself had doubts. In an article that appeared in *Foreign Affairs* in January 1969, the incoming assistant to the president warned that a standstill ceasefire could pose a number of problems for ongoing negotiations. First, it could encourage a "mad scramble" to establish predominant control over population and territory just prior to any cessation of hostilities. With time running out and the opportunity to maximize their respective political and administrative zones rapidly diminishing, the RVN and the NLF would make every effort to expand their control over rural areas before hostilities ended. Next, an in-place ceasefire could make it virtually impossible to verify the withdrawal of NVA forces from South Vietnam. The communists, he argued, would more than likely lie about the nature of forces within the areas they controlled in order to keep as many of them as possible inside the country. This would make things infinitely more difficult during the post-hostilities period. Lastly, such an arrangement would likely facilitate a continuation rather than termination of hostilities, particularly low-level conflict, guerrilla activities, and terrorism aimed at disrupting the pacification program. Although the communists would more than likely honor the ceasefire militarily, Kissinger reasoned, they would at the same time ramp up the political struggle at the local level and continue to use violence and coercion to tip the balance of forces in their favor.[62]

Some officials expressed reservations about the possibility of an in-place ceasefire and territorial accommodation. Others, however, outright rejected it. President Thieu stood among the staunchest critics of such a proposal. Throughout 1969 and most of 1970, the RVN repeatedly dismissed the idea of an agreement that would permanently partition South Vietnam and grant the NLF jurisdiction over the scattered areas it occupied, something often referred to as a "leopard spot" configuration. Although Thieu and his government had repeatedly claimed they were amenable to a supervised in-place ceasefire as a *temporary* measure, they would never accept such a plan if it led to a permanent territorial division and coalition government. Such an arrangement would not only allow the communists to rebuild their political and military strength to resume the struggle; it would gradually dissolve the Saigon government. "To accept a leopard-spot solution," Thieu informed the press in July 1970, "is to go from step to step toward a coalition government, from rear to high echelons, from local areas to the entire nation. I confirm that I will never accept it."[63] RAND analyst Guy Pauker also believed territorial accommodation should be rejected because it would create an unpopular balkanized state. Such an arrangement, he reasoned, would not last long. Pauker argued that a "leopard-spot" division of the country would be detrimental to the political stability of the RVN because it would force a "medieval pattern of authority" on a country that had long ago ceased to consist of isolated "autarchic" village communities. Should such a governing formula be imposed, he argued, it would quickly be overthrown. "Sooner rather than later," he wrote, "a new power center would have to emerge, smash through the barriers, destroy the new 'feudalism,' and unify the country." Any unifying movement, he argued, would more than likely have the backing of a substantial portion of the population because accommodation and the balkanization of South Vietnamese society it inferred would impose "intolerable restrictions on the daily life of all people."[64]

Aside from concerns about an inevitable descent into coalition government, Thieu and other Saigon officials rejected proposals for territorial accommodation because they believed that any agreement that involved political inclusion of the NLF (which in their eyes was simply an extension of the North Vietnamese government) would be tantamount to allowing Hanoi to intervene in RVN affairs.[65] Moreover, accommodation struck directly at issues related to the territorial sovereignty and political legitimacy of the Saigon government. Should the current regime accept and recognize communist control over any portion of South Vietnam, however tentative or temporary, it would signal the first step in a loss of national sovereignty. This is something Thieu was reluctant to do. After all, the RVN's legitimacy and its claim as the rightful government in South Vietnam lay at the heart of the war. Rather than accede to territorial accommodation, Thieu and others advocated national elections, particularly after 1969. This was because they believed that extensive RVN control would translate into an electoral win for the government should a referendum be held.[66] Caught between communist efforts to destroy his government and the disillusionment and growing impatience of the American people, Thieu found himself on the horns of a dilemma. Yet he had no choice but to "hold the line," as one commentator put it, in order to protect the sovereignty of his country. "The U.S. government has gone to some lengths to establish the legitimacy of the Government of South Vietnam and has a continuing interest in seeing that legitimacy maintained. To pressure Thieu into further concessions would invite chaos within his regime."[67]

There were other signs of pessimism concerning the possibility of an in-place ceasefire and the dangers it could bring. MACV believed a standstill ceasefire would do little to alter Hanoi and the National Liberation Front's goal of toppling the RVN and seizing control of South Vietnam. Nor would such an arrangement rein in communist activities, although an armistice would more than likely prompt a shift from military to political struggle. The communists, they argued, would make every effort to expand their control and attack pacification during an orchestrated cessation of hostilities. "The enemy, even after an agreement is reached, will continue to pursue his ultimate objective of controlling South Vietnam," the plan asserted. "Thus, even if he abides with the agreement with respect to his regular forces, an upsurge in terrorist and propaganda action is likely. Even a ceasefire of short duration would probably result in an increase in enemy influence and control over the countryside and the rural population."[68] Military planners were also concerned that once a ceasefire had gone into effect, the NLF would move to announce a capital for the PRG and begin administering the areas under their control through liberation committees. Combined, these developments would give the NLF de jure as well as de facto control over portions of the country and grant the Front an aura of national and international legitimacy it had not previously enjoyed. Lastly, an in-place ceasefire was dangerous because it would afford the communists the time and freedom to prepare militarily to resume main-force hostilities. "He would be able to refurbish and strengthen in-country and out-country base areas," the plan's author continued, "and develop with relative impunity his [lines of communication] in the Khmer Republic and Laos."[69]

The Vietnam Special Studies Group harbored reservations concerning a standstill ceasefire as well. In 1969, the group conducted its own ceasefire study in twelve selected provinces throughout South Vietnam. The study concluded that an in-place ceasefire would, at best, yield mixed results, producing benefits in some provinces but serious drawbacks in others. The VSSG determined that in provinces where RVN authority was strong and NLF guerrillas and political cadres relatively weak, an in-place ceasefire would be beneficial. Such an arrangement would allow Saigon to consolidate and eventually expand its control over people and territory. Conversely, in areas where the NLF was strong and the government's administrative apparatus relatively weaker, an in-place ceasefire would be harmful. Here a cessation of hostilities would be followed by a steady erosion of RVN control and by NLF ascendancy.[70] Given the impact of existing levels of control and the ebb and flow of the control war in the countryside, the timing of an in-place ceasefire would be crucial. Should such an arrangement occur when levels of government control were at their peak and those of the NLF at their nadir, it would benefit the RVN and the United States. Should it occur when levels of communist control were expanding, however, it would benefit Hanoi and the NLF. Both sides, therefore, would focus their efforts on maximizing influence and control over the operational environment in the expectation that an in-place ceasefire, should one be declared, would effectively halt opportunities for future expansion.

The VSSG study predicted a ceasefire that did not include the withdrawal of North Vietnamese units from South Vietnam would bring about "major and serious losses" in government control of territory and population at the national level. Such losses, the report continued, "are likely to be irreversible without the reinsertion of massive U.S. troops; and barring that, the enemy's prospects for a military victory would be greatly enhanced. At best, South Vietnam would be a divided country with the enemy in control of I and II Corps and the control of GVN in III and probably IV Corps. However, this situation might not be stable, making further GVN deterioration a possibility. At worst, the GVN would grow weaker and fall, by political or military means, to the communists."[71] The only conceivable way the RVN could survive, the study concluded, was if there was a withdrawal of NVA units as part of a ceasefire package. Although the removal of North Vietnamese troops did not necessarily make victory over the insurgency inevitable, the removal of the NVA from south of the seventeenth parallel would give Saigon the best chance of defending itself should Hanoi abandon political competition and once again seek to overthrow the government through military means.[72]

Debate among allied policymakers over the decision to pursue an in-place ceasefire and territorial accommodation was effectively settled in October 1970 through unilateral executive actions. In a televised address delivered from the White House, Nixon attempted to break the negotiating "logjam" by asking the communists to join the United States and the RVN in an internationally supervised ceasefire followed by an Indochina Peace Conference. "I propose that all armed forces throughout Indochina cease firing their weapons and remain in the positions they now hold," the president declared. "It would not in itself be an end to the conflict," he continued, "but it would accomplish one goal all of us have been working toward: an end to the killing. I do not minimize the difficulty of maintaining a ceasefire in a guerrilla war where there are no front lines. But an unconventional war may require an unconventional truce; our side is ready to stand still and cease firing."[73] Though relatively simple in concept, Nixon's proposal did little to dispel the uncertainty surrounding the modalities of an in-place ceasefire. "If a ceasefire were declared immediately in South Vietnam," one journalist pondered, "who would be holding what, and where?"[74] Another newspaper stated acerbically, "If there is one clearly outstanding feature of the President's 'new' proposal it is ambiguity. It could be new, or old; real or illusory; promising or not. . . . [I]n this matter of a Vietnam ceasefire everything depends on how you define the term."[75]

Despite reservations, Nixon's offer for an in-place ceasefire reflected a growing confidence in Washington that the dual strategy of pacification and Vietnamization was working, and that progress in these areas, along with the recent allied incursion into Cambodia, had made the allied position strong enough to make an offer for territorial concessions possible. In his 7 October speech Nixon stated that his new peace initiative had "been made possible in large part by the success of the Vietnamization policy."[76] Several days later, a White House spokesman informed the *New York Times* that Nixon had made the decision "because of the improved military and pacification picture." HES statistics supported such a claim. That same week, pacification officials reported that 92.8 percent of the population was living in "relatively secure" areas.[77] American media were quick to pick up on Washington's newfound optimism. "White House officials acknowledged . . . that as recently as last year U.S. military leaders opposed a ceasefire in Vietnam on grounds it would then be extremely disadvantageous to allied forces. Now the tables are turned enough for the United States and South Vietnam to propose what they once feared would be thrust at them," noted the *Washington Post*.[78] "It was a sense of military and political weakness that had hitherto caused both the Johnson and Nixon Administrations to shy away from the idea of a ceasefire," the *New York Time*s declared. "These fears have been dispelled in recent months. Even officials who continue to doubt the ability of Saigon to withstand a long guerrilla war believe that its position for the time being . . . has been made reasonably secure."[79] These assessments were valid. In fact, Washington officials were so optimistic they were able to convince Thieu and other South Vietnamese officials to agree to a standstill ceasefire. In a statement issued hours after Nixon delivered his address, Saigon announced its support for the plan. Although Thieu continued to express reservations concerning the success of such a proposal, he nonetheless reversed his previous position and reluctantly agreed to support an in-place ceasefire throughout Indochina.[80] The decision, more than likely, was based on Saigon's conviction that the regime currently held a winning hand and that Washington would continue to assist the RVN in its fight against insurgency and military aggression from the DRV.

Washington and Saigon's willingness to sanction an in-place ceasefire was not matched in the communist camp; Hanoi and the NLF immediately rejected the offer. There could be no ceasefire, they argued, without first removing all U.S. forces and deposing the Thieu-Ky-Khiem regime. Within days of the president's 7 October speech, DRV and PRG spokesmen put forth a point-by-point denunciation of the proposal. Only when the United States agreed to withdraw quickly and unconditionally and end its support for the current Saigon government, they maintained, could a ceasefire be considered.[81] "We are not opposed to a ceasefire," the DRV's chief negotiator in Paris, Xuan Thuy, stated in an interview. "The two fundamental questions—the military question and the political question—should be settled first. That is to say the United States should withdraw completely all its troops, and those of the other foreign countries, from South Vietnam. And the government of Thieu, Ky, and Khiem should be changed for the formation of a provisional coalition government in South Vietnam."[82] A Liberation Radio broadcast asked, "How can peace be achieved while more than half a million U.S. expeditionary troops and satellite servicemen and the country-selling, dictatorial Thieu-Ky-Khiem clique, which are the instruments of war and the basic cause behind the waging of this aggressive war, continue to exist?"[83] On 16 November, Thuy reiterated to American ambassador David Bruce that Hanoi and the NLF would not consent to a ceasefire until all political and military problems were settled and an agreement had been signed.[84] The communists' refusal to declare a ceasefire until a settlement had been reached and American combat forces were gone from South Vietnam was not new. In November 1969, Nguyen Thi Binh, the PRG representative at the Paris peace talks, informed the American press that Hanoi and the PRG would never agree to a ceasefire unless Washington and Saigon accepted the NLF's ten-point program for ending the war. The central demand of the program was unconditional American withdrawal. "Our position is very clear," Madame Binh informed the delegation. "A ceasefire is not enough. It must go together with withdrawal of all United States troops from South Vietnam and only by this means can an end be put to the war in Vietnam. . . . Justice requires that we must demand that the United States withdraw all troops from South Vietnam. It must not simply be a ceasefire. It must be a lasting peace."[85]

The communists rejected an in-place ceasefire and political accommodation because they were not interested in a compromise settlement; they were, instead, intent on crushing the Saigon regime and seizing power. Should an in-place ceasefire be declared while the United States and the RVN held a superior position in South Vietnam (something the allies did in fact have at the time), these objectives would be almost impossible to fulfill. Thus, a ceasefire had to be avoided because it would stop the communist revolution dead in its tracks, both on the battlefield and at the negotiating table. Moreover, if revolutionary activities were halted, it would afford the RVN and the United States an opportunity to consolidate and strengthen Saigon's position in the countryside. Hanoi and the NLF knew Washington and Saigon would use a cessation of hostilities to shrink communist influence and control. "We oppose a ceasefire," Binh declared in 1970, "because this is the way for the United States to compel the South Vietnamese people to stop fighting while U.S. aggression continues there."[86] Madame Binh and others argued that such an arrangement prior to a settlement would leave Saigon in a dominant position in regards to population and territorial control, and that the RVN would attempt to consolidate the pacification gains made since 1968.[87] "If the ceasefire means no more shooting while Americans and their allies go swarming all over South Vietnam to distribute food, build homes, and deliver the mail," one journalist noted, "it would be tantamount to an enemy surrender."[88] Many communists also feared that the allies would use the lull in fighting to press the attack against vulnerable guerrilla units and political cadres. A standstill ceasefire, an American journalist noted, would "force the NLF to reveal its positions and expose itself to being liquidated by superior firepower." A North Vietnamese official declared that any ceasefire without a formal settlement would be extremely risky. "The proposed ceasefire," he said, "is in fact a camouflaged demand for the surrender of the National Liberation Front."[89] Still others cited the violence and repression of the Diem regime after the 1954 ceasefire as proof that the RVN would not honor a cessation of hostilities but rather use it to hunt down revolutionaries and attempt to gain leverage. "With such a ceasefire there will be no more gunfire but there will still be death, imprisonment, humiliation, poverty and starvation," declared a Liberation Radio broadcast. "With such a ceasefire territorial integrity will be encroached upon, the people's right to self-determination will be trampled on, and there will be no independence, freedom, peace and democracy. Because the South Vietnamese people lived out the dark days during the six years of restored peace under the U.S.-Diem regime, they can now clearly realize what would happen, after such a ceasefire would become effective."[90] According to Wilfred Burchett, a communist sympathizer and journalist who spent years embedded with the National Liberation Front, the International Control Commission established after Geneva in 1954 failed to prevent "the physical liquidation of hundreds and thousands of patriots" at the hands of Diem's police and military units. The only guarantee against a repetition of these acts, Burchett argued, would be the "maintenance of the armed forces of liberation and the preservation of their military and political integrity."[91]

Hanoi and the NLF also rejected an in-place ceasefire in 1970 because they feared an agreement that ended hostilities while large numbers of American troops still "occupied" portions of Vietnam would legitimize the neo-imperialism of the United States and its Vietnamese ally. Moreover, it would rob the people of their sacred right to rebellion and armed resistance against "illegal occupation" by a foreign power and illegitimate "puppet" regime.[92] "A ceasefire of the Nixon type," a VWP spokesman noted bitterly, "is to deprive the South Vietnamese people of their right of self-defense, to ask the people to unconditionally surrender, to let the American troops freely occupy South Vietnam for an unlimited period of time and to legalize the Thieu-Ky-Khiem clique, the instrument of U.S. aggression in South Vietnam."[93] According to Xuan Thuy, Nixon's ceasefire proposal was "tantamount to the legalization of the U.S. aggression in South Vietnam and also the legitimization of the position of the U.S. agents in South Vietnam and other countries of Indochina."[94] Yet another justification given for rejecting a ceasefire was that it would deny Hanoi and the NLF a key bargaining chip in their efforts to force the United States to withdraw: dead GIs. As one American journalist noted, the communists' preference during the Paris peace talks had been for "a political negotiation during which American forces are being killed and held hostage in Vietnam."[95] Should hostilities cease, some argued, the communists would lose the ability to apply political pressure by inflicting casualties on American forces. In November 1969 Nguyen Thi Binh dismissed suggestions for a ceasefire based on this very reason. According to reports, Binh maintained that a ceasefire at the present time would leave more than a half million American military personnel in South Vietnam with no firm program for their withdrawal and no means to exert pressure to speed their exit. "A ceasefire when there are 500,000 American troops staying on in Vietnam," she declared in a press conference, "would mean we would be compelled to accept their presence without being able to oppose it."[96]

Concerned as they were about the possibility the allies would violate a ceasefire and attempt to use it to their advantage, the communists nonetheless had their own plans to exploit a cessation of hostilities and expand their influence and control should there be a negotiated settlement. According to the Nixon administration, Hanoi had been making contingency plans for a potential ceasefire as early as 1969. Reports from communist defectors indicated that Hanoi planned to manipulate a potential standstill ceasefire both before and after U.S. withdrawal and the installation of a coalition government. According to these sources, communist leaders anticipated not only that NLF political cadres would infiltrate cities and other urban areas to foment demonstrations and popular uprisings between a ceasefire and American withdrawal, but that the NLF/NVA would attack South Vietnamese forces once the Americans had left.[97] Here we see the General Offensive and General Uprising strategy applied to a ceasefire situation. Captured communist documents reveal similar machinations in the months leading up to the Paris Peace Accords. "When a political solution is reached," a December 1972 directive declared, "we have to initiate a general uprising which is carried out continuously on the largest scale and which will go on before, during, and after the ceasefire becomes effective, the purpose is to gain biggest victories and to liberate the bulk of the rural area."[98]

According to COSVN, the period between a ceasefire and a final political settlement was a favorable occasion to continue the struggle against the RVN and advance the revolution. "This period will be a great opportunity for revolutionary violence, for gaining power in South Vietnam, for troop and enemy proselyting, and for making great leaps in the balance of forces," it read.[99]

The RVN was no less determined to use a ceasefire to its advantage. Although the allies had no formal plans to openly violate a cessation of hostilities, Thieu's insistence that his country's sovereignty be preserved at all costs, and its territorial integrity respected, prompted him to exert every effort to expand government control just prior to a ceasefire and eliminate remaining communist areas after an armistice went into effect. Thieu's platform for negotiating with the communists between 1969 and 1972 was infamously known as the "four no's" policy: no recognition of the NLF as a legal entity in South Vietnam; no neutralization of his regime; no coalition government; and no surrender of territory to the enemy.[100] In 1972, the deputy assistant to the president for national security affairs, Major General Alexander Haig, informed Kissinger that while Thieu might be willing to show *some* flexibility concerning the "practical mechanics" of a ceasefire, he would more than likely not "back off a micromillimeter from his public claim to sovereignty and lawful jurisdiction over all of South Vietnam's people and all of its territory."[101]

The United States appears to have had every intention of honoring a ceasefire once a negotiated settlement went into effect. Nonetheless, the Nixon administration collaborated with Saigon's plan to seize as much populated territory as possible in the fall of 1972 when a settlement appeared imminent. On 8 October, Haig suggested Kissinger send a back-channel message to Ambassador Bunker instructing him to press Thieu to move rapidly and grab as much territory as possible in anticipation of a possible ceasefire offer by the communists. The following day, Bunker responded that Thieu had already issued such a command. "Concerning your reference to fact that it is essential for Thieu to instruct his commanders to move promptly to seize the maximum amount of critical territory," the ambassador informed Haig, "Thieu informed me at my meeting with him on October 6 that he had given these instructions to his Corps commanders."[102] On 12 October, Kissinger reiterated to Bunker the need to press Saigon to seize as much land as possible in light of a potential break in negotiations and Hanoi's willingness to agree to a standstill ceasefire.[103]

Like virtually every other issue discussed at the peace talks in Paris, the modalities of an in-place ceasefire and territorial accommodation remained unresolved until January 1973, when all four parties signed the treaty that would ostensibly end the war in Vietnam. Although the informal allied proposal for a standstill ceasefire first introduced by Nixon in October 1970 had become part of the allied negotiating platform the following year, when the American delegates in Paris incorporated an Indochina-wide in-place ceasefire into their final proposal for a negotiated settlement,[104] Hanoi and the NLF were unwilling to reciprocate. In 1971, the communists had codified their previously informal rejection of an in-place ceasefire as part of the DRV's Nine-Point Peace Program proposed by Xuan Thuy in Paris. Among its terms for a settlement was the oft-repeated stipulation that a ceasefire would only be implemented once a final agreement to end the war had been signed.[105] As it turned out, the issue of a standstill ceasefire and territorial accommodation, although it exerted a profound influence on the control war and the struggle for population and territory in rural South Vietnam, remained deadlocked in Paris for another eighteen months.

VIETNAMIZATION AND AMERICAN WITHDRAWAL

If negotiations and peace talks in Paris were the diplomatic leg of Nixon's strategy for ending the war, Vietnamization, along with pacification, constituted the military leg. Vietnamization was the centerpiece of allied military policy in the years following *Tet Mau Than*. Designed to allow the RVN to assume stewardship of the war effort and provide for its own defense, the program centered on efforts to modernize and develop the RVNAF and expand government control over the countryside. "Vietnamization has two principal components," Nixon informed Congress in February 1970. "The first is the strengthening of the armed forces of the South Vietnamese in numbers, equipment, leadership and combat skills, and overall capability. The second component is the extension of the pacification program in South Vietnam."[106] The president's description was appropriate. Vietnamization and pacification, while distinct programs, were part of a single overarching plan to strengthen the RVN militarily and politically so that it could stand on its own without American assistance. They were, in short, two faces of a single strategic coin. "Vietnamization is on the nonmilitary side what is usually embraced in the term 'pacification'; that is, the extension of government control, government services, and the presence of the government within the countryside," Undersecretary of State U. Alexis Johnson informed the House of Representatives Committee on Appropriations in 1970. "On the military side, Vietnamization means improving the training, improving the equipment, and improving, in general, the capabilities of the Vietnamese forces to deal with the enemy they face."[107] Vietnamization was also closely linked to the ongoing peace talks and part of Nixon's overall diplomatic strategy to compel Hanoi to negotiate. Frustrated by the stalemate in Paris, Nixon touted the program as a way to break the diplomatic logjam by applying pressure to the communists. "Vietnamization is . . . a spur to negotiations," the president declared. "In strengthening the capability of the government and people of South Vietnam to defend themselves, we provide Hanoi with an authentic incentive to negotiate seriously now."[108] The program also contained the added bonus of placating—or at least undermining—domestic critics of the administration and opponents of the war in Vietnam. "Withdrawing a number of American combat troops," Nixon recalled after the war, "would demonstrate to Hanoi that we were serious in seeking a diplomatic settlement; it might also calm domestic public opinion by graphically demonstrating that we were beginning to wind down the war."[109]

Empowering the RVN and its people to assume the burden of their own defense first meant strengthening the nation's armed forces so that the government could protect the population and handle the communist threat on its own. The overall objective was military self-sufficiency. "Vietnamization seeks as its primary objective the assumption by the South Vietnamese Armed Forces of the total responsibility for military security within South Vietnam," declared a congressional report. "In the context of a war that is part conventional and part guerrilla, this entails the development of a multi-function force that can carry out several types of security related tasks."[110] A "multi-function" force had already existed in South Vietnam. The one in existence when the Vietnamization program had been put in place, however, was inadequate for meeting the task at hand. It needed, therefore, to be expanded and improved. The problem was not that the RVNAF lacked a variety of forces geared toward tackling different security contingencies; the issue was that Saigon's multifunction military was unable to defend the nation without the help of U.S. and other allied combat forces. Vietnamization was meant to remedy this weakness. The program, therefore, worked to quantitatively and qualitatively improve Saigon's conventional and territorial forces via conscription, modernization, and training. Special attention was devoted to improving and expanding units charged with local defense, particularly the RF/PF, so that the ARVN could be freed to assume mobile tactical operations against enemy main-forces as the Americans drew down. According to South Vietnamese general Dong Van Khuyen, one-half of the total increase in force structure under the Vietnamization program went to the RF/PF. The buildup of regular army combat forces, meanwhile, also progressed, though modestly when compared to the number of U.S. forces they were replacing.[111] In the three years that followed *Tet Mau Than*, the RVNAF grew by more than 400,000 troops, swelling the ranks to 1,045,000 personnel by 1971. The following year, overall enrollment peaked at 1,090,000 men under arms.[112] RF/PF, meanwhile, jumped from 392,000 to 532,000 members during the same time period. In 1972, its membership consisted of 520,000 personnel.[113]

As originally conceived, Vietnamization was intended to build a fighting force that could provide security against the NLF insurgency alone. This was done for a number of reasons. "The present RVNAF modernization program was designed only to build up the South Vietnamese forces so that they could cope with VC insurgents," Secretary of Defense Melvin Laird informed Nixon in March 1969. "Our military authorities believe neither the . . . manpower base nor any possible modernization program would enable the RVNAF to cope alone with a threat comparable to the present level of aggression."[114] The original plan for a limited defense force also reflected the assumption that a negotiated settlement to end the war would remove all North Vietnamese troops from southern soil. General Creighton Abrams revealed this supposition in 1969. "We've tried to design them [the RVNAF] to deal with insurgency, and not with an NVA force of the magnitude and kind that is present at the moment in and around South Vietnam," Abrams stated. "The thrust of this whole thing has counted on . . . negotiating the North Vietnamese out of South Vietnam."[115]

The initial limited mission of the RVNAF as articulated by Laird and Abrams was later altered when it became obvious that North Vietnamese units could not (and most likely would not) be negotiated out of the country. By September 1969, plans for improving and modernizing the RVNAF had changed. Instead of gearing them toward fighting insurgent forces only, the program was expanded so that the RVN could meet a mixed threat of NVA regulars and NLF guerrillas. In September, Laird informed Nixon of this change in focus. The "Improvement and Modernization Program was originally intended to prepare the RVNAF to meet the residual VC insurgency threat after the North Vietnamese troops had been withdrawn," Laird said. "That residual VC threat, however, has been declining. This diminishing VC threat, coupled with RVNAF improvement, must lead us to reorient our thinking of Vietnamization goals. We are now considering the feasibility of expanding the program to prepare the RVNAF to meet a combined VC/NVA threat." According to the secretary of defense, goals for the revamped program included improvement in military leadership and esprit de corps, reduction in desertion rates, improved logistics and intelligence capabilities, increased combined operations and planning, and the development of strategy and tactics to meet military capabilities.[116]

Vietnamization was part of the planned American withdrawal from Southeast Asia. By April 1968, it was obvious that the United States was intent on reducing its combat commitments and expected the RVN to assume responsibility for its own defense if allied strategic aims were to be salvaged and the RVN have a fair chance at survival once the Americans departed. A central feature of the program, therefore, was the phased drawdown of U.S. combat forces. The plan was that as American forces "stood down," the RVNAF would "stand up." American policymakers believed that current efforts to expand pacification and strengthen the RVNAF made the American drawdown possible. "These two things, together," Undersecretary Johnson informed Congress, "mean a lessening need for a U.S. role and are thus what enable us to bring about a reduction of our forces without endangering the existence of this program."[117] Other Washington officials credited the improved security situation in the countryside. "This withdrawal program, at least in the areas of ground combat, could and should be accelerated," Congressman Howard W. Robinson informed the Select Committee on United States Involvement in Southeast Asia in July 1970. "The military situation in South Vietnam today," he continued, "is better for both the U.S. and the South Vietnamese than at any time since American units first entered the war in the summer of 1965. In addition, the South Vietnamese fighting units have undoubtedly gained new self-confidence as a result of their role in the Cambodian sanctuary operation."[118] The first round of withdrawals, which included 25,000 troops, was announced on 8 June 1969 and completed by the end of August.[119] Another 35,000 were removed by December. A third increment of 50,000 was withdrawn by the end of April 1970.[120] On 20 April, Nixon announced plans to withdraw an additional 150,000 troops over the next year. By that point, American troops had been leaving Vietnam at a rate of about 12,000 men a month.[121] As of 1 May 1971, nearly half of all U.S. forces had returned home from Southeast Asia. Of the remaining 284,000, another 257,000 were sent back to the United States over the next eighteen months. By the end of 1972, only 27,000 American military personnel remained in South Vietnam, most of them advisers.[122] The great American war machine that had been sent to Southeast Asia to save the teetering Saigon regime in 1965 and that had reached upwards of a half million men by the spring of 1969 had been dismantled. By December 1972, the conflict in South Vietnam had essentially been de-Americanized, and full custodianship for the ground war had passed to the RVNAF.

Vietnamization and the phased withdrawal of U.S. combat forces had a profound impact on pacification and RVN efforts to control territory and population. To begin with, it diminished the overall levels of American advice and support, particularly in pacification. According to historian Richard Hunt, "Vietnamization meant fewer and fewer American forces would be available to provide security for cadre teams, government officials, and territorial forces implementing pacification. There would also be fewer advisers overseeing South Vietnamese planning and program administration."[123] The American drawdown also placed added pressure on both U.S. forces and the RVNAF, demanding the latter achieve proficiency within a limited time frame. Although the extraction of American units was gradual, South Vietnamese units still had to be rapidly brought up to speed so that the transition was seamless and there were no gaps in allied combat capabilities. Once gone, the Americans would not be coming back. The RVNAF, therefore, had to be capable of fulfilling security missions even before the Americans had left. "The process of withdrawal was likely to become irreversible," Kissinger wrote after the war. "Henceforth, we would be in a race between the decline in our combat capability and the improvement of South Vietnamese forces—a race whose outcome was at best uncertain."[124] The fact that U.S. combat forces were leaving also mandated the RVN sprint to expand the government's presence and control as quickly as possible. American withdrawal elicited a need for speed not only on the military side of Vietnamization (the modernization and strengthening of the RVNAF) but on the political side as well. Dwindling numbers of U.S. combat forces meant that the RVN had to shrink the insurgency and maximize government control over population and territory within a relatively short period of time. In short, Saigon had to quickly "pacify" the countryside and create an environment that was both safe and productive if the nation was to, on its own, manage the ongoing insurgency and survive. "As the withdrawal of the U.S. presence in the country proceeds," read a report generated by a special committee of the House of Representatives, "only a secure and productive countryside can maintain in South Vietnam a society economically viable and politically stable."[125]

As with virtually all armed conflicts, the war in Southeast Asia shaped and was shaped by larger political and diplomatic currents far removed from the battlefield. After *Tet Mau Than*, both sides scaled back their main-force operations and placed greater emphasis on expanding and consolidating their hold on the countryside. The Tet attacks and the immense political changes they wrought, particularly the advent of peace talks in Paris and American withdrawal, were responsible for these adjustments. As this chapter has shown, the conflict in Southeast Asia had, beginning in 1968, moved from what Charles Joiner had called the "politics of war" to the "politics of peace." This new era of peace politics had a profound impact on efforts to control the villages and hamlets of rural South Vietnam. Once Vietnamization and the American drawdown had become reality, and the possibility of a negotiated settlement became imminent, both sides altered their approach and adjusted their tactics. The goal was to rapidly maximize control over the civilian environment in preparation for the postwar political struggle that would surely follow a brokered peace. Neither the communists nor the RVN and its allies knew what would come out of the ongoing peace talks. The only certainty was that time was now of the essence and that both sides had to move quickly to improve their situation in case an in-place ceasefire and territorial accommodation were implemented. What followed was an immense effort to expand a physical and/or psychological presence over as much of the operational environment as possible. The control war was off and running.

6 ALLIED ASCENDANCE AND THE CRISIS OF THE INSURGENT STATE, 1968–1971

Beginning in the fall of 1968, efforts to expand and consolidate control over the operational environment had assumed unprecedented urgency owing to the fallout generated by *Tet Mau Than* and the "politics of peace." Establishing political suzerainty in the villages and hamlets of rural South Vietnam and attaining hegemony over land and people had become a top priority for both sides with the advent of peace talks in Paris and the possibility of a negotiated settlement that could include an in-place ceasefire and territorial accommodation. Over the next four years, the struggle to seize control of the operational environment would dominate the war inside South Vietnam and serve as a touchstone for the ongoing diplomatic efforts in Paris.

PACIFICATION AND THE 1968 TET OFFENSIVE

At first glance, *Tet Mau Than* appeared to have damaged pacification as much as the insurgency. Less than a week after the January 30 attacks had commenced, the

American press and allied policymakers were claiming pacification had been dealt a severe if not mortal blow. "Pacification is dead," declared Ward Just of the *Washington Post*. "The [Tet] raids, audaciously conceived and executed with extraordinary ferocity, have as a practical matter killed dead the pacification program—and most of the assumptions that went with it."[1] In a 5 February brief, Thomas Hughes, the director of the Bureau of Intelligence and Research, informed Secretary of State Dean Rusk that pacification had suffered "serious setbacks" in the countryside as a result of the attacks. "Headquarters and support facilities have been disrupted or destroyed in a number of towns, and many pacification . . . cadres presumably are without any guidance and/or in a highly demoralized state. More importantly . . . the Communist attacks have almost certainly raised further doubts among the rural populace as to the ability of the government to protect the countryside when it cannot protect secure urban areas."[2] General Earle G. Wheeler, the chairman of the Joint Chiefs of Staff, made an equally grim assessment on 27 February. "The enemy is operating with relative freedom in the countryside, probably recruiting heavily and no doubt infiltrating NVA units and personnel," he proclaimed. "The defensive posture of ARVN," he continued, "is permitting the VC to make rapid inroads in the formerly pacified countryside. ARVN, in its own words, is in a dilemma as it cannot afford another enemy thrust into the cities and towns and yet if it remains in a defensive posture against this contingency, the countryside goes by default."[3] Further adding to the pall cast over the pacification program, the Central Intelligence Agency proffered a similar assessment shortly thereafter. In a memorandum dated 19 March, the agency announced that pacification had suffered greatly as a result of the Tet attacks. "The pacification program has received a severe setback in the majority of South Vietnam's 44 provinces as a result of enemy activities since the initiation of the Tet offensive on 30 January," it stated. "In the long run, the most damaging aspect of the offensive may well prove to be its adverse impact on popular attitudes toward pacification. Evidence already indicates that the enemy action has greatly increased the apathy and passivity of many rural residents toward government programs and personnel."[4] Pacification field reports generated by province senior advisers confirmed Tet had damaged pacification at the local level. In the northern province of Thua Thien the situation was dire. "The Tet offensive in late January disrupted all efforts in the countryside, and effectively destroyed all semblance of GVN rule until late February," noted the March pacification summary.[5] Down in the densely populated Mekong Delta, things were not much different. "The Viet Cong are succeeding in their strategy to contain the government within the cities while they occupy the countryside," read the March summary for Kien Hoa Province. "They are also having some success," it continued, "in discrediting the Government of Vietnam's competence and capability to protect and provide for its people."[6]

Although the initial attacks of *Tet Mau Than* had hurt pacification in rural areas, the overall damage was far from incurable. Despite its injuries, pacification remained a viable program. The initial attacks had seriously eroded security in many areas and brought numerous villages and hamlets under communist control as security forces abandoned rural areas to protect towns and provincial capitals.[7] Yet the program survived. As some of Washington's more optimistic officials indicated, pacification, contrary to early reports, had not only weathered the storm, but was poised to make a comeback. U.S. Ambassador Ellsworth Bunker informed President Johnson only weeks after the attacks began that despite many alarming reports, pacification was alive and well. "We now have a better feel of the pacification situation," he stated. "Reports are in from all province senior advisors . . . [and] their major conclusion is that there is not nearly as much damage to pacification as we earlier feared and [the] press is still reporting. Pacification is far from dead."[8] In April, General George Forsythe, acting deputy of CORDS, estimated that pacification was only "seriously interrupted" (a 3–6 months' setback, he estimated) in nineteen provinces. The remaining twenty-five provinces experienced either a minor setback (less than 3 months) or no setback at all.[9] That same month, American advisers in all four military regions reported that the situation in the countryside had stabilized and that pacification was once again moving forward. "Setbacks experienced during Tet," one American adviser declared, "have largely been restored and there are indications that the overall program has a chance to attain original goals."[10] According to another adviser, "The trend upward from the nadir of February continued through April with a definite improvement in nearly every aspect of pacification. Generally, this reporting period is the most encouraging that IV Corps has experienced in the past eight months."[11] On 18 April, Robert Komer informed the American press that pacification was "back on track."[12] Although he admitted the RVN had lost control of more than a million people, Komer claimed the overall damage to the program was far less serious than had been initially reported. "I've said before that pacification suffered a distinct setback," he declared. "But the more data that comes in, the more we get a perspective on the situation, and the more it seems as though the impact was more limited than we had earlier thought." Citing a relatively small number of hamlets that came under full communist control, the high number of personnel assigned to pacification work, and positive developments such as the reopening and clearing of roads and waterways, the ever-optimistic Komer painted an encouraging picture of pacification recovery. "The bulk of our pacification assets are intact," he proclaimed. "We are moving increasingly back into the countryside. In effect, we've . . . absorbed the Tet setback and we're on a recovery curve. I think the outlook is an encouraging one."[13]

For Komer and others, the damage Tet inflicted on pacification was more psychological than physical. The major problem with the program, they believed, was not the loss of government control and material assets, but timidity and a lack of initiative among the RVNAF and CORDS personnel. In his 20 March report to the president, Ambassador Bunker argued that the loss of territory and population was not a major issue; these things could and would be recouped. What was problematic, he asserted, was the lingering tentativeness and inactivity brought on by the communist general offensive. "The key problem is one of defensive-mindedness," Bunker wrote. "RVNAF forces and U.S. advisors alike are recovering only gradually from the shock effect of the Tet Offensive. As a result they are only slowly pushing back into the rural areas which were lost."[14] On 29 March newspaper reports confirmed this assessment, indicating that insecurity and fear of another communist attack was undermining efforts to revive pacification in the countryside. "Our biggest concern," one senior official explained to the *New York Times*, "lies in the greater fear and apathy among the rural population, some loss of confidence in the government's ability to protect them, and a greater defensive-mindedness among government security forces worried over renewed Viet Cong attacks."[15]

When combined with the urban focus of *Tet Mau Than* and erosion and disorganization of NLF guerrillas and political cadres in the countryside, the allies' timid approach to recovering territory lost during the communist offensive created a politico-military vacuum in rural areas—a void that both sides were determined to fill. Tet had denuded the countryside of military forces as both the NLF and allied forces focused on cities and provincial capitals. Once the communist onslaught had been checked in urban areas and the failure of the General Offensive and General Uprising became evident, the allies pivoted back toward the countryside in an effort to rapidly reassert control over the operational environment. "The real question now is who will fill the vacuum in the countryside," stated an American official. "It depends on how fast the South Vietnamese government moves in and how aggressive and how fast the enemy will be."[16] Richard Nixon made an identical assessment nearly two decades later. "Our defeat of the Tet Offensive had produced a political vacuum in the countryside," he wrote in retrospect. "Areas that the National Liberation Front had controlled for years were now up for grabs. We knew that whichever side won the race to take control of the hamlets would have won half the battle."[17]

To rapidly fill the politico-military void in the countryside and preempt communists' efforts to reassert their influence and control over the operational environment, allied policymakers urged immediate action. In his 20 March transmission, Ambassador Bunker informed President Johnson that the NLF, despite suffering heavy losses, was trying quickly to fill the political void in rural areas. This, Bunker argued, demanded a swift riposte. "We must get our local forces back out into the hamlets to cope with [the enemy], which means overcoming the defensive-mindedness of GVN pacification forces in the field," he stated. "The critical need is to reassert a GVN presence in the countryside as fast as possible." Bunker went on to assert that what the allies needed were "quick-fix solutions" that could produce "high impact" over the short term. The "slow, painstaking work by RD teams in hamlets as they seek through 98 tasks to accomplish everything from elimination of VC infrastructure to eradication of illiteracy," he declared, "simply cannot be afforded." To facilitate an accelerated government presence in rural areas, Bunker suggested cutting the size of RD teams from fifty-nine men to thirty- or forty-man "half-teams" so that Saigon could "show the GVN flag" in a larger number of hamlets. He also urged RD cadres to focus on security, psychological operations, self-defense, and anti-NLF activities at the expense of development activities. The latter, he reasoned, could resume once communist guerrillas and political cadres had been eliminated.[18]

The RVN and CORDS made this hasty, corner-cutting method the modus operandi of allied pacification efforts over the next two years. Later referred to by John Paul Vann as a "fast-and-thin" approach, the type of pacification advocated by Bunker privileged the rapid (albeit tenuous) extension of government control through force rather than the slower and more methodical task of nation building. "In this situation we intend to focus on short-term, high-impact programs to capitalize on [the] existing situation," the acting deputy for CORDS stated in April 1968. "This may require temporary subordination of certain of our long-term programs of economic and political development."[19] The decision to pursue politico-military control rather than politico-economic development lay at the heart of the Accelerated Pacification Campaign and subsequent operations. The objective was not to develop rural areas but to rapidly establish some degree of government presence throughout the countryside. Thus, nation building and its emphasis on social, economic, and political development took a back seat to the elimination of communist military forces and political infrastructure. "The accelerated pacification program," noted a British newspaper, "has wisely shed some overambitious attempts at rural reform. . . . It is now purely and simply an effort to wrest control of hamlets and villages from the Viet Cong."[20]

The decision to pursue fast-and-thin pacification and hurried expansion of government control rather than slow-and-steady nation building and meticulous political and economic development came in reaction to the shifting circumstances and profound changes wrought by the Tet attacks. As already discussed, *Tet Mau Than* was a "game changer" because it fundamentally altered the dynamics and trajectory of the war in Vietnam. Although the communist offensive did little to dissuade the contending parties from pursuing military victory or undermine their faith in armed force and violence to achieve political aims, it did convince both sides that current methods were largely ineffective and that alternative strategies must be pursued if an advantage was to be gained. This, as we shall see, prompted both sides to modify their behavior and pursue significant changes in policy and action.

TALK-FIGHT

Between 1968 and 1972, Hanoi and the NLF engaged in a strategy that was popularly known as "talking while fighting." The Vietnamese communists, however, were not alone in this approach; Saigon and Washington adopted a similar strategy. Thus, both sides employed a reciprocal method that used violence and combat actions to influence the ongoing negotiations on the one hand, and diplomatic pressure at the negotiating table to affect events on the ground on the other. Neither the communists nor the RVN and its allies were willing to stop fighting or halt attempts to gain advantage through military force despite the ongoing talks in Paris. Both sides believed progress at the negotiating table was only possible through sustained military pressure on the battlefield.

In December 1965, Le Duan established the principle that would later animate Hanoi's talk-fight strategy. Although the time was not yet ripe for formal talks, he argued, the goal of any future negotiations would be to facilitate the battlefield and advance Hanoi's war efforts. "At some point in time and under certain specific conditions, we may be able to fight and talk simultaneously with the goal of restricting our opponents' military actions, of winning broader sympathy and support throughout the world, and of concealing our own strategic intentions. The issue right now is the question of a favorable opportunity to employ this stratagem. The opportunity will come when our forces have won greater and more complete victories on the battlefield, when the enemy's situation has become more desperate and confused, and when the enemy's will to commit aggression has deteriorated further."[21] According to former ARVN colonel Hoang Ngoc Lung, Hanoi and the NLF's willingness to enter into negotiations in the wake of *Tet Mau Than* did not mean they had abandoned the goal of crushing the Saigon regime and seizing power through force. "Peace talks did not delude our enemy into forsaking or even diminishing his war efforts," Lung testified after the war. "On the contrary, military victory still remained the *sine qua non* for peace and the ultimate goal to be achieved regardless of the talks."[22]

Hanoi's talk-fight strategy was not only tailored to battlefield realities and current capabilities; it worked hand in glove with ongoing peace negotiations. "We must fight to win great victories with which to compel the enemy to accept our conditions," declared North Vietnamese general Nguyen Van Vinh. "If we stop fighting . . . no considerable success can be achieved in negotiations. . . . Thus, we will take advantage of the opportunity offered by the negotiations to step up further our military attacks."[23] In May 1968, Tran Van Dac, a former North Vietnamese officer who defected to the RVN, claimed that the communists would continue to fight even despite the advent of formal negotiations. "Well, you must realize," he stated in a press conference, "that the Paris talks will be like any other peace negotiations with the communists. For them, the battlefield will be supporting the talks—at any sacrifice. You can expect them to fight, even when it means heavy losses, to win successes on the battlefield that will strengthen their position in the talks."[24] A captured directive issued by COSVN in February 1969 confirmed this position. "We are attacking the enemy in the military, political, and diplomatic fields," it read. "We must positively affirm that the strong position in which we attack the enemy military and politically, and our actual military and political forces on the battlefield will determine the attitude of the enemy at the conference table and the issue of negotiations." It continued, "We should not in any way disregard the effect of the diplomatic struggle, of the art of using stratagems in foreign relations, which did and will secure success. But we must not absolutely depend on and expect too much from the Paris Conference, and nurse an illusion of peace."[25]

Saigon and Washington, despite overtures toward peace, also adopted a strategy of fighting while negotiating in the years between *Tet Mau Than* and the Paris Peace Accords. "Although the United States is trying to negotiate a settlement in Paris and the [Nixon] administration seems to have given up all hope of achieving a purely military victory in Vietnam," a May 1969 *New York Times* article observed, "in the absence of orders to let up on the enemy the military sees no choice but to continue to bear down." Media interviews with American officers confirmed that the military had been under orders to pursue the enemy relentlessly and exert "maximum pressure" on communist forces inside of South Vietnam ever since the implementation of the 1 November bombing halt against the DRV. "The message . . . is simple," one officer stated. "It is to go all out, all the time, to search out the enemy, cut off his lines of supply, disrupt his communications and block his major infiltration routes," he explained. "The fact that there are negotiations under way doesn't mean we should stop fighting," stated another. "The enemy certainly hasn't and I don't think we should unless we get something in exchange."[26]

The Nixon administration had made hard-hitting combat options a cornerstone of its strategy to end the war. Although Nixon and Kissinger had opted to maintain the Johnson administration's bombing moratorium against the DRV (at least up until May 1972), they placed no corresponding limitations on the use of force south of the seventeenth parallel or in cross-border operations into Laos and Cambodia. "The mix of actions should be something like this," Nixon declared at a 24 January meeting of the National Security Council. "We talk hard in private but with an obvious peaceful public stance, seeking to gain time, initially giving the South Vietnamese a chance to strengthen the regime to add to the pacification effort while punishing the Viet Cong."[27] According to one internal document: "Our current strategy aims at keeping two options open: negotiation of a political settlement in Paris and gradual, flexible Vietnamization of the war to permit U.S. disengagement in the absence of a settlement. Our military tactics are designed to keep 'maximum' pressure on the enemy to induce them to negotiate and to minimize our casualties to buy time at home."[28] Less than a month after taking office, Nixon had made it clear that allied military forces were to maintain an offensive position in South Vietnam given the ongoing negotiations. "In reading the January 31 news report on the Paris negotiations," the president informed Kissinger on 1 February 1969, "it seems vitally important to me at this time that we increase as much as we possibly can the military pressure on the enemy in South Vietnam." He continued, "I believe it is absolutely urgent if we are to make any kind of headway in Vietnam that we find new ways to increase the pressure militarily without going to the point that we break off negotiations."[29] After the war, both Nixon and Kissinger admitted that military force was an indispensable corollary to the Paris peace talks. "I was far less optimistic than some of my advisers about the possibility of quick progress in the negotiations unless we coupled our diplomatic efforts with irresistible military pressure," Nixon wrote in 1985.[30] Kissinger confirmed this admission. In his memoirs, the former assistant to the president stated not only that Nixon favored a military policy of "maximum pressure," but that the president was "not too eager for negotiations until some military progress had been made."[31]

COMMUNIST STRATEGY IN THE WAKE OF TET MAU THAN

Communist strategy during the Second Indochina War can largely be credited to one man: Le Duan. According to historian Lien-Hang Nguyen, Le Duan was the "architect, main strategist, and commander-in-chief of communist Vietnam's war effort."[32] Le Duan's strategy to achieve "decisive victory" in Vietnam—the General Offensive and General Uprising—had dominated communist efforts to topple the Saigon regime from the time it was first introduced at the Ninth Plenum of the Central Committee in December 1963. From 1964 to 1969, Hanoi and COSVN had attempted to win the war militarily through main-force combat. Le Duan's aggressive conventional approach was intended to crush Saigon's "puppet" army and drive the United States out of Indochina by frustrating its objectives, killing its soldiers, and exhausting its political will. The adoption of Le Duan's program by Hanoi and COSVN, however, belied a schism within the Politburo concerning strategy. Despite the image of a unified front, there was considerable opposition to Le Duan's costly "go for broke" strategy among high-ranking members of the Politburo. Critics of the strategy were divided into two camps. The first consisted of "North first" moderates, a group that favored internal development in the DRV rather than revolution in the South. Hanoi's moderate faction favored political *dau tranh* over armed *dau tranh*, and thus eschewed the general offensive and uprising strategy; instead, they advocated a negotiated settlement to the war. The second group, comprising protracted war advocates, included none other than Vo Nguyen Giap and Ho Chi Minh. Although equally committed to toppling the Saigon regime as Le Duan, this faction favored a more patient approach over the precipitous General Offensive and General Uprising. Hanoi's sponsors were equally divided over DRV strategy and offered conflicting advice as a result. Beijing advised Hanoi to adopt a protracted guerrilla war in the countryside. Moscow, meanwhile, urged the Politburo to negotiate even as it pushed the DRV to continue its main-force war against Saigon and its American ally.

By 1967, the war in Indochina was deadlocked. Two years of large-scale main-force combat had failed to move the needle in South Vietnam. Le Duan's response was to take another stab at a general offensive and uprising. According to Lien-Hang Nguyen, Le Duan and his chief deputy, Le Duc Tho, were desperate to retain control over their war in the RVN and their leadership in the DRV. The two men, she asserts, "resolved to take extreme measures both militarily and politically. As America's war of attrition reached the end of its second year and the conflict in the South entered its seventh, Le Duan drew from his revolutionary experiences in the Mekong Delta, where he had learned that in order to wage a successful revolution, one must always be on the offensive."[33] Le Duan's faith in aggressive warfare and his unshakable conviction that the general offensive and uprising was the surest means to crush the Saigon regime and achieve victory gave birth to *Tet Mau Than*.

The decision to launch *Tet Mau Than* and subsequent operations despite repeated failure and an obvious disadvantage in the main-force war reflected Le Duan's fierce unwillingness to abandon military *dau tranh* in pursuit of victory. The attacks of May and August 1968 along with the massive 1969 winter–spring offensive (often referred to as "Tet 1969" because it was synchronized, as was the previous year's offensive, to coincide with the Tet holiday) demonstrated just how tenaciously Le Duan and his faction clung to major military offensives and the general offensive and uprising as the primary means to advance the revolution in South Vietnam. This unshakable commitment to the traditional use of military force, however, was accompanied by a strategic pragmatism and flexibility that allowed Hanoi and the NLF to modify their method and transition back and forth between main-force and protracted warfare when necessary. By mid-1969, after four years of large-scale military efforts failed to break the stalemate, the need for a change was obvious. The devastating impact of the 1968 and 1969 offensives on the ranks of the southern insurgency,[34] along with the failure of these initiatives to achieve strategic objectives or tip the balance of forces in the communists' favor, had made it evident that the current military strategy was not only unsustainable, but unachievable so long as American combat units remained in force. Le Duan, therefore, had no choice but to shelve the General Offensive and General Uprising approach and resort to a protracted war strategy that was built predominantly—though not exclusively—on guerrilla tactics and politico-military struggle at the village and hamlet level.

Some had predicted a shift in the communist strategy months before any tangible changes occurred. In September 1968, Nguyen Van Thieu claimed the communists would return to low-level guerrilla war and political struggle now that the third leg of *Tet Mau Than* had failed. Hanoi and the NLF, he argued, would devote most of their resources to rural areas in an attempt to control territory and gain leverage. In a telegram sent to the White House that month, Ambassador Bunker informed Nixon of the South Vietnamese president's views. "Failing to get a major offensive off the ground," Bunker declared, "[Thieu] believes the enemy will now revert to the tactic of attempting to control more and more territory and thus divert more of his attention to rural areas. While he will exploit any opportunity he may discover to carry out attacks on cities, and will attempt to exert enough pressure to hold allied troops near the urban areas, his real objective will be the countryside." According to Bunker, Thieu was certain that the communists would attempt to seize control of as much of the operational environment as possible before calling for a ceasefire followed by elections. Douglas Pike, a government information officer and authority on Vietnamese communism, also predicted the communists would emphasize protracted war strategy after *Tet Mau Than*. Pike informed Bunker that the communists would likely abandon their main-force effort, albeit temporarily, to pursue a protracted war aimed at gaining control of rural areas. The Politburo, he surmised, was currently experiencing a period of "great doctrinal indecision" and that given current circumstances, the faction that was advocating a return to protracted war strategy would probably win out over those that favored a negotiated settlement or a continuation of maximum offensive military pressure. Indications of a shift in communist strategy, Pike informed the ambassador, would include a reemphasis on training and indoctrination in guerrilla warfare, avoidance of contact with large allied military formations, and greater attention to seizing and expanding control in the countryside. Should such a change occur, however, it would likely be only temporary; the communists would once again return to the general offensive and uprising after the Americans had left.[35]

Evidence that the communists were reemphasizing low-level protracted war soon followed these predictions. In November, American advisers in all four military regions reported a sharp decline in overall levels of enemy activity. According to pacification officials, communist main-forces were deliberately avoiding major engagements with allied units while stepping up harassment, terrorism, small-scale operations against territorial forces and pacification personnel, and other guerrilla activities. "The enemy continued to avoid large unit engagements but significantly increased his indirect fire attacks on outposts, RD teams, and district and provincial towns," declared the senior adviser in MR 4.[36] "The overall low level of enemy activity," noted the November assessment for MR 2, "is primarily in the form of standoff attacks against military installations and isolated ground attacks against vulnerable villages and RF/PF units, coupled with increased propaganda activity (especially flag raising and leaflet distribution), assassinations of civilians, tax collection, and hamlet intrusions."[37] In MR 3, CORDS personnel informed MACV that captured documents and other indicators had revealed that the communists were conducting extensive refitting and political reorientation, possibly in preparation for a future offensive.[38] Meanwhile, in MR 1 it was reported that most enemy activities, aside from some coordinated attacks by fire and ground assaults against RVNAF/FWMAF and select villages and hamlets, involved guerrilla activity combined with low-level political agitation and organization.[39]

MACV also sensed a change. "What we've seen over the past few weeks leaves little doubt that the enemy has made some fundamental changes in his prosecution of the war," an intelligence officer declared at a 15 November briefing. These changes, he informed Abrams, were "possibly as far-reaching and important as his July 1967 decision to escalate to the general offensive and general uprising." According to intelligence estimates, Hanoi and the NLF had begun to alter their tactics as a result of a "fundamental reappraisal" of current efforts, one that urged military, political, and proselytizing efforts be "radically redirected" in light of the failure of *Tet Mau Than*. These modifications included an emphasis on small-unit and guerrilla-sized actions; standoff and sapper attacks against outposts, allied war facilities, and lines of communication; and a shift toward insurgent activities in populated areas to increase the effectiveness of the political infrastructure and expand communist control over the people. "His political tactics are aimed at the immediate goal of bringing about U.S. withdrawal," MACV intelligence concluded, "while his long-term goal is to so weaken the GVN that it will be unable to survive alone."[40]

By December, reports began trickling in that NVA main-forces were being deliberately and systematically broken down into roving companies and that NLF local forces were being splintered into platoons and squads. Allied intelligence believed the communists were dividing insurgent forces into smaller units to help them recruit labor and guerrilla fighters, collect taxes, eliminate government officials, and make the revolution's presence felt in as many villages and hamlets as possible. Others believed these efforts were meant to offset the APC, which began 1 November.[41] According to sources within MACV, the breakdown of larger NLF/NVA units into smaller entities was meant to advance protracted guerrilla warfare by facilitating a faster and more flexible maneuver capability, reducing the need for logistical support from the DRV, easing the ability of communist units to gain local support from the populace, and increasing both the geographical scope and frequency of attacks against allied units. "Starting in November . . . all [MR] observers noted that battalion-sized or larger NVA units were being systematically broken down and, in several cases, combined with guerrilla units and [NLF] cadre[s]," MACV intelligence determined. "It now seems clear that the first step was then being taken in accordance with . . . COSVN directives . . . to construct a mini-war army composed of small units comparatively well equipped, well trained, well led, and largely locally victualized. This type of unit is well suited to carry out the mobile, well-coordinated, centrally conceived and directed tactics of protracted war."[42] MACV also indicated that the purpose of subdividing larger units into smaller ones was not to break down communist main-forces, but to build up local guerrilla forces.

The construction and use of a "mini-war army" comprising an ad hoc mixture of NLF local guerrillas and NVA regulars[43] signaled the emergence of a new and unique form of armed *dau tranh*, one never before seen in South Vietnam. Douglas Pike labeled this phenomenon the "super guerrilla" concept. This was essentially a return to protracted guerrilla war albeit with a new twist: well-trained sapper or commando teams that were designed to undermine the enemy's advantage in firepower and mobility, and whose purpose was to harass and wear out enemy military forces rather than decimate them. "Gone was the guerrilla in black pajamas, fighting in his home region with homemade or captured weapons," Pike wrote. "In his place came a new fighting man, organized into a highly skilled team as well knit as a professional football squad, proficient in the best weapons and explosive devices the communist world could produce, linked to higher headquarters with sophisticated communication equipment, striking sudden deadly blows that would neatly destroy medium-sized enemy installations."[44] Hanoi and COSVN considered an economy-of-force strategy built around the extensive use of "sapper" tactics by guerrilla forces to be the most effective measure for inflicting damage with minimum expenditure. Such an approach was intended to not only preserve communist forces, but to maintain constant military pressure on South Vietnamese and American units, inflict costly military and political damage on the allies, tighten insurgent control over the countryside, disrupt the economic and political stability of the RVN, and influence ongoing diplomatic negotiations in Paris. In short, the "super guerrilla" concept was implemented to maximize communist control over the operational environment at a minimal cost, and to place the revolution in an optimal position to resume the political struggle should a negotiated settlement be reached.[45] According to MACV, the communists' newfound emphasis on small-unit tactics, sapper activities, and guerrilla warfare was not simply a defensive response to allied initiatives, nor was it suggestive of an inability to win the main-force war against conventional RVNAF and American units; it was "a positive response, a strategy," for defeating the pacification program, which itself hinged on the work of relatively small politico-military "units" spread over a large geographical area.[46]

Ground-level observations along with other intelligence convinced allied leaders that a shift in communist strategy was under way. On 24 November, General Abrams, then head of MACV, issued a report estimating that enemy activity had undergone a paradigm shift in the wake of *Tet Mau Than*. Abrams maintained that while the "ultimate objective" of the communists remained "the complete domination" of the RVN through military, political, and psychological actions, Hanoi and the NLF were shifting their emphasis from "the military to the political." This included changes in strategy as well as in the deployment of forces. The decision to return to protracted warfare, Abrams asserted, was based on the communists' "rapidly deteriorating military posture" and the emergence of a new set of dynamics that included a possible ceasefire, the formation of a coalition government, and the potential withdrawal of American combat forces. "In preparation for these," Abrams insisted, "the enemy realizes that he must strengthen his control over the population. There is evidence of his plans to attain these near term objectives and counter our [pacification and Vietnamization] strategy." According to Abrams, Hanoi and the NLF were attempting to contest the pacification and Vietnamization programs by shifting their tactical and operational focus toward attacking outposts, territorial forces, and pacification personnel; destroying war facilities in order to weaken RVNAF capabilities; and maintaining and expanding politico-military control over the population.[47]

Abrams's estimation of the situation was shared by others. In his 30 November telegram to the president, Bunker informed Nixon that the communists would likely pursue a new strategy and step up efforts to strengthen control over the countryside. Insurgent forces, the ambassador declared, were expected to increase terrorism and employ a three-pronged counterstrategy aimed at inflicting heavy casualties on allied units, wrecking government outposts and military facilities, and, most importantly, strengthening their control over the rural population. "Enemy activity," Bunker stated, "seems to be directed primarily toward the third factor." The ambassador also informed Nixon that should this assessment be correct, it would reinforce the need for a vigorous pacification offensive to "frustrate and undercut enemy efforts to extend his control over the population in anticipation of a ceasefire."[48]

Allied suppositions concerning the shift in communist strategy turned out to be accurate. By the summer of 1969, a series of resolutions and directives disseminated to NLF/NVA units in the field were urging the type of low-level attack anticipated by Saigon and the Americans. Issued in response to the failed 1968 and 1969 offensives and reflecting what MACV called a "new realism," COSVN and VWP edicts clearly advocated a shift in strategy, from an emphasis on swift victory and aggressive all-out offensives to a more prescribed and pragmatic approach that called for protracted warfare and the careful husbanding of human and material resources. Pike was correct: the faction within the Politburo that had advocated an adjustment in the current approach and a return to protracted warfare had won out—at least temporarily. In April, VWP Resolution "C" was issued, quickly followed by COSVN Directives 81 and 88. Unlike prior directives, these instructions were notable both for their caution and their circumscribed objectives. Each recognized the immense power of the allied armies and the unlikelihood of defeating RVNAF and U.S. military forces on the battlefield. As such they omitted hopes of a rapid conventional military victory and advocated a more modest set of objectives, including inflicting maximum casualties and destroying as much war matériel as possible. Such actions, communist leaders believed, would influence U.S. domestic politics and exert pressure on the Nixon administration to rapidly withdraw America combat forces, acquiesce in the formation of a coalition government with NLF participation, or both.[49] "Our basic objective is to gain the decisive victory," read an addendum to Directive 81. "Our steadfast viewpoint is that the U.S. government must withdraw all U.S. and satellite troops from [South Vietnam] to achieve peace, independence, and territorial integrity. The coalition government must stand for democracy, peace, and neutrality. . . . We must achieve partial successes before advancing toward the final victory."[50]

As we can see, Hanoi and the NLF advocated a gradual, piecemeal political victory founded on protracted warfare rather than the impossible objective of total military victory in a relatively short period. "We do not intend to attain the victory overnight or in a single phase, nor do we plan to obtain the victory in any predetermined phase after conducting many phases," COSVN announced. "The victory will not come to us in such an easy way, but it will come in a difficult and complicated way. It will be a limited victory and not a clear-cut, complete victory."[51] Other directives supported the idea of a long and circuitous road to winning the war. "We secure victory not through a one-blow offensive, not through a phase of attack, not even through a series of attacks culminating in a final kill," read one COSVN study. "Victory will come to us," it continued, "not suddenly but in a complicated and tortuous way."[52] According to another document, the decisive victory looked for by Hanoi and the NLF would only come "step by step, bit by bit."[53] As these passages illustrate, communist leaders had embraced, albeit reluctantly, the reality that the communist revolution could only come to fruition incrementally and over time.

This reality was likely rooted in a firm understanding that the NLF had been severely weakened during *Tet Mau Than* and that the strategic position of the revolution vis-à-vis the RVN and the United States was now retrograde. The substance as well as the timing of victory had also changed. Although the definition of a complete victory had not been altered—the overthrow of the RVN and unification of Vietnam under communist rule remained the raison d'être of the revolution—the means to get there had been modified. The method for seizing political power would no longer follow Le Duan's prescription for a swift, overwhelming military victory, but rather a negotiated political settlement that included American withdrawal, coalition government, and a favorable balance of forces that would later be used to seize power in the postsettlement period. "A decisive victory is not a complete victory," a captured NLF document declared. "It is a victory gained by one step; but it is a very basic one, creating favorable conditions for us to move one step toward gaining complete victory. This decisive step means that the Americans will accept to withdraw their troops, a coalition government will be established, and the revolutionary forces will be stronger than the counterrevolutionary forces."[54]

In July, COSVN released Resolution 9, the most comprehensive pronouncement to date on the new communist strategy. Although it was in many ways a basic review and restatement of earlier directives, the resolution was, nonetheless, a unique, complex, and highly detailed strategic blueprint for advancing the revolution in the wake of *Tet Mau Than*. According to one former ARVN officer, Resolution 9 was a "watershed policy directive" because it "set the communist conduct of the war in South Vietnam on an entirely new course."[55] According to historian Ang Cheng Guan, Resolution 9 "directed the communist forces to be relentless in developing their strategic offensive. But it also required them to consolidate their strength in anticipation of a prolonged war. Large units were to be broken down into smaller ones, while provincial local force units were to split into smaller cells. The aims were twofold: 1) to hit US troops and installations in order to pressure US troops to withdraw; 2) to disrupt pacification and destroy Saigon's control of the population."[56] As Guan illustrates, the heart of the new communist approach was to attack the allies' pacification-Vietnamization strategy. Its primary goals were to force a rapid and complete American withdrawal, defeat Vietnamization by shortening the time available for U.S advisers to devote to training the RVNAF, disrupt and roll back pacification gains and allied efforts to expand influence and/or control over the countryside, and lay the groundwork for exploiting a negotiated settlement. These were to be achieved primarily through armed *dau tranh* and violence, although political *dau tranh*, particularly at the local level, was also urged. Besides advocating greater emphasis on terrorism and guerrilla activity aimed at destroying the RVN's rural administrative structure, Resolution 9 called for mobilizing the population; strengthening the insurgent base; improving the revolution's economic position; and expanding political proselytizing, particularly among the RVNAF, to spur defections and increase penetration into government organizations.[57]

In accordance with the new communist strategy, Resolution 9 called for a long-term approach based on a partial military victory (an American withdrawal) followed by a complete political victory (the overthrow of the Saigon regime via a coalition government stacked in the communists' favor).[58] According to COSVN, the objective was to "smash the Americans' will of aggression; force them to give up their intention of ending the war in a strong position and end the war quickly and withdraw troops while the puppet army and administration are still too weak to take over the responsibility of the Americans; [and] force the Americans to accept a political solution and recognize an independent, democratic, peaceful, and neutral South Vietnam with a national, democratic coalition government working toward reunifying Vietnam."[59] MACV took note of communist intentions: "COSVN Resolution 9 . . . clearly specifies the objectives for the 1969–1970 Winter-Spring Campaign are a limited military victory, followed by a coalition government leading to final political victory. In stating how it plans to achieve these objectives, the enemy leadership is clearly oriented toward planning for a post-war political struggle."[60]

Defeating the allies' pacification-Vietnamization strategy and winning the control war in the countryside was the centerpiece of the new communist strategy. Resolution 9 maintained that the objective in South Vietnam was to thwart the allies' plan to "de-Americanize and de-escalate the war step by step . . . and to compete with us in territory and population control so as they can end the war on a definite strong position." COSVN leadership recognized that in addition to consolidating and strengthening the RVNAF and government administration, the allies were expanding "clear and hold" operations in rural areas and accelerating the pacification program "in order to gain control of the people and territory . . . weaken our military and political forces, cause difficulties for us, [and] create conditions in which they could gradually withdraw American troops while keeping the puppet troops strong."[61] Thus, revolutionary forces had no choice but to confront pacification-Vietnamization by stepping up antipacification activities and by maintaining and expanding areas under communist control. "We must defeat the enemy's rural pacification campaign, dominate the major part of the rural areas, and control the important strategic positions in the jungle-and-mountain area and on strategic communication lines, and especially in areas bordering the cities and some parts right in the cities," COSVN maintained. Communist guerrillas and political cadres, it continued, must "defeat the enemy's pacification plan; wipe out the major part of the puppet administration and its control grip at the infrastructure level; . . . establish and strongly consolidate the people's revolutionary administration at various levels, especially at the infrastructure level, and promote the role of the Provisional Revolutionary Government."[62]

Subsequent COSVN resolutions and directives show that Hanoi and the NLF's key objective after *Tet Mau Than* was to thwart the pacification and Vietnamization programs and expand communist influence and/or control over the operational environment. According to Resolution 10, the aim was to frustrate allied efforts to strengthen Saigon's military and administrative structure, intensify the pacification program, and maximize government control over rural areas. Thus, revolutionary forces were ordered to "expand our control to all villages and hamlets, break the enemy defensive and oppressive control system in rural areas, and expand and improve the revolutionary movement in cities by creating strong political organizations." These tasks, COSVN argued, would thwart the enemy and help the revolution advance toward a "new strategic offensive position." In time, this new position would destroy a significant portion of the RVNAF and overthrow the Saigon regime. "In this way," Resolution 10 declared, "we will basically frustrate the enemy strategy of Vietnamizing and prolonging the war and shift the situation to our favor to facilitate an eventual decisive victory."[63]

Resolution 14, issued in October 1969, stated that the three pillars of communist strategy (armed struggle, political struggle, and military proselytizing) were meant to expand revolutionary government, consolidate revolutionary organizations, and protect and enlarge liberated areas. Although the resolution discussed the importance of the communist main-force effort and urged close coordination between conventional and guerrilla units, it largely emphasized small-unit tactics and protracted warfare. "At the present time, guerrilla warfare holds a position of particular importance," it read.[64] According to Resolution 14, guerrilla warfare helped advance the current communist strategy in three ways. First, it assisted the revolution in coping with an enemy who was much stronger. By waging a protracted war throughout the three strategic areas of South Vietnam (mountains, lowlands, and cities), NLF/NVA guerrillas stretched out and exhausted the more numerous allied forces as they dispersed to counter the guerrilla threat across a wide geographical scope. Next, guerrilla warfare helped roll back pacification by weakening Saigon's hold on rural areas and by expanding communist control in the countryside, thus tipping the balance of forces in the control war in their favor. Finally, guerrilla tactics helped provide the necessary conditions for success in the main-force war. Allied forces that were dispersed and stretched thin in response to widespread guerrilla action, Resolution 14 reasoned, were vulnerable to a coordinated main-force attack. Vo Nguyen Giap, the foremost proponent of a protracted war strategy, recognized the utility of meshing guerrilla warfare with NVA main-force actions. "In the new situation, we must further improve the local military tasks and develop the great capabilities of local armed forces," he declared in July 1969. "Under the present war conditions, it is necessary to develop guerrilla warfare to a very high level, thus creating favorable conditions for the regular war to defeat the enemy. Local armed forces must be developed vigorously and widely, thereby enabling regular troops to acquire high mobility and great offensive strength."[65] Together, these three developments—stretching RVNAF/U.S. forces thin, defeating pacification, and meshing guerrilla actions with main-force offensives—were intended to help the revolution regain a strong presence in the countryside and tighten its hold over local villages and hamlets.

Throughout 1970 and 1971, Hanoi and the NLF focused on rebuilding their shattered politico-military infrastructure and disrupting the allies' pacification-Vietnamization strategy. "The Vietnamization of the war is an overall plan consisting of many objectives, but the two fundamental objectives are pacification and the development of the Puppet troops and Puppet Government," COSVN declared on 10 March 1970. "Pacification is considered the most fundamental objective. . . . To frustrate the pacification program is to deal a deadly blow to the enemy's scheme of Vietnamization of the war, to win great victories and to achieve the decisive victory in the last period of the resistance war against the U.S."[66] Efforts to attack the pacification-Vietnamization strategy were meant to shift the balance of forces in the countryside to the communists' favor through a combination of political and military means. In VC Military Region 5, VC/NVA forces undertook two interdependent tasks: to "foil [the enemy's] pacification plan, break his oppressive control, and expand our control over the population"; and to "destroy and disrupt a great part of the puppet army and cut off his source of troop replacements." Units were ordered to intensify military operations and expand their area of operations; frustrate pacification and sweep operations; relocate the population from government- to communist-controlled areas; attack "weak" areas including hamlets, refugee camps, and isolated outposts; consolidate liberated and base areas; step up political activities in cities and in district and provincial capitals; intensify proselytizing activities among the population and RVNAF; and strengthen local guerrilla and self-defense forces and their leadership. In addition, they were instructed to "concentrate all efforts on the immediate and primary missions . . . to win and hold the people."[67] In April 1971, another directive stated that antipacification activities remained "the principal mission" of communist forces operating in South Vietnam. "Make all echelons and specialized agencies thoroughly understand the strategic significance of the counter-pacification task in the current situation," it read. "The entire Party . . . must concentrate on frustrating the enemy pacification and Vietnamization plans. At present, get rid of subjectivism, impatience, pessimism, and passiveness, and grasp missions and operation and combat procedures, and apply them appropriately with the situation of different areas, villages, and hamlets."[68]

Applying the tactics and techniques of protracted warfare appropriately within different contexts meant tailoring the method and level of activity to the local environment. In liberated areas under firm communist control, guerrillas and political cadres were ordered to hold the ground and defend the villages. They were also responsible for maintaining a "revolutionary spirit" among the population, increasing agricultural production, and conscripting the population to contribute manpower and material resources to the revolution. "In the liberated zone . . . we must motivate and unify the peasant bloc . . . push strongly toward guerrilla warfare, the construction of combat villages, the movement of killing the enemy to defend our villages, and defeating all enemy maneuvers of pacification, sabotage, espionage, psywar and appeals to rally and surrender." In contested villages and hamlets as well as those under predominant government control, insurgents were instructed to develop local guerrillas and political cadres; disrupt the RVN administrative apparatus; launch attacks against territorial forces, PSDF, and RD cadres; assassinate government officials; proselytize and attempt to strengthen the political base; and disrupt pacification activities. These actions were assigned to special operation teams trained in guerrilla warfare, armed propaganda, terrorism and repression, and the building of revolutionary bases. "At present the enemy is still maintaining his control grip on a number of densely-populated rural areas," Resolution 9 pronounced. "It is necessary to . . . transform these areas into areas of struggle against enemy oppression, for the purpose of winning political control, liberating the people from enemy schemes to take control . . . destroying the enemy sources of manpower and material resources and turning these into ours, broadening the liberated areas and areas where we have political control, and creating a new position of encirclement and pressure against the enemy."[69]

The protracted war prescriptions found within COSVN and VWP resolutions were intended to mesh with ongoing political initiatives, particularly the formation of governing and administrative bodies within liberated areas in 1968–69. After the introduction of American combat forces in 1965 and subsequent escalation of the main-force war, Hanoi and the NLF shelved efforts to establish government-style administrative councils and committees in rural areas as they pursued rapid victory through main-force warfare. The push to establish local administrative bodies was revived in the run-up to the 1968 Tet Offensive. In the fall of 1967, communist leaders instructed NLF cadres in Long An Province to elect "People's Liberation Committees" in areas that were under predominant communist control. In villages and hamlets that were either contested or controlled by the South Vietnamese, communist cadres were ordered to appoint committees with the purpose of forming an insurgent governmental structure that would parallel the RVN governmental structure at the local level. These early initiatives would later be repeated throughout

South Vietnam. VWP leaders believed "democratically elected" local governmental bodies were necessary to consolidate gains made during Tet and prepare for future campaigns.[70] "In the present situation, with the momentum of our offensive against the enemy in the military, political, and troop proselyting fields we will certainly liberate many large parts of the remaining rural areas, as well as a number of urban sections and towns," COSVN declared on 5 March 1968. "Such a situation requires revolutionary government organizations in order to firmly establish security and order and rapidly strengthen various revolutionary forces. We must continue to step up the struggle, firmly hold and protect newly liberated areas, resist every U.S. and puppet counterattack and stand victorious in every circumstance."[71]

The purpose of liberation committees, also called revolutionary committees, was to form the foundation for a rural administrative apparatus at the local level and legitimize NLF control over territory and population. Liberation committees laid the foundation for communist claims to a de facto NLF "government" within South Vietnam. "Politically, the communists are engaged in a major effort to weaken the GVN and to create the appearance, if not the substance, of an ongoing administrative apparatus 'governing' as much of South Vietnam as possible. Their aim is to boost the prestige and image of the National Liberation Front and its claims of control over territory and people," Richard Helms, then director of the CIA, declared.[72] According to MACV, liberation committees were "an apparent attempt to establish, at least on paper, a viable alternative to the GVN."[73] Moreover, they were meant to serve as transitory organs for establishing politico-military control at the local level, and were intended not only to mobilize the population, but to perform civil administrative functions much like any legitimate local government.[74] To COSVN, liberation committees were a vital mechanism for growing the revolution at the grass roots in preparation for the seizure of political power at the national level. According to a party directive dated 5 March 1968, "The problem of building state power is, above all, the problem of building village and hamlet government in liberated areas . . . an indispensable matter which we should strive to achieve positively. To achieve organizational uniformity in the local areas . . . COSVN has decided to proceed with the election of . . . People's Liberation Committees in the villages."[75]

Another function of liberation committees was to provide the NLF with leverage should a ceasefire be declared and a coalition government formed. "To regain . . . power we should at first build an administrative system in the villages and hamlets, in liberated areas," declared one directive. "This is our immediate task which requires our urgent and greatest efforts and helps us build the solid foundation for the establishment of a high level coalition government."[76] Liberation committees were instruments for expanding and consolidating local control over the civilian environment vis-à-vis the ongoing peace talks in Paris and the possibility of an in-place ceasefire. "The communists now have a legitimized administrative apparatus covering much . . . of the population under their control," noted one MACV study. "Should the talks in Paris turn to more serious negotiations or should the war take a dramatic turn in South Vietnam," it continued, "the communists [would] have, in this apparatus, a formidable card to play."[77] According to a report generated by the Combined Intelligence Center, Vietnam, liberation committees would play a critical role in the communists' bid for control of the countryside should negotiations result in a ceasefire. "If during a state of ceasefire the GVN did not immediately consolidate its control over contested villages," it read, "the VC/NVA, because of decreased military pressure, would be in a good position to establish more Liberation Committees." Greater numbers of these committees, the report continued, would "heighten the degree of population control by the VC/NVA. This increase in . . . population control could lend credence to any future claim by the VC/NVA of a *de facto* government in South Vietnam, and thus would constitute a strategic element in negotiations toward the establishment of any coalition government."[78] Ambassador Bunker shared similar reservations. Although he considered liberation committees to be simply "existing bodies under a new name," and pointed out that more than half of them were in areas already under the control of communist forces, he felt, nonetheless, that these organizations could pose problems for the allies should a ceasefire be declared. "Given some kind of international ceasefire," he informed the White House on 19 October, "Liberation Committees could lend some credibility of control over wide areas of the countryside."[79] ARVN officer Nguyen Be also believed liberation committees posed a threat. "The communists will bide their time until the last weapon falls mute and the ceasefire begins," he warned President Thieu in October 1968. "At this time, the communists will surface as the legitimate hamlet governments, with a claim to the trust and affection of the people. They will fly the communist flag and organize the people very easily according to their plan, for they will have the approval of the people."[80]

Be's warning held a great deal of truth. Many South Vietnamese citizens believed liberation committees did in fact provide an opportunity for the communists to legitimize their claim to be "the authentic representative" of the Vietnamese people. Should the current Saigon regime collapse, it was argued, these local elected committees and councils would be the closest thing to a popularly elected government in hundreds of rural villages and hamlets.[81] By the end of October 1968, sources reported that the communists had conducted elections for revolutionary committees in more than 1,200 villages and hamlets and established administrative bodies in seventeen provinces, five cities, and thirty-one districts. The NLF claimed revolutionary forces controlled four-fifths of South Vietnam's territory and 10 million of its people. The Front also claimed revolutionary administrations had been established throughout the national territory. American officials, however, refuted these claims. Some argued that the communists had held elections in, at best, 10 percent of those areas under NLF control. The United States Mission estimated that only 15.3 percent of the population was under NLF control (14.9 percent were allegedly living in contested areas while 69.8 percent lived under government control). Thus, the election of liberation committees was not viewed as a major threat. Some American analysts believed the rise of revolutionary committees was tied to the launch of *Tet Mau Than*. Hanoi and the NLF, they maintained, had given them high priority in the spring of 1968 because they assumed the General Offensive and General Uprising would succeed and bring the war to a precipitous end. Hanoi and the NLF made similar claims to legitimacy just prior to American intervention in late 1964 and early 1965, another period when the communists were on the offensive and believed a final victory was imminent.[82] Nonetheless, the proliferation of liberation committees caused concern among allied policymakers.

On 10 June 1969, the NLF announced the formation of the Provisional Revolutionary Government, an organization that assumed the mantle of "national" leadership and was immediately recognized as such by a number of communist nations. A de facto insurgent government within South Vietnam had—in the eyes of the communists at least—become a reality.[83] At a press conference held on 11 June, the president of the PRG, Huynh Tan Phat, declared, "The establishment of the Provisional Revolutionary Government of the Republic of South Vietnam is the development of a necessary step in the process of perfecting the network of the revolutionary government which has been set up over half the territory in South Vietnam. It is also aimed at meeting the great requirements of the fighting—these consist in striving to step up the general offensive on the military, political, and diplomatic fronts so as to advance toward complete victory."[84] According to the National Congress of People's Delegates, "The Provisional Revolutionary Government of the Republic of South Vietnam is the highest executive body expressing the will and aspirations of the entire people, and having as its tasks to lead the various branches of activity and the various echelons, to mobilize the entire people to step up the resistance and to liberate South Vietnam."[85] Former NLF operative Truong Nhu Tang indicated in his memoirs that the PRG was meant to establish a formal governmental structure at the regional and local levels and to create a "central political entity" that could compete with Saigon on a formal, legal footing. Among its most important benefits, the PRG was used to influence public perceptions and garner political legitimacy as well as sympathy for the revolution both inside and outside South Vietnam. "Our goal was to influence public opinion," Tang wrote, "domestically, where a noncommunist government would give us added credibility with the South Vietnamese populace; internationally, where we would be able to compete with Saigon for formal recognition . . . and in the United States, where we would enhance our claim of representing the Southern people, giving the peace movement additional ammunition. The administration's effort to portray the Saigon regime as an autonomous, legitimate government would now be answered by another Southern government fighting hard in every international forum to establish its own claim to legitimacy."[86]

Unlike the NLF and the Alliance of National, Democratic and Peace Forces of Vietnam, a front organization composed of a network of variously named organizations founded in April 1968 and committed to working with the NLF to overthrow the Saigon government, the PRG was introduced as a fully functioning government replete with a president, ministries, and an advisory council. This was an important development given that both the NLF and the alliance were never portrayed as anything other than front organizations. Once established, the PRG became the "chief public organ" for the communist revolution in South Vietnam as well as the national directing body in charge of the liberation committees. The PRG also replaced the NLF as the negotiating arm of the southern revolution at the Paris peace talks, and assumed primary responsibility for conducting diplomacy with foreign countries.[87] The appointment of Madame Nguyen Thi Binh as the PRG's representative at Paris and as its foreign minister gave the communists a powerful diplomatic tool in the negotiations.[88] To some, the arrival of the PRG signaled the communists had in fact constructed a rival, alternative government to the Saigon regime. As stated by the *New York Times*, the formation of the PRG was seen by many as "an important step in an intensive effort to establish a credible alternative to the Thieu government."[89] To others, however, the PRG was an illegitimate pretender. Thieu and the Saigon regime naturally rejected the PRG and disavowed its claims to political legitimacy. "We will never recognize a coalition or any other sort of government arbitrarily imposed by the communists," blurted one South Vietnamese official. "This new so-called government is not recognized by us in any way at all."[90]

Immediately following its formation, the PRG released a twelve-point action plan. Among the plan's provisions, the first four covered issues related to the control war, particularly communist efforts to force an American withdrawal, topple the Saigon regime, and establish a coalition government. The first point vowed to lead the people and the revolutionary armed forces and intensify the military and political struggle. The stated objective was to force an unconditional U.S. withdrawal, frustrate the allies' pacification-Vietnamization strategy, and end the war by negotiating a peace that comported with the conditions laid out in the 1954 Geneva Accords. The second point called for the eradication of the "camouflaged neocolonialist regime" set up by the "U.S. imperialists" in South Vietnam and the construction of a "truly democratic and free republican regime," one not only based on equality, freedom, and democracy but free from outside intervention. The third point mandated the formation of a provisional coalition government made up of groups reflecting the various social strata and political tendencies of the South Vietnamese people and based on "the principle of equality, democracy, and mutual respect." The coalition government would be responsible for holding local elections, electing a national assembly, and building a constitution that fully reflected the interests and aspirations of the people. The fourth point vowed to consolidate and develop a revolutionary administration, strengthen revolutionary armed forces, and expand and consolidate liberated areas.[91]

In February 1970, the People's Revolutionary Party Committee of VC Subregion 2 issued a directive that sheds light on how the PRG sought to expand and consolidate its hold on the operational environment. According to the directive, guerrillas and political cadres were ordered to establish liberation committees in contested areas using the methods of protracted war. "The mission of the . . . village committees is to intensify military and political attacks, frustrate the enemy's pacification and land encroachment plans, use our governmental authority to annihilate tyrants, break the enemy's oppressive control of the population, strengthen our forces, and expand our control over the areas by gradually consolidating our installations toward the establishment of our local government."[92] In areas under total or predominant insurgent control, the means to establish PRG administrative machinery were quite different. Here party members were required to reorganize and consolidate committees at the village, hamlet, and district level, and then assign individuals to carry out specific functions laid out in party resolutions. These included promoting political awareness among the local population; mobilizing manpower and material resources for the frontline by recruiting civilian laborers and securing financial support; increasing agricultural production; and developing public health, cultural, and education activities. Committee members were also responsible for security missions, including "the annihilation of the enemy" and the suppression of "counterrevolutionaries." Once established, PRG administrators were expected to sustain revolutionary activity among the population and to "ensure democracy for the people" by means of propaganda and population control measures.[93] "Through indoctrination, we must heighten the revolutionary vigilance and security maintenance concerns of the people, troops, and agencies and closely control our personnel and the population."[94] Stepped-up activities in cities and in provincial and district capitals were also required. Urban guerrillas and political cadres, much like their counterparts in rural areas, were expected to play "a key role" in the struggle to disintegrate the RVN and expand communist control. "Breaking the enemy's oppressive control and killing tyrants in cities, province capitals, district seats and city bordering areas, are the missions of the armed reconnaissance, military, political, and military proselyting forces, especially the local secret force, in annihilating the enemy tyrants and ringleaders who directly destroy our movements." Besides assassinating "U.S. and puppet intelligence espionage forces and public security agencies in cities, district seats, and province capitals," urban security units were expected to break the RVN's "oppressive control" through political agitation.[95]

Liberation committees and the PRG played a fundamental role in the control war and the struggle to establish dominion over the operational environment in the years immediately following *Tet Mau Than*. This is because local administration at the grass roots served as the foundation to claims of political legitimacy—a concept that was particularly urgent once the Paris peace talks had begun and the liklihood of an in-place ceasefire and coalition government increased. Much like protracted warfare and efforts to expand influence and control over the civilian environment, liberation committees and the PRG were a crucial part of communist efforts to leverage power and increase the revolution's standing before the clock ran out and a settlement was reached in Paris.

ALLIED STRATEGY IN THE WAKE OF *TET MAU THAN*

Tet Mau Than revealed the allied main-force war had yielded little in the way of bringing the conflict to a successful conclusion. Although Westmoreland's "big unit" war had advanced pacification and elevated Saigon's level of control over the countryside to peak levels,[96] the nationwide surprise attacks demonstrated that the RVN and the United States had not crushed the insurgency, blunted the communists' ability to wage war, or deterred Hanoi and the NLF from attacking the Saigon regime. More importantly, Tet showed that new initiatives were in order if the United States was to salvage its ongoing efforts to preserve the RVN. The decision to pivot toward an accelerated pacification program and rapid expansion of government control over rural areas was based on this sobering assessment of allied capabilities. It was also rooted in a number of developments and opportunities that had arisen as a result of the enemy's failed offensive.

To begin with, *Tet Mau Than* generated extremely high numbers of communist casualties. This, along with the offensive's focus on urban areas and district capitals devastated the insurgency's fighting capacity and denuded much of the countryside of military units and political infrastructure. The resultant power vacuum provided the allies with a unique opportunity to expand Saigon's control over the operational environment and strengthen the South Vietnamese military. Progress in these areas would, in turn, allow for a more rapid extraction of U.S. combat forces—a development that had become increasingly likely by the summer of 1968. In July, Ambassador Komer postulated this scenario to Ambassador Bunker and General Abrams. Should the communists launch another offensive and it was crushed, Komer suggested, Hanoi and the NLF would either have to make significant political concessions in Paris or revert back to a lower level of fighting. Should the latter occur, he asserted, the RVN and the United States would gain a unique opportunity to make rapid and substantial gains in pacification and nation building. Combined with the continued attrition of communist units via main-force combat and improvement in the RVNAF's combat capabilities, gains in the control war would put the allies in a winning position and facilitate the impending drawdown of U.S. troops. "We can gradually redeploy," Komer declared. "We do not need any more U.S. forces. What we do need is time—enough for the . . . process to bear fruit. If we can only hold the negotiating line long enough to defeat the next enemy offensive effort and let the RVN/RVNAF grow in relative power, we can reasonably expect to be in a much greater position of strength than is visible at present."[97]

Next, Saigon and Washington's emphasis on rapidly expanding government control over rural areas was meant to counter communist efforts to establish political and administrative structures in the countryside. After Tet, the allies were faced with a political imperative to "establish legitimate local authority to counter the claims of the communists and their fronts."[98] According to William Colby, the founding of the PRG and formation of liberation committees demanded an appropriate commensurate counterstrategy. Saigon needed to blunt the expansion of communist administrative apparatuses in rural areas, Colby argued. Therefore, what was needed was to implement "a vigorous extension of security and political presence by the government, with American support, in order to preempt the areas not yet penetrated by the communists and to spread the government presence into contested areas."[99] In 1968, Colby underscored the importance of defeating communist efforts to establish alternative administrative bodies in rural areas. In a memorandum to Tran Thien Khiem, then South Vietnam's minister of the interior, Colby asserted that local elections were one of the most critical aspects of pacification. Electing community leaders as soon as villages and hamlets had been secured, Colby asserted, was an "urgent objective" that must be met if the allies were to counter the recent proliferation of liberation committees. "This should be a high priority task in view of the stress the enemy has been placing upon legitimizing his political structure in the countryside," Colby stated. "The GVN," he continued, "is in a much better position to extend its influence through elections than is the enemy. Yet, the enemy seems to be moving ahead at a faster pace. In my view, this imbalance must be reversed rapidly."[100]

Rapid politico-military expansion into the countryside was also critical given the ongoing negotiations in Paris. The level of control each side held over the operational environment was a central factor in the deliberations. Washington and Saigon's ability to win concessions and secure conditions that would advance political objectives was directly linked to their capacity to shrink the insurgency and grow the RVN presence in rural areas. "Relative control in the countryside is going to be one of the things that will largely determine the shape of any settlement reached between the North and South Vietnamese," Komer declared in his final press conference as the civilian head of CORDS.[101] Bunker agreed with this argument. "Success in the pacification and military sphere will have a direct effect on the negotiations," Bunker informed the White House on 19 December 1968. "By extending territorial control and driving enemy forces across the border into Laos and Cambodia," he continued, "the South Vietnamese greatly strengthen their position at Paris. This is obviously a strong incentive, and Thieu is pushing his people to get on with the war effort and pacification faster and with better effect than at any other time since I arrived."[102] Thieu himself was more succinct. "At the conference table in Paris," he stated in December 1968, "there will be long discussion as to who controls the land and the people . . . as to who has governments in the villages and hamlets."[103] Less than two weeks after Thieu made this statement, an American journalist placed the resurgence of pacification at the center of allied efforts in the wake of *Tet Mau Than* and the commencement of formal peace talks in Paris. "No one speaks of conventional 'military victory' over the 110,000-man North Vietnamese army, with its secure Laotian and Cambodian rear base areas. Such talk died among working-level folk here long before it vanished from U.S. official rhetoric in the wake of the communist Tet Offensive last January," he wrote. "The goal . . . is more modest: rapid consolidation of South Vietnamese government authority over as many of Vietnam's 17.5 million people as possible in 1969, before any ceasefire agreement is reached in Paris."[104]

Finally, rapid pacification and a hurried expansion of government control over territory and population were critical because—owing to the advent of Vietnamization and American withdrawal—time was now of the essence. The war's unpopularity along with a growing sense of frustration and impatience for an end to the fighting made tangible results on the ground an imperative for the allied governments and their armed forces. The political pressures brought to bear by the international community and, especially, the American public and its elected representatives required Washington and Saigon to attain measurable success within a limited time frame. "The realities of the American political situation indicate a need to consider time limitations in developing a strategy to 'win.' The time available to achieve U.S. objectives in Vietnam has definite upper limits," stated the 1969 MACV Objectives Plan. "The reality of the long-term consequences of failure has not adequately impressed the majority of the American people. The U.S. public does not see the issues in Vietnam . . . as being worth the cost. Therefore, support in terms of personnel and dollars for U.S involvement in Vietnam has been given reluctantly, and related to continued promises that 'the war is just about over.' Time then, is running out." Should the allied military effort not achieve "measureable success" by July 1970, the report continued, "widespread disillusionment with the Congress" and a subsequent reduction in funding and personnel would follow.[105] According to the 1971 Combined Strategic Objectives Plan, "The length of time that the war has gone on . . . has led to a loss of enthusiasm as the people become more war weary. Therefore, the significance of time itself cannot be over-emphasized. The need for time in which to further develop the already improving RVN leadership and to show positive tangible results, as opposed to the amount of time remaining before the withdrawal of Free World support, is a very real and serious limitation."[106]

The push to gain "positive tangible results" in South Vietnam began shortly after the communists launched the 1968 General Offensive and General Uprising. Three months after Tet's January 30 kickoff, Ambassador Colby announced that efforts were under way to advance pacification and reassert government control in rural areas. "A military counteroffensive has been launched to push Viet Cong forces away from important areas," he stated on 1 May. "However, the VC political structure, looking toward the possibility of negotiations and coalition, will make every effort to retain influence and control over the rural population. It is important that the GVN assert its authority throughout the country to prevent the . . . NLF claiming that it represents any area or percentage of the people. The pacification program then must regain its earlier momentum and develop a widespread effect in the next three to six months."[107] The momentum and widespread effect Colby and others called for would have to wait, however. In May, the NLF launched the second phase of *Tet Mau Than*; three months later, they initiated a third wave of attacks. Although the May and August offensives were much smaller and less effective than the January attacks, they were disruptive enough to impede pacification and delay a counteroffensive until fall.

By October, U.S. and RVN officials were back on track and had developed a plan for rapidly expanding government control over the countryside. At its heart was the pacification counteroffensive Colby called for in May. Known initially as the Le Loi Pacification Plan and later as the Accelerated Pacification Campaign, the operation was built on the improved allied military position and growing political strength of the RVN. Unlike previous campaigns, the new pacification plan targeted contested hamlets rather than those controlled by the NLF. Should the security status of targeted hamlets be raised and added to the 5,000 or so already listed in the "relatively secure" category, it would bring an overwhelming majority of the population under a "reasonable" level of RVN control. Moreover, the campaign would, if successful, "forestall any VC effort to partition the country or justify any claim to coalition government." Besides improving security, targeting the insurgent infrastructure, and intensifying efforts to induce communist guerrillas and political cadres to "rally" (defect) to the RVN, the new pacification plan focused on strengthening hamlet and village administration to compete with NLF liberation committees at the local level.[108] American officials believed the new pacification counteroffensive took on new significance given the heavy emphasis the communists were placing on these administrative bodies. In a telegram dated 30 October, Ambassador Bunker tied the proliferation of communist liberation committees directly to pacification and urged allied leaders to push the new counteroffensive more vigorously. "Since the spread of VC 'liberation committees' and the enemy's waning military fortunes signaled the possibility of a VC switch to a more political phase," he stated, "we urged that the GVN preempt this policy by vigorously expanding its influence in the rural areas."[109]

The APC was intended to wrest South Vietnam's contested areas away from the NLF by rapidly extending RVN authority over territory and population. The plan was to upgrade approximately 1,000 of the 3,600 hamlets in the "contested" category (D, E) to the "relatively secure" (C) category. Slated to begin on 1 November, the campaign was designed to "extend GVN influence over as much presently contested population as feasible by 31 January 1969." Its primary components included "an intensive military spoiling campaign" to destroy communist forces and drive them away from populated areas; an accelerated Phung Hoang campaign to "neutralize" at least three thousand Viet Cong Infrastructure per month; and a stepped-up Chieu Hoi program to "rally" five thousand communists to the RVN by the 31 January deadline. In addition, it promoted a program to organize roughly 1 million PSDF and arm at least one in five of them. The APC also called for an "information campaign" to convince the rural population that the RVN had seized the initiative. Should it succeed, the APC would reclaim much of the civilian environment lost during Tet and bring approximately 75 percent of the total population into the "relatively secure" category.[110] According to allied planners, the APC was vital to combating the communists' resumption of protracted guerrilla warfare. "If there is any validity to estimates of the enemy's strategy," an MACV memorandum declared, "it reinforces the need for the GVN, the RVNAF, and the Free World Forces to press the Accelerated Pacification Campaign to the maximum." The one thousand target hamlets, the Phung Hoang campaign, the PSDF program, the Chieu Hoi effort—all were designed to extend Saigon's authority well beyond what many officials thought possible. "The word must be to extend the authority of the GVN," the memo continued, "and to stamp out every last vestige of VC authority throughout the countryside."[111]

Although it fell far short of stamping out "every last vestige" of the communist insurgency, the APC succeeded in rapidly extending government authority throughout the countryside. By the end of November, one month into the offensive, MACV had reported that 73.3 percent of the population was living in relatively secure areas—a new high. The population living in contested and VC-controlled areas, meanwhile, had dropped to 13.3 percent and 13.4 percent, respectively. The 3.5 percent increase in people living in relatively secure hamlets from the previous month represented an additional 636,800 people inhabiting areas considered to be under reasonably good RVN control. The largest increase in government control occurred among the rural population, which increased 5.2 percent to 60.7 percent overall, or 586,000 additional people. "I am pleased with the increased activity which has resulted [from the APC]," Colby declared, "and I believe this campaign has initiated a real momentum in pacification which is proving contagious in many quarters hitherto included to the cautious and defensive. It is an important counterpart to Paris, as it builds Vietnamese confidence in what they can do and at the same time strengthens the machinery of constitutional government for the political contest to come."[112] Colby was correct: the momentum of pacification was contagious. By the end of the APC's first month, three out of four military regions had reported a marked improvement in territorial security. Only MR 2 reported no change in security levels.[113] The lack of progress in central South Vietnam, however, did little to dampen the enthusiasm shown by American advisers in the other three zones. "The Accelerated Pacification Campaign has been, thus far, the most successful GVN program attempted in this [Military Region] in the last three years," asserted John Paul Vann, then deputy for CORDS in MR 3.[114] According to the senior adviser in the Mekong Delta, "The IV Corps pacification program in November reflected more tangible indications of progress than any previous month of 1968. Phase one of the Accelerated Pacification Campaign received enthusiastic GVN support and produced important gains in territorial security."[115] Up north in MR 1, CORDS personnel were reporting similar advances. "Significant measureable progress in pacification occurred during November. The Accelerated Pacification Campaign proved to be the vehicle around which the civil/military teams of the GVN, the U.S., and the FWMAF united in a coordinated program to improve territorial security and to attack the enemy's political infrastructure," an adviser reported. "Highly successful encounters with the enemy, encouraging progress with VCI neutralization, and optimistic plans for continued programs in all fronts," he continued, "have created a sense of purpose and optimism among U.S. and GVN officials that should cause the pacification program in I Corps to move with renewed vigor and purpose into calendar year 1969."[116]

A new sense of purpose and renewed vigor in the pacification program did in fact occur in 1969. When the APC concluded on 31 January, allied policymakers were elated by the outcome. "Results of the Accelerated Pacification Campaign generally exceeded expectations," Colby trumpeted. "The GVN has seized the initiative in the countryside and the enemy has yet to react positively. Thus, the ambitious pacification plan for 1969 starts from a greatly expanded base . . . and with considerable momentum."[117] Colby's excitement was well founded. By nearly all measures, the "1,000 Hamlet Campaign" (as some called it) had been a success. In addition to upgrading 1,055 out of 1,338 targeted hamlets to at least a C status, the campaign exceeded its goals for the Chieu Hoi and PSDF programs. As for the former, more than 3,600 additional *Hoi Chanh* (the name given to communist guerrillas and political cadres who defected to the RVN) had rallied on top of the original target number of 5,000. The goals for organizing and training the PSDF were also exceeded although the number of firearms issued to local militia fell below expectations. As of 31 January, CORDS reported 1.1 million PSDF organized, 660,000 trained, and 170,000 armed. Only area the Phung Hoang program failed to meet expectations, falling short of the stated goal of 3,000 VCI neutralizations per month. Even here, though, there was cause for optimism. The number of VCI neutralized during the APC averaged 2,400 per month, roughly double the number of the 1,200 per month prior to the launch of the campaign. Military support for the APC, particularly the contribution of territorial forces, was also deemed a success. "The RF/PF support of the APC was outstanding: they continually improved their combat performance and provided most of the protection around the target hamlets. ARVN/FWMAF were thus able to continue large scale operations."[118]

The APC was succeeded by the 1969 Pacification and Development Plan. With some minor modifications, the new plan was identical to the accelerated campaign. "During 1969 we are determined to destroy the supporting structures the enemy has created among the people by the use of violence and terrorism," the CPDC declared. "Owing to repeated military victories of ARVN and Allied Forces, the enemy has been obliged to change tactics. He has shifted from military efforts to political ones and is attempting to expand his political and guerrilla machinery in order to control rural areas. . . . In this situation, our essential task . . . is to act quickly to liberate the people from the coercion and control of the enemy. The 1969 Pacification and Development policy is thus a continuation and extension of the APC."[119] Much like its predecessor, the 1969 plan had as its centerpiece a military campaign designed to eliminate communist insurgents and expand RVN authority. The overall security goal was to bring 90 percent of the population into the relatively secure category and extend national sovereignty throughout the country. Other goals included eliminating 33,000 VCI, rallying 20,000 *Hoi Chanh*, establishing local government in all villages under RVN control, and raising the number of PSDF to more than 2 million, including 400,000 armed. The plan also included efforts to decrease the number of refugees and displaced persons to less than 1 million, increase the information and propaganda effort in the countryside, and stimulate the rural economy primarily through stepped-up rice production.[120]

Like the APC, the 1969 pacification plan was a success. Although it too had fallen short of stated goals in a number of areas, particularly *Hoi Chanh* and VCI eliminations, the program exceeded its objectives for organizing PSDF and, most importantly, solidifying government control. As of December the percentage of the population living in relatively secure areas climbed to 92.7 percent. This roughly translated into 16.3 million out of 17.6 million total population. As these numbers indicate, 1969 had turned out to be a banner year for pacification and allied progress in the control war. Although momentum began to slow in November and would continue to stagnate into the early part of 1970, the year still witnessed the largest and most rapid control gains of the war, rising from approximately 65 percent of the rural population living in relatively secure hamlets in January to more than 90 percent by year's end.[121]

In December 1969, MACV concluded that the overall progress in security over the previous thirteen months was largely the result of the allies' increasingly effective use of military resources to clear target areas along with a strong emphasis on developing local government and local defense.[122] A number of key American officials agreed that the improved conditions in the countryside were the product of conventional combat actions. In September 1968, General Abrams informed the White House National Security Council, "Where we are in Vietnam is due to the application of raw power. That is why the enemy is where he is, why pacification has moved. . . . When you turn off the power you have got an entirely new ball game."[123] The following year, Abrams again made this argument: "The accelerated pacification program, which we feel is progressing quite favorably, is made possible largely by friendly military initiative which keeps the enemy from concentrating his forces against our pacification program."[124] Abrams's predecessor concurred. "At the time we entered, the government held only a few enclaves," Westmoreland informed the House of Representatives Committee on Appropriations in February 1970. "It took a great deal of effort, sweat and blood to expand the enclaves to the point that the countryside opened up. In the process of that, we have had to grind down the guerrilla force, which has been done. Pacification has been extended bringing more people and area under government control. The indigenous enemy in the south has been reduced. The enemy now has been required to take refuge in remote areas of South Vietnam, in Cambodia, in Laos, and in the DMZ area."[125] Robert Komer made a similar assessment. "Without the regular ARVN and Americans pushing back the enemy's conventional units and in effect acting as a shield for the pacifiers," Komer stated in 1970, "the enemy's big units would have come in and messed up pacification before it could even get off the ground. In sum, you cannot talk about winning in Vietnam post 1964 by pacification alone without the big unit effort."[126] The Vietnam Special Studies Group also shared this sentiment. Conventional military operations, it argued, were the foundation for unprecedented expansion of RVN control in the wake of the Tet Offensive. "Throughout late 1968 and 1969, enemy main-forces were destroyed, dispersed, and forced to withdraw from populated areas of the countryside," one VSSG report maintained. "In most provinces studied, this change in the main-force environment was a prerequisite to the RVN control gains of late 1968 and 1969."[127]

The momentum of the 1969 pacification program was sustained, albeit at a much slower pace, throughout 1970 and 1971—this despite Vietnamization and the gradual withdrawal of American combat forces. In part, the encouraging security situation in the countryside was made possible by the Nixon administration's decision to expand the war into Laos and Cambodia. In the spring of 1969, just as the APC was beginning to yield substantial security gains in the countryside, MACV began attacking formerly inviolable communist sanctuaries in Cambodia. In March, the Nixon administration ordered the first of what would become a series of offensive cross-border operations against communist sanctuaries. For years, Hanoi and the NLF had used Laos and Cambodia to launch attacks against the RVN and provide shelter for battered NLF/NVA main-force units. Despite such provocations, however, these areas had been untouched by invasion. Sanctuaries in Laos and Cambodia were immune from allied ground attack due to the proclaimed neutrality of these two countries and the political turmoil it would have provoked, both in the United States and abroad. From the perspective of U.S. and South Vietnamese officials, the situation was akin to, in the words of one former officer, "a loaded and cocked pistol . . . being held to the head of South Vietnam."[128] This situation changed when, in March 1969, the Nixon administration sanctioned Operation Menu, a covert bombing campaign against communist base areas in eastern Cambodia. The action ended allied military abstinence in the region and precipitated two major cross-border operations in the following twenty-four months.

In May 1970, the Nixon administration upped the ante by approving a joint U.S./RVNAF cross-border ground attack into Cambodia. The "incursion," which witnessed offensive thrusts into two strategic areas known as the "Fishhook" and "Parrot's Beak," had several key operational objectives: (1) to rescue Cambodia's new proallied regime, headed by Lon Nol, from military defeat at the hands of communist forces; (2) to destroy NLF/NVA main-force units, supply caches, and base areas; and (3) to locate and neutralize COSVN headquarters. Strategically, the operation was intended to buy time so that Vietnamization and pacification could progress. Moreover, it was hoped that it would break the diplomatic stalemate in Paris.[129] Overall, the Cambodian incursion was deemed a military success. Although the allies failed to capture COSVN leadership and the political fallout in the United States was considerable (the raids were vehemently denounced by antiwar activists and members of Congress), the cross-border invasion achieved most of what it set out to do. "The operation gained much-needed time for the allies," historian James Willbanks maintains. "The Communists were unable to launch any significant operations from the Fishhook and Parrot's Beak for the next two years. Despite having provoked controversy, the president's decision to go into Cambodia had lessened the pressure in South Vietnam."[130]

The following year, the allies launched another cross-border invasion against NLF/NVA sanctuaries, this time into Laos. Known as Lam Son 719, the operation was conducted by the RVNAF without the assistance of U.S. ground forces. MACV provided logistical, artillery, aerial, and advisory support. The objective of Lam Son 719 was to sever the Ho Chi Minh Trail by occupying and destroying a network of logistical facilities and base areas in southern Laos. This region was crucial to Hanoi's support to the southern insurgency and had been so for years. U.S. intelligence estimates calculated that between 1966 and 1971 the DRV had moved 630,000 NVA troops, 400,000 weapons, 50,000 tons of ammunition, and 100,000 tons of foodstuffs down the Ho Chi Minh Trail. The strategic purpose of Lam Son 719 was to disrupt this flow and upset a potential communist offensive sometime in 1971 or 1972 so that Vietnamization and pacification could progress. Lam Son 719 was deemed a military failure largely due to a dismal combat performance by the RVNAF. Nonetheless, the operation helped buy time for Saigon to strengthen its strategic position by forcing the NVA to expend resources in a pyrrhic victory. The operation, though profoundly disappointing in many respects, was, according to Willbanks, "an important factor in delaying the next major Communist offensive; the North Vietnamese took nearly a year to replace the resources lost in the effort to destroy the South Vietnamese along Route 9. Thus, the South Vietnamese and the Vietnamization effort gained a brief, but needed respite from NVA pressure."[131]

Overall, Operation Menu, the Cambodian incursion, and Lam Son 719 did little to alter the strategic balance of forces in Indochina. They did, however, provide the RVN an opportunity to consolidate its gains in the countryside and continue to prepare its armed forces for a future without American combat forces. In Cambodia, joint U.S./RVN operations cut a vital North Vietnamese supply line and preoccupied NLF/NVA main-force units that otherwise would have threatened Saigon and the Mekong Delta. In Laos, Lam Son 719 reduced the flow of men and matériel down the Ho Chi Minh Trail and delayed a possible North Vietnamese invasion for a year. "From the viewpoint of the Military Assistance Command, Vietnam," historian Graham Cosmas contends, "three years of cross-border operations kept the enemy at bay and won time in South Vietnam for Vietnamization and pacification to run their course."[132] In South Vietnam, low levels of combat along with reports of steady progress in security levels, economic stabilization, and political development appear to have validated such claims.

By December 1970, the population living in relatively secure areas had jumped to nearly 95 percent. In the countryside, RVN control of the rural population was equally noteworthy, increasing more than 25 percentage points since the beginning of the American drawdown. The VSSG figure for December showed Saigon in control of more than 66 percent of rural inhabitants. "The war in South Vietnam has wound down to a point well below the levels of previous years and has entered a period of relative quiescence," the office of the Deputy Assistant Secretary of Defense (ODASD) declared in its end-of-year report. "Pacification progress and activity has exceeded expectations, despite reduced U.S. forces and activity," the report continued. "Experienced observers returning to Vietnam after long period out of the country unanimously agree that security conditions in the countryside are better than ever before, and that the allied main-force military campaign has achieved most of its objectives in MR's III and IV."[133] The year 1971 witnessed peak levels of control over territory and population. In December, ODASD reported that the number of people living in relatively secure areas was up to 96.8 percent. RVN control of the rural population, meanwhile, climbed to 76 percent, a 10-point increase from the previous year.

Critics have long questioned allied security ratings because HES statistics were often inaccurate, unreliable, and subject to inflation by officials who manipulated the numbers to show progress. In addition, data that exhibited large percentages of the population living in "relatively secure" areas failed to reflect the high levels of sustained communist activity inside South Vietnam. HES was an admittedly imperfect instrument that could not capture the war in its entirety or measure its intangibles—the social and political intricacies that defined the conflict at the local level. Yet despite its flaws and the fact that security ratings did not perfectly reflect the political realities of South Vietnam, the program did provide a reasonably accurate picture of the degree of government/communist control over population and territory. HES delivered useful insights into pacification trends; and after March 1965, the trend was clear: area and population control had progressed. Statistics compiled by ODASD show that 4.2 million people had been brought under government control between December 1964 and June 1967. During the same period, the number of people living under communist control had dropped by nearly 1 million. This trend continued in the years that followed. Data show that an additional 8.2 million South Vietnamese were brought under government control between December 1967 and December 1972. Thus, 80 percent of the population was living in "secure" areas on the eve of American withdrawal, and an additional 13 percent were living in areas designated "relatively secure." By the end of 1972, 18 million out of a total population of 19.3 million South Vietnamese were living in areas under predominant or full government control.[134] Although evidence indicates that communist activity was much more prevalent than the official numbers suggest, and that a large percentage of the overall increase in government control—particularly during the 1964–67 period—was a product of urbanization and increased numbers of people moving into secure areas rather than the expansion of territory secured by allied forces, the trend toward greater RVN control over the civilian population was real. Thomas Thayer was correct: the secure population was often overstated and the VC/NVA control was almost certainly greater than HES data recognized. Nonetheless, the fundamental trends in pacification were reasonably sound.[135]

What critics fail to mention is that HES statistics were not the only measure demonstrating progress in South Vietnam; other indicators showed pacification had made genuine forward movement in the countryside after *Tet Mau Than*. Beginning in 1968, marked improvements were reported in key economic indicators, including animal husbandry, fisheries, and agricultural output.[136] According to USAID, poultry, swine, and fish production had increased between 10 and 15 percent during 1970. Moreover, rice production in South Vietnam had climbed from 4.3 metric tons in 1967 to 5.7 metric tons three years later.[137] This was due, in part, to the advent of the "Green Revolution" and the proliferation of high-yield rice varieties—the highly celebrated IR-8 "miracle" rice first introduced by USAID in 1966. Higher rice production was also the result of increased acreage under cultivation, which in turn was attributed to improved security in the countryside.[138] By December 1971, it was estimated that 80 percent of the essential roads and 75 percent of the essential waterways in South Vietnam were deemed safe. Polls taken in the summer of 1972 indicated that 75 percent of the rural population had expressed no problems getting themselves or their products to local markets. HES statistics, meanwhile, showed 70 percent of the rural population were able to move their produce to and from local markets without harassment or taxation by local NLF cadres. Increased security also allowed South Vietnamese president Nguyen Van Thieu to implement sweeping land reform legislation in 1970. Known as Land to the Tiller (LTT), the program distributed 2.5 million acres of land to 800,000 tenant farmers and their families between 1970 and 1973, thus reducing the land tenancy rate from 60 to 10 percent of the rural population.[139]

The improved security situation in the countryside and the economic benefits this brought were evident by early 1971. In April, a *Los Angeles Times* reporter captured South Vietnam's new modus vivendi as witnessed in the Mekong Delta. "The real business of the Vietnamese, looking with some confidence toward the future for the first time in decades," he wrote, "is business."

> Local capital is investing heavily in construction, while Saigon is beginning to get badly needed daubs of paint. In the countryside, the emphasis lies upon reconstruction. Sixty-eight sawmills are now

> operating in one western border province against 15 a year ago. Loggers' bright yellow tractors contrast with drab military convoys. Greatly increased security provided by the 1970's Cambodian operation has opened notorious War Zones to logging. The most striking visual impact of Vietnam's new preoccupations is felt driving to My Tho, chief city of the upper Mekong Delta 50 miles south of Saigon. For the first time, buildings under construction greatly outnumber those gutted by war. Billboards touting aluminum roofing alternate with billboards for small machine-plows. Raising new roofs with government aid is as much a political as an economic gesture. The Viet Cong cannot enforce their "strict prohibition" on such "collaboration." Food trucks, rather than tanks or troop trucks, now dominate Route Four from the fertile delta to Saigon. Market gardening is thriving. New strains, fertilizer, irrigation, and techniques have given rice production a tremendous fillip. But the chief factor is security. Vietnam imported 800,000 tons of rice in the worst years. Imports in 1970 were 200,000 and the government is waiting till September 1971 before deciding this year's needs. They will probably not exceed 150,000 tons. By 1972, self-sufficiency, inconceivable a few years ago, is confidently expected.[140]

As we can see, improved security in the countryside not only boosted food production and reduced the RVN's reliance on rice imports; it brought a new prosperity to farmers and new way of living in rural areas that could not be denied.

The improved security situation in rural areas also facilitated refugee resettlement, as increased government control allowed the RVN to help peasants displaced by the war. Beginning in late 1967, the RVN was providing emergency assistance to refugees, enabling them to resettle or return to their homes villages, and preparing for large-scale rehabilitation programs. By 1970, considerable progress had been made in these areas. According to testimony delivered before Congress, nearly 600,000 refugees had been returned to their homes or paid resettlement allowances between November 1968 and February 1970. In addition to financial assistance, the allies, under the auspices of the Ministry of Social Welfare, provided refugees with roofing, cement, blankets, mosquito netting, and foodstuffs.[141] The number of individuals living in refugee camps, meanwhile, dropped from approximately 700,000 in January 1969 to about 150,000 by year's end.[142] More than any other factor, the expansion of security and government control over rural areas begun under the Accelerated Pacification Campaign enabled the South Vietnamese and

American governments to devote time and resources to assisting refugees. "The extension of security over a large part of the countryside, plus the increase in [RVN] self-assurance, were the main reasons for the favorable developments in the refugee program in 1969, in which it was possible to give largely undiverted attention to the overarching refugee program," William Hitchcock, director of the Refugee Directorate for CORDS, declared in February 1970.[143] The following year, William Colby also made the connection between increased security in the countryside and refugee relief. "The political, economic, and psychological atmosphere today is markedly changed from earlier days," Colby informed Congress in 1971. "Thus, while a refugee problem remains and new refugees appear, the government is undertaking programs to rebuild a war-torn society rather than meeting only the daily crises of refugee handling and care."[144]

American officials, particularly those invested in the pacification program, were heartened by the security and development trends in South Vietnam. "The signs of progress on many fronts since my previous visit to South Vietnam in August of 1969 are clear," Congressman Orval Hansen informed a House of Representative select committee in July 1970. "The country is far more secure now than it was at that time. This greater security from hostile enemy action has been a stimulus to expanded agricultural production and increased commercial activity generally. There are also encouraging signs of progress in the development of viable political processes and institutions at the local level."[145] In February 1970, Colby relayed to Congress the qualitative improvements found in rural areas. "The key development of 1969 was the further [government] expansion in the new areas throughout the countryside," he stated. "In early 1970 the change . . . is there to be seen. Except in one or two areas, the large enemy battalions, regiments and divisions are in the border sanctuaries. The roads are open to many markets and, from the air, tin roofs sparkle throughout the countryside where families are once again tilling their long-abandoned farms."[146] Lieutenant General A. S. Collins made a similar assessment. "Progress of pacification has been marked," he stated in his final report as the commander of I Field Force Vietnam. "Rice production . . . has increased . . . so that the region as a whole is verging on self-sufficiency. Village and hamlet chiefs have been elected and I consider this to be a major step forward. . . . Traffic on the roads and highways is heavy and ever increasing. Many other pacification objectives have been achieved. In the past, all too often the VC have had the upper hand. Now it is obvious that the [RVN] is in the driver's seat."[147] Historian Ronald Spector argues that it was virtually impossible after 1969 to deny genuine forward movement in pacification. "Even those who had come to regard all Saigon reports and statistics with deep skepticism," he writes, "could not deny the physical evidence of improved security."[148]

The gains in allied control over territory and population naturally came at the expense of the communist insurgency. As early as 1969, intelligence reports indicated pacification had "constricted" communist control over the civilian environment and limited the insurgency's ability to secure material resources and recruit personnel. "Allied military activities have imposed major limitations on communist military operations," read a July 1969 CIA national security estimate. "Allied military action has resulted in the capture and destruction of large quantities of communist supplies and, along with the GVN's pacification program, has constricted communist controlled territory and hampered their ability to recruit personnel; it has also limited communist access to local funds and supplies. In short, allied forces have not only denied the communist forces in the field any military progress but have also reduced their overall military capabilities during the past year and a half."[149] Other reports indicated a serious decline in insurgent manpower. "Since the Tet Offensives of 1968, VC guerrilla and local forces have faced increased frustration in their efforts to maintain effective strength levels," MACV stated. Decreased support from the rural population along with the APC, it continued, had made it exceedingly difficult for the communists to maintain a robust guerrilla movement. These developments not only reduced the NLF's overall numbers, but weakened the insurgency's ability to maintain itself at the local level. "The VC guerrilla and local forces are considered the mainstay of the enemy's efforts in the Republic of Vietnam. To be effective, they must control the population, extort taxes, gather supplies, and conduct proselyting activities among the South Vietnamese. Failure in these areas will hinder all efforts to gain ultimate control of the south."[150]

According to a special national security intelligence estimate issued in February 1970, the NLF was "in trouble." The report listed three positive developments: (1) NLF casualties had exceeded infiltration; (2) the quality of combat units had declined and they were heavily dependent on NVA support; and (3) the morale of soldiers and political cadres had dropped. Large areas of the South Vietnamese countryside, it continued, were now denied to communist forces, which reduced access to crucial manpower and economic resources. The loss of access to local populations also reduced the Front's ability to maintain an effective political apparatus, proselytize, and gain support.[151] "During the past year and one-half, the GVN has sharply reduced the population controlled or influenced by the communists," the intelligence brief read. "As a result, communist access to local manpower and sources of supply is being restricted, the mobility of many communist units limited, and local cadres exposed and demoralized." The estimate continued, "Hanoi may well believe that GVN pacification progress is essentially fragile and can be rolled back once U.S. forces are out of the way. However, it may now feel compelled to recognize that further contraction of communist-controlled areas, or even a prolonged continuation of the current situation, will limit the ability of its forces to confront the RVNAF, to launch major attacks on population centers, or even to wage a prolonged struggle."[152]

Hanoi and the NLF acknowledged that pacification was not only taking a heavy toll on the insurgency but had degraded their ability to wage protracted war. Between 1969 and 1971, scores of directives and interviews with captured guerrillas and political cadres revealed allied pacification efforts had in fact damaged the NLF. "The enemy's pacification program will bring great danger to friendly forces because their pacification activities are steadily increasing," one cadre lamented. "The enemy pacification program is very dangerous," he continued. "This program can affect the potential of the revolution, the morale of the people, and even the morale of the friendly soldiers. . . . Due to fierce enemy attacks and poor cadre leadership, the guerrilla movements in a number of local areas have suffered a setback. This setback was even felt in areas where the guerrilla movement had previously been strong."[153] After defecting to the RVN in 1970, one former NLF member stated, "In 1969 . . . pacification accelerated. From then on, the Front has been going downhill." When asked if the program had caused difficulties to the NLF cadres in his area, he replied, "Yes—a lot of difficulties. . . . Life has become ten times more difficult for the cadres now. The cadres at present only see imminent death wherever they turn."[154] According to a captured document dated May 1970, communist units were becoming increasingly weak and demoralized as a result of expanded RVN activities. "The enemy has gradually expanded his control over our areas," the author noted bitterly. "The enemy has succeeded in his pacification program. Our areas of activities have been narrowed considerably. . . . [A]t present . . . we are exhausted."[155] Postwar analyses also admit the reinvigorated pacification program imposed major difficulties on the revolution. "The enemy's horrible, insidious pacification program and his acts of destruction created immeasurable difficulties and complications for our armed forces and civilian population," the official history of the People's Army of Vietnam (PAVN) read. "Our liberated areas were shrinking, our bases were under pressure, and both our local and strategic lines of supply were under ferocious enemy attack. . . . In the meantime, the enemy gained control of many heavily populated areas."[156]

The deterioration of the NLF and declining strategic position of the communists had given rise to a new sense of optimism in Saigon and Washington. According to some observers, the pacification-Vietnamization strategy, when coupled with progress in the main-force war, had "carved out a new set of realities on the ground in Vietnam."[157] By 1970, the allies' ascendant position in the control war had given a number of officials cause to celebrate. In the fall of 1969, General Abrams declared, "The pacification program is no longer a laughing matter from his [the enemy's] viewpoint. It's after people, and that's what he's after—that's what the whole thing's about."[158] He continued, "I must say when you take the whole thing . . . the progress in pacification . . . no one has to buy the 90 percent or 86 percent, but the fact that security of the people is improving every month, I think is clear. I feel that he has to do something because the march of events seems to me to be inexorable."[159] Robert Thompson, a staunch critic of the U.S. war effort until 1969, was equally optimistic. "I was very impressed by the improvement in the military and political situation in Vietnam, as compared with all previous visits, and especially in the security situation both in Saigon and the rural areas," Thompson wrote of a trip he made to South Vietnam in the fall of 1969. "The position of the GVN is undoubtedly more stable and its performance increasingly effective. This has best been demonstrated in the 1969 pacification program which has resulted in greatly increased government control in rural areas."[160] John Paul Vann, another longtime skeptic, underwent a similar transformation after 1968. Once an acerbic critic of South Vietnamese and American performance in South Vietnam, Vann joined the ranks of the "new optimists" by 1970. In testimony delivered before the Senate Committee on Foreign Relations in 1970, Vann spoke frankly about his metamorphosis. "Up until 1968," he declared, "I was highly dissatisfied with the manner in which the war was being conducted in Vietnam, and I did not anticipate that it was going to be successful. Since 1968 I have become increasingly convinced that, with the changes that have been made not only by our side but by the enemy side, our objectives in Vietnam and, coincidentally, the objectives of the majority of the Vietnamese people, will be achieved."[161] Vann continued, "We have essentially stolen the enemy's thunder by engaging in a people's war on our side. This is what has made such a difference in Vietnam. That is why for the last eighteen months I have been called an optimist in Washington."[162] The positivity surrounding pacification and the control war in 1969 and 1970 was so strong that it persisted long after the war had been lost. "Pacification worked wonders in South Vietnam," Nixon retrospectively declared in 1985. "We were in control of areas that we had previously not dared to enter. . . . [O]ur pacification program had extended Saigon's control throughout the country down to the hamlet level."[163]

Despite overall high levels of government control and the heady optimism of American officials, there were dark spots in the overall pacification picture. As 1971 drew to a close, a number of provinces remained beset by persistent insurgent activities. In addition, American intelligence analysts again determined that many of the 1970 and 1971 pacification gains had resulted as much from reduced enemy activity as allied initiatives.[164] Moreover, it was known that many of the 1968–71 pacification gains were a product of the devastating losses suffered by VC units during Tet, a significant decline in communist main-force activity, and a large demographic shift in South Vietnam from contested to government-controlled areas. Thus, there were many who were less sanguine about the state of pacification and the control war than Vann, Thompson, and Abrams.

As we will see next, concerns over the durability of pacification gains were not only prevalent but well founded. The program and the progress it made between November 1968 and December 1971 were much more fragile than many realized. The communist insurgency, despite unrelenting pressure from the allies, had managed to hang on and survive to varying degrees throughout South Vietnam. The RVNAF, meanwhile, was becoming increasingly overburdened by territorial defense and stretched thin as Saigon's armed forces attempted to hold on to more territory and population with diminishing American assistance. South Vietnamese military units in support of pacification were also beginning to suffer from a number of systemic problems, including poor leadership, low morale, and desertion. These weaknesses undermined Vietnamization and the RVN's efforts to maintain its hold over populated areas. By the spring of 1972, the unique spatiality of the control war and a protracted guerrilla war without frontlines had left the RVN and its armed forces overstretched and unable to adequately defend the national territory without the help of U.S. combat forces. This became increasingly evident beginning in March, when a new communist offensive laid bare the hollowness of the RVN's 90-plus percent security ratings and the inherent weaknesses of the fast-and-thin approach to pacification. The outcome of the offensive was considered by some to be a victory for the RVN and a vindication of the pacification-Vietnamization strategy. Nonetheless, the communist general offensive of 1972 would in time bear bitter fruit for the RVN. Although beaten back, the Easter Offensive established a firm NVA foothold in South Vietnam and set in motion a series of events that led to the decline and eventual collapse of the Saigon regime in April 1975.

7 THE TENACIOUS REVOLUTION AND REVIVAL OF THE INSURGENT STATE, 1969–1972

Pacification had achieved notable gains in the control war between November 1968 and December 1971, bringing a reasonable degree of security to roughly 95 percent of the South Vietnamese population. Despite the unprecedented level of RVN control over territory and population and historically low levels of communist insurgent activity in the countryside, there was some cause for concern. Security levels in a number of provinces were extremely fragile and susceptible to regression should the communists step up antipacification activities or launch another large-scale offensive. While things may have appeared relatively quiet on the surface by 1971, there existed putrefaction underneath that belied the RVN's quiescent facade. Not only had the communist insurgent state survived, leaving pockets of rebellion and sustained politico-military conflict throughout South Vietnam, but the pacification and Vietnamization programs were beset by a host of debilitating problems that grossly undermined Saigon's ability to maintain control over the national territory and its population with diminishing assistance from U.S. forces. In February

1970, a report on the pacification program generated by a subcommittee of the House Armed Services Committee expressed lingering suspicion regarding the war in Vietnam. Given the history of that conflict, the report argued, there were grounds for reservation. "Anyone who has followed developments in and reports on Vietnam over the years has learned the value of skepticism," the subcommittee report declared. "No other war, no other great national undertaking has suffered so much from overly optimistic estimates. We are constrained to inject a note of caution, to remember past occasions when indices of progress proved to be illusions."[1]

THE PARADOX OF FAST-AND-THIN PACIFICATION

Pacification progress was not an illusion; yet high levels of security and government control were much more precarious than the official numbers suggested. As early as January 1969, before the Accelerated Pacification Campaign had even concluded, a number of analysts and advisers expressed concern regarding the rapid pace of the program. Some argued it would spread available forces too thin and weaken territorial security and government control.[2] Others believed the speed of security gains left them vulnerable to regression. "A principal conclusion about the entire pacification program . . . is that its gains are tentative and can be adversely affected by military setbacks; this is especially true of the gains registered during the APC," declared a January 16 national security intelligence estimate. "The VC . . . will attempt to counter gains in pacification . . . and they will probably have some success; such an intensification of fighting would . . . set back the pacification program."[3] Two months later, a national security study commissioned by the Nixon administration revealed that the Central Intelligence Agency and State Department both believed the security situation in the countryside was fragile, and that a single shock, such as an attack similar to *Tet Mau Than*, could wipe out the gains that were made during 1968. According to National Security Study Memorandum One, there was cause for skepticism. Security improvements made during the APC, it argued, were (a) thin and tenuous because they were achieved through a "hard-and-fast" approach that spread military and administrative resources thinly over contested areas; (b) made only because the insurgents did not vigorously oppose pacification; (c) as much the product of urbanization and a massive demographic shift to the cities due to pacification operations. Intelligence analysts acknowledged recent progress but deemed it largely untested. "The GVN has yet to succeed in the countryside," the study concluded. "The rural population security has not changed significantly and certainly not at a rate which will free us of noticeable burdens within 2–5 years. We may even be overextended in the rural areas and open to a damaging VC counterattack. . . . The real test of how solid recent gains in pacification have been will come when the VC initiate serious counter-pacification activity."[4] In April, John Paul Vann expressed similar reservations. Now that Vietnamization and the withdrawal of American combat forces had become a reality, Vann argued, the risks inherent in fast-and-thin pacification were becoming apparent. "I am questioning the wisdom of taking what might be the remaining one to two years of U.S. presence in Vietnam and using this period in an attempt to gain major additional population control," he declared.

> It is obvious that there is no way to "wipe out" the enemy during this period and our actions may well be setting the stage to permit him, particularly after we have forced him into a vastly more mobile role, to gobble up our newly acquired hamlets in the same manner that he did the strategic hamlets [program under Diem]. If the de-Americanization of this war is to be accomplished within one or two years, then I am certain our present pacification policy is wrong. Even if the de-Americanization were to be over an extended period wherein we would not be below 200,000 Americans by 1972, our pacification policy may still be too ambitious. I believe we should re-examine the direction in which we are now moving for the express purpose of determining whether the GVN position in later years would not be stronger if it had firmer control over less population than the predictably modest to weak control it may have over 90 percent of the population if our 1969 program is successful.[5]

The following September, Kissinger, who had himself advanced a "less is more" argument the previous January,[6] informed Nixon that while large numbers of the rural population had come under government control over the past ten months, the durability of recent pacification gains remained in doubt. This was particularly true in contested hamlets. Much of the progress in government control, he informed the president, stemmed from mass emigration from rural areas to the cities rather than the efficacy of the pacification program. Kissinger also noted that officials from the top agencies continued to demonstrate major disagreements over pacification progress. As usual, the Joint Chiefs and MACV offered more optimistic assessments and argued substantial progress had been made. The CIA and State Department, meanwhile, were much more pessimistic, declaring that pacification had not made sufficient gains and that the program would continue to founder unless "radical changes" were put into place.[7]

Despite its detractors, fast-and-thin pacification still had its supporters, who accepted the inherent risks of the approach. When asked in 1970 if pacification was built on "numbers and mass" rather than high quality security and development, Ambassador Komer rushed to the program's defense. He admitted that while the allies had employed a qualitative approach whenever possible, the overall low quality of available resources and the need for a quick impact had forced Saigon and the Americans to adopt a predominantly quantitative approach. Pacification, he argued, received only a fraction of available material resources and quality personnel despite its massive territorial obligations (the main-force war received the lion's share of both). Sacrifices in quality, therefore, had to be made. "We needed a massive across-the-board effort on a countrywide basis if we were to achieve the necessary impact in the village war even though this necessarily meant some degradation of quality," Komer admitted. Contradicting Vann and Kissinger, Komer claimed a 10 percent improvement in every province was better than a 60 percent increase in only a few. This, he reasoned, was because every province in South Vietnam possessed at least some administrative structure and resources. "Why confine ourselves to a few [provinces]?" he queried; "the enemy didn't." According to the ambassador, the need to deliver at least a modicum of security to approximately 12,000 hamlets, 2,500 villages, and 236 districts within forty-four provinces dictated that the allies use "simple mass approaches" for pacification. "To provide some permanent in-depth security presence took a lot of people," Komer explained. "We had to provide continuous local security to thousands of hamlets, not just for a few."

Komer defended the timing as well as the scope of the current pacification effort. It was urgent, he argued, that the allies expand pacification into the countryside quickly because—after years of neglect—the program had finally received the attention it required. "If we had started way back in the late Fifties or even the early Sixties," he declared, "we could have had a much more careful buildup; we could have used much more sophisticated techniques; could have had much more time to learn and improve. By 1967 we didn't have time . . . to work on it village by village. There are 2,000 villages in Vietnam, and it'd be 1984 before we even got a quarter of them garrisoned, much less pacified." Komer concluded with the argument that the allies simply did not have the time or resources to conduct a slow and methodical pacification campaign. Substituting quantity for quality and running a simplified program, then, was the only effective way the RVN and United States could pacify the countryside and maximize government control in a relatively short period of time.[8]

DOWN BUT NOT OUT: THE DURABILITY OF COMMUNIST REVOLUTION AND THE INSURGENT STATE

In the summer of 1969, intelligence analysts estimated that while communist forces had taken a beating and suffered a severe setback in rural areas over the past sixteen months, they were far from defeated. Overall, the insurgency and its position in South Vietnam had eroded: the political infrastructure was being squeezed by growing government control over rural areas, and guerrilla units were experiencing a decline in morale and military effectiveness under heavy casualties and local logistical problems. Yet despite these difficulties and in spite of persistent allied military pressure and an accelerated pacification campaign, the communist insurgency had managed to survive. In July, a CIA national security estimate verified that the insurgency was gravely hurt but far from dead. "The military capabilities of communist forces in the field in South Vietnam have declined over the past year, and the overall intensity of their military effort probably fell below intended levels," it stated. "Nonetheless, the communists retain a substantial ability to sustain military operations."[9] According to the estimate, the political infrastructure showed no signs of weakening; it maintained a presence in nearly every village and hamlet. Moreover, it continued to operate effectively despite a loss of personnel and diminished contact with the rural population. Local and main-force guerrillas also remained viable thanks to an abundant manpower base and strong logistical system that ran from the DRV through Laos and Cambodia. The flow of men and materials down the Ho Chi Minh Trail, the CIA estimate concluded, would enable communist fighters to absorb casualties and continue to apply military pressure at "1969 levels" for several years.

The insurgency's ability to survive in the years following *Tet Mau Than* resulted largely from infiltration. According to figures compiled by MACV, more than 104,000 communist troops had slipped into South Vietnam in 1969. This was in addition to the 57,300 individuals who were recruited by the NLF within the RVN that year. Although the overall number of infiltrators had suffered a sharp decline in 1970 (57,200), the number rose again to approximately 73,000 in 1971 and more than 150,000 in 1972. In-country recruitment by the NLF during these years, meanwhile, topped more than 68,000 men.[10] As for war matériel, the communists moved more than 26,000 short tons down the Ho Chi Minh Trail in 1970–71.[11] This infusion of manpower and matériel had allowed Hanoi and the NLF to sustain their hybrid warfare approach and preserve the revolution. "It seems likely," one CIA report concluded, "that communist strategy will call for them to stay in the field with forces which will enable them to conduct periodic offensives, some of which may be fairly large-scale and costly, and to maintain pressure on the GVN presence in the countryside."[12]

Infiltration revealed both the strengths and weaknesses of allied efforts in the control war. On the one hand, the need to ferry manpower and material resources into the RVN from outside the country stood as a testament to the success of pacification and allied military operations. It was also an indicator of how far the revolution had fallen from favor among the rural population. After *Tet Mau Than*, communist revolutionary forces inside South Vietnam simply could not sustain the war from within; only with generous support from Hanoi could they survive. On the other hand, successful communist infiltration into South Vietnam revealed the allies' inability to police the borders or control the national territory. Had the constellation of guerrilla base areas and liberated zones inside the RVN been eliminated, then the infiltration of men and supplies would have been moot. This was not the case, however. Although greatly reduced, the insurgent state was able to preserve the insurgency and sustain the revolution by exercising control over the land and establishing a revolutionary space, with troops and supplies from the DRV, to sustain attacks against the Saigon regime and its American ally.

In South Vietnam, Hanoi and the NLF maintained a relatively high level of military activity because they managed to preserve their network of liberated zones and base areas inside and outside the country—despite the immense allied effort to destroy them. The 1970 Combined Strategic Objectives Plan confirmed this observation. "Although faced with obstacles and the allies' continued battle victories, the VC have not been immediately defeated," it declared. "With their capability to initiate military operations from base areas beyond the borders, the enemy may exert significant pressure at the time and place of his choosing. . . . Until these hideouts or base areas are denied the enemy, particularly the NVA, he will continue to represent an immediate threat to the achievement of our combined objectives."[13] Although base areas had been a consistent target of allied main-force operations throughout the war, they proved incredibly resilient and impervious to destruction. Even when allied forces effectively destroyed an insurgent stronghold by dispersing forces, wrecking fortifications and infrastructure, and liquidating weapons caches and foodstuffs, these areas were quickly reconstituted once major combat operations concluded.[14] "We were able to rebuild many of our 'leopard spot' bases within the enemy-controlled areas," declared the PAVN official history. "Many areas in the . . . lowlands that the enemy had 'pacified' during the 1969–1970 period were again liberated."[15] Massive B-52 bombing campaigns and large-scale search-and-destroy operations against cross-border sanctuaries and weapons caches, including the 1970 Cambodian incursion and Lam

Son 719, also failed to eliminate communist redoubts. Base areas and liberated zones remained a persistent and ubiquitous threat despite years of allied efforts to destroy them. "In contested and even in GVN controlled areas," CORDS asserted in 1972, "the enemy may still be found in 'mini-bases.' The enemy can train in these bases and can conduct operations, taxation, and recruiting out of them."[16] The following October, it was reported that communist forces were reestablishing base areas inside South Vietnam and making their presence felt in rural areas. According to one newspaper account, "The communists have returned to many interior base areas from which they were painfully pushed over the past four years. From these base areas they are seeking to expand and re-exert their presence in the hamlets and to return as much of the countryside as possible to 'contested' status."[17]

The resiliency of the communist administrative apparatus at the village and hamlet level also helps explain how the insurgency managed to survive the 1969–71 allied onslaught. According to the CIA, the overall durability of the VCI was a major element in the communists' ability to sustain the revolution and "stay the course" in South Vietnam after *Tet Mau Than*. "The Viet Cong infrastructure . . . continues to function effectively despite some attrition and reduced access to the population and resources of the South," a 1969 study declared.[18] Individuals who personally fought insurgents in the villages and hamlets confirmed the tenacity of the communist shadow government. Bing West, a RAND Corporation analyst and former participant in the U.S. Marine Corps' Combined Action Program, was one. "The Viet Cong have the stronger organizational cement in the countryside and the main reason it has not hardened and held has been the jarring efforts of American armed forces," West stated in 1969. "But this defensive sparring has not destroyed the enemy. The order of battle for the Viet Cong shows their infrastructure and local forces to be relatively intact (with their main-forces heavily reliant on inputs from the NVA resource pool)."[19] West possessed intimate knowledge of the control war at the grassroots level. He therefore knew that the only way the Saigon government and its American ally could defeat the communist insurgency was by dismantling its political infrastructure. Many agreed with his assessment. As one American newspaper indicated, the absence of NLF/NVA main-forces did not translate into absolute government control over the countryside. "The fact that communist combat forces have been pushed out of most South Vietnamese populated areas does not really mean the country is pacified," two senior Vietnam correspondents declared in the *Washington Post*. The VCI, they argued, had merely gone underground and was biding time until its strength was rebuilt and a favorable strategic situation emerged. Until the infrastructure was eliminated, the authors reasoned, the RVN would be at war—a fact that explained the residual pessimism among segments of the allied camp despite pacification progress and in spite of incessant setbacks suffered by communist fighters. "Although the VC guerrillas are cut to shreds," the article concluded, "the VCI, in large part, endures."[20]

The allies' inability to eliminate the VCI was touted by the communists as an indicator that pacification had failed. "During the past two years, the U.S. and puppet focused their efforts on pacifying and encroaching upon rural areas," declared a 1971 COSVN directive. "As a result, they caused many difficulties to and inflicted losses on friendly forces. Generally however, they were unable to attain their basic objectives. They failed to destroy or wipe out the revolutionary infrastructure or our local and guerrilla forces which continued to remain in their areas of operation. In some areas, we were even able to increase our forces. In spite of his oppressive control, the enemy failed to subdue our people."[21]

The VCI had a profound impact on the control war. Its ability to survive, therefore, became a major factor in the struggle to establish dominion over rural areas during the communist revolution's touch-and-go years in South Vietnam. According to a report generated by the Pacification Studies Group in 1970, the infrastructure's ability to survive, despite extreme allied pressure, allowed the revolution to maintain influence and control over portions of the civilian environment. "The enemy . . . has been able to maintain a degree of control in the face of opposing military activity of greater intensity than that threatening the GVN control apparatus," the report declared. "This is because the enemy has a well-developed doctrine and organizational structure for the extension and maintenance of control through covert means (especially the VCI), while the GVN relies on more overt control mechanisms (RF/PF forces, RD teams, etc.) which are more easily targeted and attacked. Thus the enemy has a significant doctrinal and organizational advantage in the control war." The report concluded that while the government improved its performance in recent years, the communists' control apparatus remained "inherently more effective, in relation to its size and overt strength, than the GVN's."[22] David Elliott, who conducted hundreds of interviews during the war with communist defectors and prisoners of war, verified that the communists had indeed maintained a superior control apparatus in the countryside, one that allowed the infrastructure (and by extension the revolution itself) to survive an accelerated pacification program and relentless allied combat operations. As long as the VCI retained a presence among the population, however secretive or clandestine, Elliott argued, it was able to elicit support from a hard core of sympathizers and influence the attitudes and behavior of the remaining peasantry in the villages and hamlets. "Although the village cadres were largely passive, holed up in their 'mini-bases,' they were able to maintain a presence in almost every village, and they were able to come out and appear suddenly in populated areas often enough to remind people that they were still around," Elliott wrote. "Even in villages that had been extensively 'pacified' . . . cadres could still operate."[23] Overall, the ability of the infrastructure to maintain its presence in the countryside was critical because it permitted taxation and recruitment—two fundamental activities that provided a lifeline to the battered insurgency by providing the manpower and supplies it needed to continue low-level protracted warfare.

The communist insurgency's ability to survive allowed it to sustain an antipacification campaign in the countryside. Although Hanoi and the NLF were unable to drive government forces permanently from the vast majority of villages and hamlets under RVN control after *Tet Mau Than*, they were able to disrupt pacification by incessantly harassing local militia and territorial forces, conducting acts of terror against local populations, and occasionally overrunning populated areas and government outposts. Given the diversity of the communist force structure, its hybrid method of warfare, and the broad geographical scope of protracted guerrilla warfare, no area in South Vietnam was exempt from attack. Even in areas considered secure, NLF local forces and political cadres were able to initiate indirect fire and terrorist attacks.[24] As monthly province reports and military region overviews generated by CORDS in 1970 and 1971 indicate, practically every province experienced some level of insurgent activity. Insurgent combat actions varied from place to place and included attacks by fire, sapper attacks, and small-scale ground assaults against critical targets such as lines of communication, contested villages and hamlets, military and government installations, and pacification personnel, particularly RF/PF, PSDF, and RD cadres. Nonmilitary activities involved political proselytizing and terrorist activities aimed at civilians, refugees, government officials, and those tied to the pacification program. Indiscriminate rocket and mortar attacks against refugee camps, cities, and district and provincial capitals were also prevalent. For the most part, communist guerrillas and main-forces continued to avoid contact with allied forces in accordance with COSVN Resolution 9 and the protracted war strategy put in place in 1969. Nonetheless, NLF/NVA units took advantage of every opportunity to attack allied main-force units and territorial forces in an effort to inflict heavy casualties, disrupt the pacification program, and corrode the government's hold on the operational environment.

Regardless of form or intensity, sustained insurgent activity meant the revolution's presence, however small, was felt in nearly every district and hamlet throughout

South Vietnam. It would be impossible within the confines of this study to give an exhaustive account of communist activity in each of the country's provinces during this period. Even so, a small sample of select provinces adequately captures what had occurred each month in nearly every district and province between 1969 and 1971. Here are some examples.

Thua Thien Province, located in MR 1, reported moderate enemy activity in May 1970, including attacks against civilians, booby traps, sabotage against government installations, and landmine incidents against lines of communication. The estimated casualties caused by these operations included eleven government personnel and thirteen civilians killed and another sixty wounded. On 6 and 31 May, insurgents launched rocket attacks against the provincial capital of Hue.[25] According to newspaper reports, one rocket struck a military hospital, killing twelve people and wounding seven. Another four people were injured elsewhere in the city.[26] The following month, CORDS advisers reported thirty-eight enemy-initiated incidents. Although this number was relatively low when compared to other provinces (Thua Thien was rated third-most secure overall in South Vietnam according to HES ratings at the time), it demonstrates the insurgency's ability to launch attacks even in the most secure areas. Local and main-force guerrillas reinforced with NVA regulars were responsible for most of the insurgent activities in Thua Thien. In addition, large concentrations of NVA forces were reported operating in the mountainous western region of the province along the Laotian border. Although these units were not engaged in the guerrilla activities prevalent in the populated coastal regions, they were still considered a major threat to pacification. "Enemy lowland activity should continue with small unit reinforcements to the VCI and harassment of GVN pacification efforts. The enemy has the capability of company to battalion size attacks in the western portion of the province," the CORDS report concluded.[27] Also in June, the senior adviser in Binh Dinh Province, located in MR 2, reported that communist activity had disrupted the pacification program, slowing and in some areas reversing recent progress. "The communists' use of selective terror," he stated, "has been facilitated by the apparent assignment of small elements of NVA units to act in the role of village guerrillas and district units—providing muscle to the local insurgent apparatus." Binh Dinh reported 142 enemy incidents that month, predominantly terrorism and attacks by fire by small enemy units against territorial forces and government outposts.[28]

Things were not much different in the southern provinces. Binh Duong, located in MR 3, also reported moderate insurgent activity during the month of June. According to official accounts, local insurgent forces, aided by elements of the 101st

NVA Regiment, had mined or ambushed lines of communication, planted booby traps, assassinated RVN officials and civilians, and launched attacks by fire against allied military installations. These actions killed several civilians and pacification personnel. "The growing boldness and aggressiveness of enemy forces," declared the senior province adviser, "indicated that the enemy is relatively confident of his ability to avoid detection and to initiate day and night action at the time and place of his own choosing."[29] Meanwhile down in the Mekong Delta, CORDS advisers in Kien Hoa Province reported 143 communist-initiated incidents in May, including a battalion-strength attack against a village complex in Ba Tri District. Also that month an RVN district chief was killed while leading troops against an NLF/NVA main-force battalion. The ARVN attack was intended to recapture a populated area that had been seized by communist units. According to reports, the battle marked the sixth time since the 1st of April that the communists attacked secure areas in an attempt to disrupt pacification and reverse recent improvements in the local security situation. "This long-anticipated tactic," the senior province adviser declared, "places a premium on finding and destroying these battalions in their base areas, on improved security measures to detect their movement prior to entering the villages, and on sound tactics designed to eject them from the village without inordinate destruction and civilian casualties when they do succeed in gaining access."[30]

The ability to sustain revolutionary violence throughout the countryside via guerrilla warfare and terrorism had a tremendous impact on the control war. Insurgent activity in the provinces discussed above indicates that Hanoi and the NLF had largely managed to implement the antipacification, protracted war strategy laid out in COSVN Resolution 9 and other directives. As previously discussed, the communist insurgency neared the verge of extinction in 1971 owing to *Tet Mau Than*, the APC, and unrelenting military pressure by allied military forces. Despite extreme difficulties and severe attrition, the revolution survived. Thanks to the infiltration of NVA regulars into depleted and battered NLF main-force and guerrilla units, a constellation of base areas and safe havens both inside and outside the RVN, and the VCI's ability to maintain a presence among the population, the communist insurgency maintained enough influence and control over the operational environment to sustain its protracted war strategy and attack the pacification program. In spite of strenuous efforts by the RVN and the United States to expel the communist presence, the insurgent state endured in rural areas. Its presence corrupted the national body politic, provided a platform for attack across countless points of conflict, and created a landscape honeycombed with liberated zones, base areas, and other pockets of communist control. These facts on the ground severely weakened Saigon's strategic position by stretching its armed forces dangerously thin and leaving populated areas highly susceptible to communist attack and subversion. As noted by one communist source, "The dispersal of a portion of our main-force troops to operate in local areas contributed to the continuation of the people's war at the local level and compelled the enemy to disperse his forces to counter our troops. This effort enabled us to defend our base areas and preserve our main-force army. It was also a realistic measure that allowed our main-force units to expand and provide a basis from which we could launch future attacks."[31] As will be seen next, the protracted war strategy used by the communists to maintain pressure in the control war and prepare for future offensives began paying dividends in the spring of 1972, when a new communist offensive laid bare the weaknesses of the pacification-Vietnamization strategy.

THE 1972 NGUYEN HUE ("EASTER") OFFENSIVE AND EXPANSION OF THE INSURGENT STATE

In the spring of 1971, Le Duan determined it was time to abandon the low-level protracted warfare approach that had dominated the revolution since the summer of 1969 and return to the General Offensive and General Uprising strategy shelved two years earlier. In May, the Politburo's Central Military Committee issued a resolution that called for "winning a decisive victory in 1972." The resolution advocated the close coordination of all assets with a combination of military and political offensives, military proselytizing, guerrilla warfare, and mass uprisings in all three strategic areas. It also laid out a number of specific missions meant to collapse the RVN's military and political forces and expand liberated areas. According to the committee, the specific missions of the 1972 offensive would be to annihilate the RVNAF, defeat the pacification program and destroy government administration at the local level, draw in and tie down enemy forces, and seize "all areas we are capable of liberating and, when there is a favorable opportunity . . . expand our liberated area."[32] The following month, the committee met with members of the Politburo to draw up plans for a strategic general offensive that would take place in early 1972. Shortly thereafter, Le Duan detailed the plan's objectives in a letter to COSVN. They were, first, to destroy the pacification program and regain popular support in the countryside; second, to step up the political struggle in the cities; and third, to redress the balance of forces and regain the initiative in South Vietnam.[33] Historian Lien-Hang Nguyen has shown that plans for the 1972 were inspired by Le Duan's belief that Vietnamization had been defeated, largely as a result of the failed ARVN incursion into Laos (Lam Son 719). Although many within the Politburo believed pacification was still a threat, the losses suffered by ARVN units in Laos along with the withdrawal of U.S. combat forces had "reached a level where communist forces possessed a clear superiority on the battlefield." Communist leadership believed the RVN was now vulnerable to a concerted conventional attack. According to the PAVN official history, the purpose of the 1972 spring offensive was to defeat Vietnamization and gain a "decisive victory" in 1972, one that would "force the U.S. imperialists to negotiate an end to the war from a position of defeat." The immediate objectives established by the Politburo in May 1971, the history continues, were "to launch large offensive campaigns using our main-force units in the important strategic theaters. We would simultaneously mount wide-ranging military attacks coordinated with mass popular uprisings aimed at destroying the enemy's 'pacification' program in the rural lowlands. These actions would totally change the character of the war in South Vietnam."[34]

The offensive began on March 30, when three lavishly supported NVA divisions crossed the seventeenth parallel into South Vietnam. These actions marked the beginning of what was at the time the largest offensive of the war. Known as the Nguyen Hue Offensive (or "Easter" Offensive, as the Americans and South Vietnamese called it), the operation involved four separate thrusts against key strategic regions throughout the RVN. The first took place in Quang Tri and Thua Thien Provinces, located just below the DMZ in MR 1. The second phase, directed against Binh Long Province just northeast of the capital of Saigon in MR 3, commenced the following week. The third, which unfolded in the western highlands provinces of Kontum and Binh Dinh, began in mid-April. The final front, smaller and less intense than the other three, was launched in MR 4 against outposts and other government-controlled areas throughout the Mekong Delta. Overall, Hanoi committed 14 divisions, 26 separate regiments, and supporting armor and artillery units. Within weeks, these forces achieved substantial gains in territory and population control. After three months of steady progress, however, the offensive ground to a halt and the RVNAF began to retake what it had lost.[35] By September, the South Vietnamese counteroffensive had reclaimed a significant portion of the territory and population that had been seized by the communists during the first few weeks of the offensive.

As indicated in Le Duan's letter to COSVN, the aim of the Nguyen Hue Offensive was to disrupt pacification, grow the insurgent state, and gain leverage in the control war. The allies were quick to pick up on the intent of the attack. "The clear purpose of the North Vietnamese invasion," Abrams informed the

White House in April, "is to destroy the Armed Forces of South Vietnam, to seize, occupy and hold territory in South Vietnam, control as many South Vietnamese citizens by military force as possible, and cause the downfall of the present South Vietnamese Government."[36] Lieutenant Colonel Doan Cong Hau, a province chief in MR 3, supported Abrams's claim with his own observations. "After several months it is apparent that the communists' Nguyen Hue offensive is a campaign to gain control of the people," he informed CORDS.[37] The PAVN official history later corroborated these claims. "Our goals for 1972," it declared, "were to annihilate a number of enemy regimental task forces and brigades; to render entire puppet regular divisions combat ineffective; to liberate a number of areas; to expand our base areas; to move our main-force units back into the various battlefields of South Vietnam and provide them a firm foothold there; and to provide direct support for mass popular movements that would conduct attacks and uprisings to destroy the pacification program in the rural lowlands."[38] Former ARVN commander Ngo Quang Truong also asserted that seizing control of territory and population was a fundamental aspect of the resurrected general offensive and uprising. According to Truong, PRG representatives in Paris had disclosed Hanoi's plan to seize control of as much RVN territory as possible in order to obtain leverage in the negotiations and strengthen the demand for a coalition government. "North Vietnam's principal goal for 1972," he wrote, "seemed to be a standstill cease-fire as soon as it had achieved some territorial gains. From this improved bargaining position, our enemy felt certain he could force negotiators into accepting a solution to the war that would be most favorable for his side."[39]

Documents captured at the time of the offensive reinforce the political impetus behind the invasion. COSVN Directive 43, for instance, indicated that the communists would demand an in-place ceasefire once they had seized significant land areas. The purpose, the directive claimed, would be to negotiate a settlement on terms favorable to the revolution.[40] COSVN Directive 51 showed that recent communist attacks were intended to disperse the RVNAF, use up ARVN manpower, and roll back pacification and government control in rural areas. "Once these attacks had widely spread the ARVN forces and reduced their reserves," Abrams declared in an assessment of the resolution, "the enemy would then demand a ceasefire in place and attempt to install a coalition government."[41] Saigon and its military agreed with Abrams's analysis. In April, Thieu informed the press that the purpose of the offensive was to capture the northern provinces of Quang Tri and Thua Thien to establish a capital for the PRG in South Vietnam. Should the communists succeed in this task, the South Vietnamese president exclaimed, Hanoi and the NLF would demand a political solution that would include either a coalition government or territorial concession—demands, he warned, that were designed to ultimately take over the country.[42] In July, an ARVN officer in Hau Nghia Province posited that the communists had planned to "obtain a ceasefire and then negotiate for a place in [the] government. In order to strengthen their position for negotiations, they must show that they control part of the population." It was for this reason, he argued, that Saigon could expect to see sustained military pressure throughout the countryside.[43]

Historical scholarship supports wartime arguments that the purpose of the 1972 offensive was to expand the insurgent state to gain political and diplomatic leverage. According to a number of specialists, the overall objective of the NVA invasion was to strengthen the communists' territorial position and enable them to negotiate from a position of greater strength. "A resounding defeat of Saigon's army would demonstrate the futility of Vietnamization and weaken the American position at the Paris peace talks," Dale Andrade asserts. "Before a settlement was reached, Hanoi wanted to be in possession of as much South Vietnamese territory as possible. . . . As the communists were always quick to point out, military success dictates diplomatic success."[44] Ang Cheng Guan maintains the offensive was intended to "complement and support the foreign policy offensive."[45] Lien-Hang Nguyen has made similar claims. Unlike *Tet Mau Than* and other previous offensives, she contends, the 1972 invasion was planned with the recognition that military force alone could not achieve Hanoi's goals, and that the success of the general offensive and uprising was dependent on the success of the political and diplomatic struggles that were currently under way on the ground and at the negotiating table. "In order to carry out a lengthy military campaign and to translate battlefield success into total victory," Nguyen writes, "the Party needed to coordinate the military offensive with the political movement in the cities and the countryside as well as the diplomatic struggle in Paris."[46]

The Nguyen Hue Offensive was largely a conventional fight between communist and South Vietnamese main-force units. Despite the VCI's resilience and a revival of guerrilla activities in many areas, particularly the Mekong Delta, the overwhelming majority of attacks were conducted by the NVA. As a number of observers indicated at the time, the 1972 offensive was qualitatively different than *Tet Mau Than*. The objectives of these two attacks may have been similar—a precipitous American withdrawal followed by a favorable settlement that would benefit Hanoi and the NLF—but the composition of forces was not. "In 1968, the Viet Cong in South Vietnam were the ones who were called upon for total effort, with the North Vietnamese army units in a supporting role," veteran correspondent Joseph Alsop stated in April 1972, "whereas this time, the North Vietnamese army itself is being called upon for total effort, without any very striking assistance from the indigenous VC."[47] Allied leaders agreed with this assessment. "This is a great difference between 1968 and this year," Ambassador Bunker declared. "It was a largely VC operation in 1968."[48] Other high-ranking officials underscored the conspicuous lack of participation by NLF guerrillas and local forces. In April, a summary of Abrams's assessment of the offensive stated, "One of the most significant features of the current general situation is the absence of any widespread uprising throughout the Republic by local force guerrillas."[49] Kissinger echoed this finding the following month. "The enemy has had only minimal assistance from local force units and guerrillas," he informed Nixon on 19 May, "evidenced by the fact that since the beginning of the offensive there has been significant activity in only 16 of the 44 provinces."[50] Province reports and pacification overviews from all four military regions confirm, largely through omission, the supremacy of NVA participation in the offensive. With the exception of sporadic sapper activities, terrorism, harassment, and small-unit attacks against government outposts and isolated villages and hamlets by local NLF forces, the preponderance of reported enemy activity involved engagements between RVNAF and NVA units. Intelligence reports also revealed that most of the gains made in population and territorial control were made by NVA main-forces rather than NLF guerrillas. "Invading North Vietnamese regulars have replaced local Viet Cong forces as the enemy's main strike force," one CIA estimate declared. "In four major areas . . . the communists have caused significant damage to pacification in the last three weeks: in Quang Tri Province, along the coast from Quang Nam to Binh Dinh Provinces, in Binh Long Province north of Saigon, and in Chuong Thien Province in the Mekong Delta. In three of these areas . . . the damage appears to have been the work of main-force units, largely made up of North Vietnamese troops."[51]

The communists' 1972 spring offensive had a significant detrimental impact on the control war and the rural pacification program. "While South Vietnam's regular army has been all too busy, half a million people and hundreds of tiny hamlets have slipped back under communist control," declared an American newspaper on 31 May. "The Viet Cong flag now flies over the capitals in 14 of South Vietnam's [236] districts. Nine weeks after Hanoi launched its current offensive the damage to South Vietnam's pacification program is evident."[52] Reports generated by MACV in the fall of 1972 reveal that the damage caused to pacification by the Nguyen Hue Offensive was widespread but uneven. Some areas had suffered a deep decline in security ratings and levels of disruption, while others fared better. The impact of the offensive was felt most acutely in MR 1, where communist forces had completely overrun Quang Tri and inflicted severe losses in population and territory control in the remaining provinces: Quang Ngai (–56%), Thua Thien (–39%), Quang Nam (–38%), and Quang Tin (–35%). The damage done to pacification in MR 2 was more ambiguous. Here there were moderate to serious regressions in four of the northernmost provinces and only minor regression, even some gains, in the southern provinces. Of those seriously hit, the worst losses in control were in Kontum (–96%), Binh Dinh (–35%), and Phu Yen (–36%). In MR 3 four provinces sustained moderate to serious regression, while six either made gains or sustained a regression of less than 7 percent. Here the worst control losses were in Binh Long (–100%), Phuoc Long (–38%), Long Khanh (–24%), and Hau Nghia (–18%). Down in the Mekong Delta, nearly every province in MR 4 suffered regression in population and territory control with the exception of three that maintained security ratings of 99 percent or better. Some of the more severe losses were found in Chuong Thien (–45%) and Vinh Binh (–32%). Regression in the remaining Delta provinces varied from 5 percent to 19 percent.[53]

The Nguyen Hue Offensive increased and solidified the revolution's presence inside South Vietnam. Four months before the offensive, the communists exercised predominant or total control over 7.3 percent of all hamlets nationwide. This translated into approximately 880 out of 12,032 total hamlets. By December 1972, the communists' share had climbed to 21 percent, or 2,554 hamlets out of 12,207. The actual number of hamlets that communist forces fully controlled and administered rose from a mere 34 in November 1971 to 885 twelve months later. This, however, was a high point. By January 1973, the communists exercised predominant or complete control over just 18 percent of hamlets (2,176) as a result of the RVNAF counteroffensive. Nonetheless, communist military forces and political cadres retained significant influence and/or control over approximately 2,250 hamlets throughout South Vietnam at the time the Paris Peace Accords went into effect.[54]

MACV and CORDS recognized that the Nguyen Hue Offensive had caused substantial injury to the rural pacification program in select areas. However, many believed the damage was reversible. "The impact of the 1972 enemy offensive on pacification has not been insignificant," a 16 September report declared. "Both security and development efforts have suffered, although the amount of damage varies widely from area to area, and the enemy has not yet exhausted his offensive potential. In some areas recovery has already begun, while in other areas the most which can be said at present is that the worst may be over."[55] The report concluded optimistically. Despite pacification setbacks, it argued, the overall security situation in South Vietnam remained "far more favorable" than after *Tet Mau Than*. Officials at CORDS were equally sanguine. According to a briefing on the effects of the NVA offensive, pacification had only been "temporarily set back" in several areas; overall, the program remained strong. "In the large bulk of the country," it stated, "the pacification programs go on as before."[56] This briefing also ended on a high note. "With the exception of one or two hard-hit provinces in each of the four military regions," it concluded, "the government has been able to maintain some degree of pacification momentum."[57]

Given the politico-military reality in South Vietnam and the disposition of opposing forces on the ground, the optimism expressed by MACV and CORDS in the wake of the Nguyen Hue Offensive is perplexing. Although the RVNAF counteroffensive successfully beat back the communist invasion and recovered most of the territory lost in April and May, it was unable to dislodge NVA forces from southern soil. Moreover, as before, communist base areas and liberated zones continued to dot the landscape. In October, the CIA estimated that there were close to 200,000 NVA troops inside South Vietnam.[58] That same month, the agency reported "heavy concentrations" of communist units in all five provinces of MR 1 and in the central highlands, including northern Binh Dinh, Kontum, and Pleiku Provinces. Multiple NLF/NVA units were also reported north and northwest of Saigon in MR 3 and throughout the Mekong Delta.[59] According to historian William Le Gro, there were sixteen NVA divisions inside South Vietnam at the time of the ceasefire in January 1973, including eight in MR 1, three each in MR 2 and MR 4, and two in MR 3. Overall, Le Gro's figures show there were close to 220,000 communist forces operating to varying degrees throughout South Vietnam, including 123,000 NVA combat troops, 25,000 NLF guerrillas, and more than 71,000 administrative and logistical troops.[60] These figures were only slightly less than the number of communist troops reported by MACV to be operating in South Vietnam in December 1965 (230,000) and December 1968 (259,000).[61] Former ARVN officer Nguyen Duy Hinh put the number of communist forces in South Vietnam at the time of the ceasefire closer to 300,000 (293,000), including 167,000 NVA, 41,000 guerrillas/VCI, and 85,000 administration and rear service troops.[62] The RVN, meanwhile, possessed more than 1.1 million men under arms at the time of the peace agreement, although the actual strength of South Vietnam's armed forces was probably less than a million. Total numbers for the RVNAF in January 1973 included 450,000 ARVN personnel, 525,000 in the RF/PF, and 96,000 in the Vietnamese naval and air forces.[63]

The communists also had a substantial amount of real estate under their control. Because of the Nguyen Hue Offensive, North Vietnamese troops had made significant territorial gains and were occupying, in addition to an increased number of villages and hamlets, significant swaths of unpopulated or sparsely populated territory, predominantly in the northern and western portions of South Vietnam. In the five northernmost provinces of South Vietnam, the NVA controlled nearly all of the rugged terrain from the Laotian border to approximately twenty miles west of Route 1, the main coastal road running from the DMZ to Saigon. In addition, they occupied a small sliver of area on the coast southeast of the Demilitarized Zone and just north of the provincial capital of Quang Tri. In the central highlands, communist forces were in control of huge areas of Kontum and Pleiku Provinces along the Laotian and Cambodian borders. NVA units also retained significant portions of land in northern Binh Dinh Province along with territory northwest of the capital and in the Mekong Delta.[64] "Below the Demilitarized Zone and down 500 miles of South Vietnam's western edge, nearly all the outlying bases in the thinly populated areas had fallen, leaving the communist army emplaced closer than before to the populous, prosperous coastal lowlands," journalist Arnold Isaacs noted.[65] Although most of these areas were largely unpopulated and the overwhelming majority of the population living in the narrow coastal plain remained under government control, there was still cause for alarm. "Though most of the population remains under government control with enemy units outside the heavily populated areas, the North Vietnamese and Viet Cong hold a substantial part of the countryside," read a 27 October newspaper article.[66] "The other side is building a nation in the hills," an American officer noted bitterly. "The way they're going, they'll soon have two-thirds if not three-quarters of the physical geography of the region."[67] Another high-ranking American officer argued that it would be impossible to say how much territory the communists controlled. "It's just a guess," he stated. "If you say 25 [percent], nobody in the world can show that you are wrong."[68] The CIA was equally concerned with the communists' territorial disposition. "Present enemy force deployments would support a communist claim to de facto control over considerable unpopulated territory in western MRs 1 and 2, plus a more arguable claim to enclaves in the populated lowlands of southern MR 1 and MR 2," one analyst informed Kissinger. "In MRs 3 and 4 . . . the communists have moved significant forces out of relatively isolated border sanctuaries and/or base areas and committed them to action in relatively close proximity to urban population concentrations. These most recent moves, coupled with previous enemy deployments, could easily be presented as a loose encirclement of Saigon."[69]

The 1972 Nguyen Hue Offensive can in many ways be characterized as a tactical failure. The invasion, after all, was defeated and its larger objective of crippling the

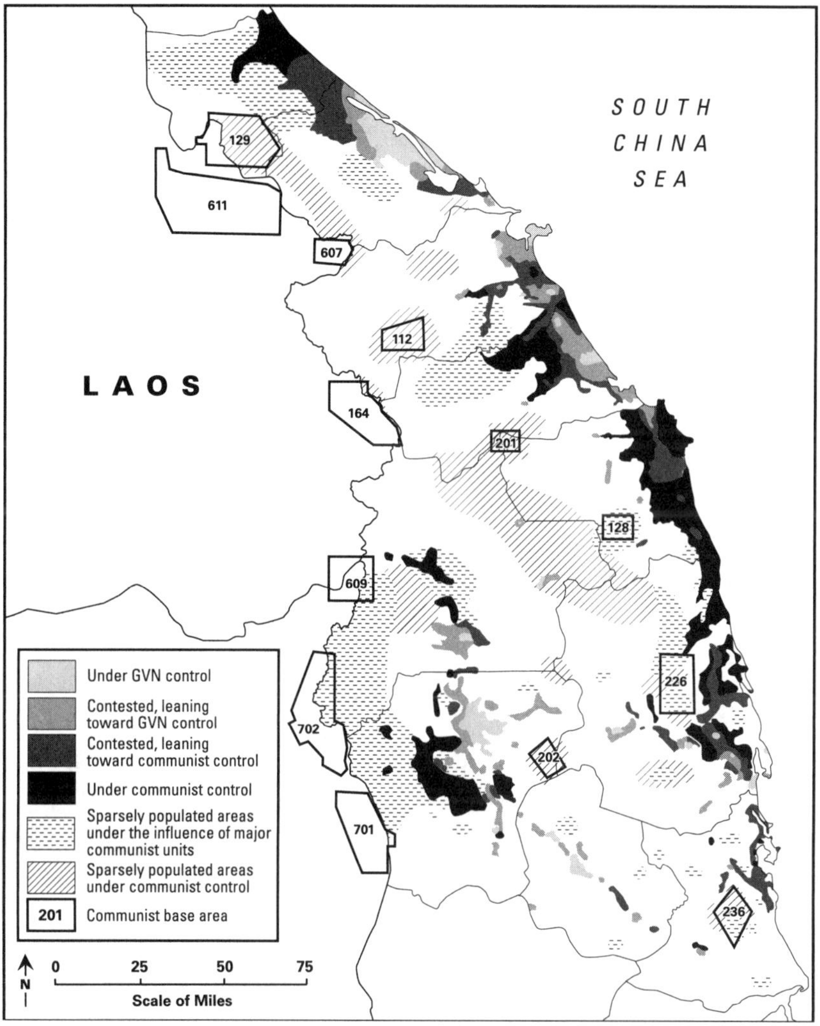

Military Region 1, showing control by differing HES ratings.
Map by Erin Greb Cartography. Copyright © 2018 by the University of Oklahoma Press.

RVNAF went largely unfulfilled. Moreover, the expected general uprising by the civilian population once again failed to materialize. As in 1968, the population did not greet communist units as liberators or rise up against the Saigon regime as Le Duan and others in the Politburo had hoped. Nonetheless, the attacks—much like *Tet Mau Than*—turned out to be a long-term strategic success. The spring offensive not only seized and held additional territory inside South Vietnam, it established a permanent

NVA presence within the national boundaries of the RVN. Although beaten back almost to the point of departure, the NVA offensive created a permanent toehold and expanded the size and scope of the insurgent state. "It can be argued with considerable logic that the North Vietnamese position at present is considerably improved over that of 1970–71," a September CIA intelligence memorandum stated. "Base areas and logistical routes through southern Laos and eastern Cambodia, for example, are more secure than ever before. Serious ARVN incursions into these areas are increasingly unlikely. . . . [I]ts operations now constitute little more than an occasional nuisance. NVA troops are present in South Vietnam in greater strength than a year ago and to date have successfully resisted ARVN efforts to push them back over the borders."[70]

Contemporary maps confirm Hanoi's success in seizing large swaths of territory and vastly expanding the insurgent antistate within South Vietnam. Moreover, they illustrate the unique spatiality of the Second Indochina War. Rather than produce clearly demarcated and well-defined spatial divisions like those found in conventional wars, the complex politico-military struggle to control South Vietnam's operational environment generated a series of messy patchworks and nebulous territorial boundaries that defied easy definition. The interlaced deployment of friendly and enemy forces fragmented the battle space and produced irregular and interspersed patterns of control. This made it nearly impossible to determine what area was under whose control. According to Isaacs, any attempt to identify combatant zones of control was like asking someone to distinguish "the whites and yokes in a dish of scrambled eggs."[71] Another American journalist observed that the landscape of South Vietnam did not look "so much like Viet Cong islands in a Saigon lake, nor like Saigon islands in a communist lake but rather like marshland—with land and water pretty much indistinguishable and fluctuating with the tides and storms."[72]

Allied policymakers understood that the fragmentation of space, state of mixed control, and presence of communist "leopard spots" meant a continuation rather than an end to hostilities. Kissinger declared in 1969,

> If there existed a front line with unchallenged control behind it, as in Korea the solution would be traditional and relatively simple: the two sides could stop shooting at each other and the ceasefire line could follow the front line. But there are no front lines in Vietnam; control . . . depends on who has forces in a given area and on the time of day. If a ceasefire permits the government to move without challenge, day or night, it will amount to a Saigon victory. If Saigon is prevented from entering certain areas, it means in effect partition which . . . tends

> towards permanency. . . . [T]he pattern would be a crazy quilt, with enclaves of conflicting loyalties all over the country.[73]

The policymakers also knew such a situation was politically unacceptable. Thieu feared mixed control would undermine national sovereignty and Saigon's political legitimacy by establishing a de facto communist state within the geographical borders of South Vietnam. "A leopard-skin solution," he lamented, "would mean the communists control certain areas by themselves, including political organization, with their own flags, their own military organization, their own government—tiny communist nations situated in [South] Vietnam."[74] Although American officials (including Kissinger, who later reversed his reservations concerning the permanent "partition" of the RVN into opposing zones of control) had repeatedly tried to assuage such fears by assuring Saigon that areas under communist domination were not only relatively small but were located predominantly in remote and sparsely populated regions, Thieu and other officials continued to fear communist "leopard spots" as a "cancer" that would grow and metastasize through sustained guerrilla activity, political struggle, and accommodation between the population and insurgents at the local level.[75] Many feared the scenario posited by Thieu would amount to a "third Vietnam"—that is, a situation where zones of communist control were not only delineated but legitimated through the establishment of a communist capital and formal recognition by the international community that the Provisional Revolutionary Government was the proper authority in areas of South Vietnam under communist control. Such a condition, needless to say, was fiercely rejected by the RVN. "A third Vietnam . . . is more than we can afford to give," roared one Saigon official. "We dread it."[76]

On the eve of the Paris Peace Agreement and full American withdrawal from Southeast Asia, communist forces held a considerable percentage of South Vietnamese territory. Although most of the areas under NLF/NVA control were largely unpopulated (more than 90 percent of the population continued to live in RVN-controlled areas), the communists had expanded their political and military footprint inside of South Vietnam and solidified communist control over the operational environment. This not only gave Hanoi and the NLF tremendous leverage in the ongoing peace talks in Paris, which were quickly hurtling toward conclusion in the fall of 1972; it provided significant advantage in the control war, an advantage Hanoi later capitalized on. As will be seen next, the failure of the RVN and the United States to negotiate NVA units out of South Vietnam or fully eliminate territory under communist control by January 1973 proved fatal to the Saigon regime in the years that followed.

8 TRIUMPH OF THE INSURGENT STATE, 1972–1975

As the Saigon regime feared, a "third Vietnam" had been created and maintained after the 1972 Nguyen Hue Offensive had secured an expansive communist base of operations within the geographical confines of South Vietnam. The Paris Peace Accords sanctified this situation; and in doing so, it perpetuated rather than ended the war in Southeast Asia. The 1973 Agreement was highly problematic for the Saigon regime (much more so than the 1954 Geneva Accords) because it sanctioned a complex and dangerous "standstill modus vivendi" where enemy and friendly zones were interspersed. Unlike the latter agreement, which authorized two well-defined regroupment areas and mandated troop withdrawals, the Paris Accords locked opposing forces in place within the territory both were fighting to control. This produced a highly volatile situation that enhanced rather than suppressed the state of armed conflict in South Vietnam. According to South Vietnamese general Nguyen Duy Hinh, the territorial control situation agreed on in Paris was a confusing, war-perpetuating mess. "How could the contending adversaries arrive at a satisfactory agreement as to where one's control was supposed to extend if they both claimed the same areas?" he asked. "The standstill idea was the basis for the land grab that occurred before, during, and after the ceasefire, with both sides resorting to mutual denunciation. The territorial problem proved to be the thorniest of the problems and probably could never be settled." Hinh was correct; the standstill ceasefire imposed by the Paris Peace Accords simply did not work in a war of innumerable fronts. "Troops of both sides could be everywhere and nowhere at any given time and where the adversaries had built up so great a hatred for each other," he lamented. "It was understandable that either side would want to settle for no less than a bigger piece of the pie than the other."[1] The final phase of the Second Indochina War was in fact a contest to gain control over an expanded slice of the operational environment.

TO THE PARIS PEACE AGREEMENT AND BEYOND

The short-term failures of the Nguyen Hue Offensive convinced Le Duan and other high-ranking members of the Politburo that the time had come to consider a negotiated settlement. The pacification-Vietnamization strategy, they reluctantly admitted, had established conditions on the ground that made it impossible to overthrow the RVN by military means alone. There was simply no choice, then, but to accept a brokered peace and wait until the Americans had left and a more favorable balance of forces could be developed without the risk of outside interference. Although the U.S. withdrawal was already well under way and the Nixon administration had made the vital concession in Paris that would allow North Vietnamese units to remain inside South Vietnam as part of a settlement, the current and short-term politico-military situation still favored Saigon and its armed forces.[2] Hanoi's long-term prospects, however, looked promising. To begin with, its "great rear area" north of the seventeenth parallel, though severely battered by MACV's intensive bombing campaign, was still inviolate and functional. Moreover, the DRV continued to receive robust military and economic aid from China and, increasingly, the Soviet Union. Nonetheless, the time had come to postpone the pursuit of total victory via military means until a more opportune moment arrived. "The North Vietnamese leadership decided to switch from a strategy of war to a strategy of peace," writes Ang Cheng Guan.[3] According to Lien-Hang Nguyen, Le Duan no longer thought it prudent to attempt a violent overthrow of the Saigon regime as a means to force Washington's hand in the negotiations after the invasion. Rather, the general secretary "strove for more minimal terms in Paris and shelved his military plans until the Americans vacated Vietnam. For the remainder of the war against the

United States, then, Le Duan and the Politburo relied primarily on diplomatic and political struggles and downgraded the military offensive."[4] Before brokering a deal and agreeing to a negotiated settlement and ceasefire, however, the communists were determined to expand their grip on rural areas in preparation for the postceasefire period. Hanoi's decision to downgrade its military offensive did not mean that it had abandoned armed *dau tranh* to expand the insurgent state and gain advantage in the control war. The Politburo and the NLF, now that they had accepted the modality of a negotiated settlement, looked forward to the period between the signing of a peace agreement and its implementation. This, they believed, would be "the decisive phase and the high point" of future communist efforts. Often described as the "twilight period" by communist leaders, the preimplementation phase was looked on as a golden opportunity to rapidly expand control over the operational environment in preparation for a future political contest at the local level. "The aim was to take control of the bulk of the countryside," Guan argues. "After the peace agreement became effective, they were to use political struggle to consolidate all the areas under the PRG and to work towards weakening the Thieu administration."[5]

For the remainder of 1972 and into 1973, communist guerrillas and political cadres made a violent attempt to expand their influence and/or control over territory before an agreement was reached and an in-place ceasefire went into effect. In June, the deputy assistant to the president for national security affairs, Major General Alexander Haig, informed Kissinger that COSVN had recently issued instructions to its cadres to seize as much land as they could to expand their base and prepare for the grassroots political struggle that would take place once an armistice had been signed. According to Haig, the cadres were instructed to "lay low until just before the ceasefire is to be accepted, at which time they will attempt to overrun as much territory as possible."[6] In October, COSVN released a number of directives that ordered units to rapidly expand their hold over territory and population in anticipation of an imminent ceasefire. As one directive read:

> This is the period of transition between two struggle phases of the revolution in South Vietnam. It is a decisive race between us and the enemy. We must work against time to seize new victories each day and each hour to hit the enemy so hard that he cannot react or recover. . . . Before the ceasefire agreement goes into effect, our armed forces are to work against time to strive hard to intensify military attacks in coordination with political and military proselyting attacks so as to

> win great victories on the battlefield, chiefly in the counter-pacification field. At the same time, they are to resolutely defeat all enemy attempts to reoccupy [lost territory].[7]

According to another directive, the "central task" of guerrilla units and political cadres was to increase military attacks and promote uprisings to expand and protect communist-controlled villages and hamlets and other liberated areas. "The task of winning and keeping the people, seizing and holding land, [and] disputing and occupying the important territories is a strategic matter, a pressing requirement for the present time and for the long future, and also a matter of survival for the revolution containing a historical meaning of utmost importance. For these reasons, we have to dispute decisively with the enemy for every citizen, every inch of land, to plant our flags, encroach deeply on enemy territory and enlarge our area."[8]

In response to these orders, NLF/NVA units launched widespread low-intensity attacks throughout South Vietnam's rural areas. Such efforts aimed to spread revolutionary forces over a wide geographical area and maximize communist control over territory and population. "It appears that the enemy has recently broken down into small groups," declared a senior American adviser in the Mekong Delta. "The small size unit may be part of his plan to position himself in as many places as possible in preparation for a possible ceasefire."[9] In Binh Duong Province, just northwest of Saigon, reports indicated that communist units had occupied a number of hamlets and were preparing for additional actions in anticipation of a truce. "At this time, it appears that the VC/NVA forces are awaiting new instructions, possibly those based on the occurrence of a ceasefire," read a monthly pacification report. "Should a ceasefire materialize, greatly increased activity on the part of the VC/NVA can be expected if they attempt a last minute expansion of their control of the population and land."[10] This estimate turned out to be accurate. The following month, CORDS advisers reported that NVA main-force units, guided by local guerrillas, had seized forty hamlets in Binh Duong. The general enemy tactic, it was reported, was to infiltrate hamlets and then "bunker in heavily" to make it difficult for ARVN forces to retake these areas.[11] According to U.S. officials in Military Region 2, statements from POWs, and captured documents, communist units were "determined to occupy currently contested territory and control the population in these contested areas after the declaration of a ceasefire."[12] Meanwhile, in MR 1, it was reported that a "battle of the flags" was occurring, as communist units attempted to take down, tear up, or scrape off RVN flags in villages and hamlets and replace them with their own. A sharp increase in insurgent political activity also occurred. "These incidents," one account concluded, "are expected to continue as the ceasefire talk increases in activity."[13]

The RVN and the Americans were no less determined to rapidly expand their hold over the operational environment in the final months of the war. In July, Bunker informed Kissinger that there was a general feeling that "peace was in the air." Many people in South Vietnam, the ambassador argued, believed a ceasefire would go into effect and the conflict would end sometime soon. The sense that the war was ending or entering a new phase, he continued, raised hopes and prompted government efforts to maximize control over territory and population before a truce ended the fighting and froze political and military forces in place. "This gut feeling regarding an imminent ceasefire," Bunker wrote, "is spurring the GVN to recapture lost ground. It is clear that the GVN does not intend to be caught napping in what may be a fast moving situation and will be resourceful in utilizing all its assets to prevent a last-minute [land] grab by the enemy if a ceasefire is declared."[14] Anticipating a ceasefire proposal by the communists during the upcoming October talks, Haig urged Kissinger to deliver a back-channel message instructing Bunker to pressure Thieu to move quickly and grab as much territory as it could. "You should tell Thieu as soon as possible that at today's meeting the other side pressed very softly on political issues and major concentration was on military and security arrangements. This means that the other side may surface a ceasefire proposal during [the] meetings. While we certainly will not agree without further consultations, it is essential that Thieu instruct his commanders to move promptly and seize the maximum amount of critical territory." The following day, Bunker responded that Thieu had in fact already issued such an order to his military commanders.[15] On 12 October, Kissinger reiterated to Bunker the need to implore the Saigon regime to seize as much land as it could in light of a potential break in the negotiations and Hanoi's willingness to agree to a standstill ceasefire. "My judgment at this juncture would be that they appear ready to accept a ceasefire in place in the near future," Kissinger wrote. "This, of course is corroborated by field intelligence and it is for this reason that you cannot overemphasize upon Thieu the need to regain as much territory as possible."[16]

As it turned out, there was no settlement that October. Shortly after a possible break in the negotiations appeared, the peace talks broke down and the frustrated American delegation returned to Washington. This collapse essentially scuttled any chance for an armistice for the remainder of the year. There was, however, one major diplomatic development in the autumn of 1972. Hanoi and the PRG had, after years of fierce opposition, finally agreed to an in-place ceasefire. The concession was made for two reasons, both of which were rooted in the Nguyen Hue Offensive. First, as previously discussed, the tactical failures of the offensive revealed that it was impossible under current circumstances to achieve a military victory and to overthrow the RVN with force. Given the unexpected resiliency of the RVNAF, the severe beating suffered by communist forces in the Nguyen Hue Offensive, and Washington's continued use of airpower to defend the RVN, it was obvious that a period of respite to rebuild and reconstitute battered NVA units was necessary. Second, Hanoi believed that the disposition of ground forces made possible by the offensive was a long-term, strategic victory for the revolution. The net gain in real estate, particularly in the northern and western areas of South Vietnam, along with the presence of nearly 200,000 NVA soldiers on southern soil, convinced communist leaders that the revolution's strategic position had been strengthened to the point where they could temporize and agree to a negotiated settlement and in-place ceasefire. "A political settlement and ceasefire represent a basic victory for us and a fundamental defeat for the enemy," the People's Revolutionary Party trumpeted in October 1972. "Today, in South Vietnam, we have large liberated areas, a people's administration, and strong people's liberation armed forces, especially the main-forces; we have a political force, a complete system of leadership from high to low levels and a time-tested infrastructure; we have the National Front for the Liberation of South Vietnam and the Provisional Revolutionary Government of the Republic of South Vietnam which enjoy great prestige on the international scene; and we will occupy a position of equality in the administration of national accord."[17] Although Hanoi had retained its demand that all American and allied troops be withdrawn and the bombing and mining of the DRV terminated as part of a final settlement, it dropped the longstanding requirement that an armistice be reached before a standstill ceasefire went into effect. Le Duan and others had agreed to the condition that all Vietnamese armed forces, including their own, would be "frozen in the positions they currently held"—there would be no withdrawal of North Vietnamese units or a regroupment of forces as there had been in 1954. Hanoi recognized that, in addition to its military benefits, such a development was politically advantageous because it forced Washington and Saigon to acknowledge, even if only tacitly, that there were in fact "two governments, two armies, and two zones of control" in South Vietnam. Such recognition, they reasoned, was extremely favorable because it allowed the revolution to continue the struggle and work to overthrow the RVN in the months and years ahead.[18]

In contrast to Le Duan and the Politburo, Thieu and other Saigon officials remained adamantly opposed to a ceasefire that would allow NVA units to remain in South Vietnam. Such an arrangement and the "leopard spot" territorial configuration it produced, they argued, not only would threaten the future security of the nation, but would grant Hanoi and the NLF sovereignty over the territory they held at the time of a ceasefire. In order to allay Saigon's fears and garner its support for a settlement, U.S. policymakers pointed out that the areas under NVA control were mostly the remote, sparsely populated hinterlands of South Vietnam. They also gave assurances that communist enclaves would likely shrink over time. Nonetheless, Thieu continued to warn of the dangers territorial accommodation posed to national security. "Saigon," one newspaper declared, "tends to see any leopard spot as a cancer that will gradually expand through guerrilla activity, political struggle and local accommodation."[19] Thieu also opposed from a legal standpoint any territorial accommodation or agreement that would leave North Vietnamese troops in the South. In October, the South Vietnamese president informed Bunker that the DMZ was not merely a military boundary meant to protect the RVN against armed aggression from the DRV; it was, he argued, a legally constituted political boundary that divided two independent and autonomous nations and protected the sovereignty of the Republic of Vietnam. Thus, Saigon could not and would not accept the presence of North Vietnamese soldiers on its soil or accept the current ceasefire proposal.[20] "Thieu makes clear that he cannot accept the Draft Peace Agreement unless it includes a parallel withdrawal of North Vietnamese troops from the south under effective supervision and removes any references to the Provisional Revolutionary Government," Kissinger informed Nixon on 20 December. "Although his letter is couched in terms of making further concessions by accepting the political provisions of the Agreement," Kissinger concluded, "Thieu's demand for North Vietnamese withdrawal, of course, makes a comprehensive settlement with Hanoi impossible, as he well knows."[21]

Thieu's intransigence was eventually overcome and the South Vietnamese president buckled under American pressure. Despite months of bitter debate between Saigon and Washington and accusations that the Americans had sold out the Saigon regime,[22] Thieu finally conceded to the settlement proposed by Nixon and Kissinger and agreed to sign the Paris agreement in January 1973. Although the South Vietnamese leader still held serious reservations about the treaty, he assented after Nixon presented him with a written ultimatum that included a mixture of threats and assurances. In a personal letter sent to the Presidential Palace on 16 January, Nixon informed Thieu that should Saigon refuse to sign the treaty or agree to the settlement proposed by the White House, the United States would have no choice but to sign a bilateral agreement on its own, without the participation of the RVN, and to terminate all U.S. economic and military assistance. Should Thieu agree to the plan, however, Saigon could count on the U.S. government to sustain economic and military aid and, if necessary, to reintervene in the conflict should Hanoi violate the ceasefire and launch another major attack against the RVN. Nixon wrote,

> I hope . . . that after all our two countries have shared and suffered together in conflict, we will stay together to preserve peace and reap its benefits. To this end I want to repeat to you the assurances that I have already conveyed. At the time of signing the agreement I will make emphatically clear that the United States recognizes your government as the only legal government of South Vietnam; that we do not recognize the right of any foreign troops to be present on South Vietnamese territory; and that we will react strongly in the event the agreement is violated. Finally, I want to emphasize my continued commitment to the freedom and progress of the Republic of Vietnam. It is my firm intention to continue full economic and military aid.[23]

When combined with the RVN's absolute dependence on American assistance, the existential nature of these conditions convinced Thieu he had no choice but to let the chips fall where they may and accept the proposal put forth by Kissinger and Nixon.

In January, when the possibility of a negotiated settlement once again became imminent, COSVN replicated orders to seize and occupy as much territory as possible before a ceasefire went into effect. Guerrillas and political cadres were also instructed to capitalize on the initial period of confusion after the cessation of hostilities by hiding their motives and falsely attributing their actions to communication failures and misunderstandings on the exact time and circumstances of the armistice. Intelligence analysts predicted that these last-minute efforts to expand control would be more intense than those of the previous October, and that while such efforts would take place throughout South Vietnam, the main thrust would likely occur in the more densely populated areas around Saigon and in the Mekong Delta.[24] A week before the Paris Peace Accords took effect, COSVN issued a final directive to its cadres to prepare for the upcoming political struggle. In particular, they were told to develop political and military forces, sustain the violence program against the "puppet" regime, disintegrate and collapse South Vietnamese military and pacification personnel, and consolidate and expand liberated areas. "Our immediate objective," the directive read, "is to seize control at the base level and to disrupt the enemy."[25]

The Agreement on Ending the War and Restoring Peace in Vietnam, also known as the Paris Peace Accords, went into effect on 27 January 1973. The treaty called for an in-place ceasefire and stipulated that the use of military force and other "hostile acts," including terrorism and reprisals, would be prohibited. In addition, it required the Vietnamese parties to "respect the ceasefire and maintain peace in South Vietnam, settle all matters of contention through negotiations, and avoid all armed conflict." The agreement also banned the introduction of combat troops, military personnel, arms, and war matériel into the RVN, although it did afford for the replacement of these items after the ceasefire should they be destroyed, damaged, worn out, or depleted. In the political realm, the treaty called on the two Vietnamese parties to establish a National Council of National Reconciliation and Concord, an organization that would be responsible for implementing the agreement, ensuring the democratic liberties of all people within their respective zones of control, and organizing elections to reconcile political divisions and mend the social fabric of the nation. "The South Vietnamese people shall decide themselves the political future of South Vietnam through genuinely free and democratic general elections under international supervision," declared Article 9. According to the treaty, the reunification of Vietnam was to be determined at a later date through "peaceful means" and based on "discussion and agreement" between the DRV and the RVN "without coercion or annexation by either party, and without foreign interference." Finally, the Paris Accords called for the dismantlement of all foreign military bases and the complete removal of all American and Free World military forces, including advisers and support personnel, within sixty days.[26]

In an attempt to manage the territorial disposition of opposing armies after what had essentially been a war without front lines, the Paris agreement called for a Four-Party Joint Military Commission (JMC) to implement the modalities of the in-place ceasefire for sixty days. The Four-Party JMC included members from the United States, DRV, RVN, and PRG. Once all foreign troops had been removed and their military installations deactivated, the commission would be replaced by a Two-Party JMC with just the RVN and PRG. Both the Four-Party and the Two-Party JMCs were responsible for enforcing the ceasefire and determining the areas controlled by each party.[27] Scattered across twenty-six locations throughout South Vietnam, JMC teams were responsible for drawing the boundary lines that separated the territory controlled by the RVN from the "leopard spots" and sanctuaries controlled by the PRG and DRV.[28] To oversee the process and to report RVN and PRG activities in implementing the modalities of the ceasefire, the Paris agreement established the International Commission of Control and Supervision (ICCS), a multinational body that would be activated once the ceasefire was effective and disbanded once the new government was formed by the general elections called for in Article 9. It was the job of the ICCS to supervise the ceasefire by investigating and reporting violations to the Four-Party and, later, the Two-Party JMC. The ICCS was also responsible for helping control the movement of manpower and military supplies through the respective zones of control and helping supervise national elections.[29] ICSS teams comprised members from four nations: two communist and two noncommunist. The original membership included Poland, Hungary, Canada, and Indonesia, but after becoming frustrated by operational gridlock, Canada decided to withdraw and was replaced by Iran.[30]

Within weeks, it became apparent that the primary instruments for implementing the treaty were ineffectual. While on paper the JMC and the ICCS appeared sound, in practice they proved unworkable. Both institutions were crippled by their own operating principles. To begin with, they lacked teeth and contained no protocols that would put into force the provisions of the agreement. The opposing camps, therefore, were reliant on a process that lacked an enforcement mechanism and had no tangible way to punish violators. As such, each side had no other recourse than to verbally denounce the ceasefire violations of the other. Over time, the competition over territory and population control devolved into what one former South Vietnamese officer characterized as "a propaganda battle in which each side strived to demonstrate the ill will and belligerency of the other."[31] Another fundamental flaw of the ICSS was that it operated on the principle of unanimity—no investigations of ceasefire violations could be conducted unless all four members agreed. Predictably, there was never any agreement between member states and a persistent impasse quickly paralyzed the organization. According to Stuart Herrington, a former CORDS adviser who returned to the RVN in 1973 to work for the American Defense Attaché Office (DAO), the ICCS was essentially useless. "When the communists were the apparent violators, the Poles or the Hungarians vetoed the investigation. If the Saigon government's forces appeared to be guilty, then the Iranians or the Indonesians would nix the probe." Such a situation, Herrington stated, led some observers to declare that ICCS actually stood for "I can't control shit."[32] In the end, the dysfunction of the JMC and the ICCS, particularly the lack of enforcement capability and the unworkability of the unanimity principle, doomed these institutions—and by extension the ceasefire and entire peace treaty itself—to failure. Ironically, the mechanisms put in place to regulate the Paris Peace Accords and adjudicate zones of control had rendered the treaty inoperable. Within a year, both institutions had collapsed and the facade of compromise had fallen away, revealing a state of open conflict in South Vietnam that had been neither interrupted nor dampened by the Paris agreement.

While it is true that the Paris Accords had officially brokered a de jure settlement between the RVN, DRV, and NLF, and extricated the United States from its long and agonizing intervention in the conflict, the treaty nonetheless failed to bring a de facto compromise or true reconciliation between Hanoi and Saigon. Contrary to its name, the Agreement on Ending the War and Restoring Peace in Vietnam neither ended the war nor restored peace. Before the ink on the treaty had even dried, both Vietnamese parties resumed their bid to expand their presence throughout the countryside and control the operational environment. South Vietnamese general Cao Van Vien was correct when he remarked that the Paris treaty did not bring an end to the fighting but rather initiated a period of "half war—half peace."[33] Shortly after the Paris Accords were signed, Vietnam scholar Douglas Pike pointed out that it was obvious to most observers that the conflict would continue. The communists, he argued, were not going to leave the RVN alone or abandon their objective of unifying Vietnam under communist rule. Nor were they going to give up the goal to defend, consolidate, and expand areas under their control. As for Thieu and the South Vietnamese government, they too were not going to stop fighting to defend the nation's sovereignty or abandon their claim to dominion over all of the national territory. "Both sides," Pike wrote, "jockeyed for position, fought bitterly against encroachment of territory, [and] loudly appealed to their respective allies for support in the 'war of flags.' Each probed the other, searching for some fundamental weakness."[34] Former ARVN officer Tran Van Don reaffirmed Pike's observation years after the war. "The National Security Council was called together by Thieu at the presidential palace with the task of determining the best ways to thwart the enemy's efforts to make its presence felt everywhere in order to profit from the ceasefire," Don wrote in reference to the situation that existed in the early months of 1973. "In fairness it must be said that the ARVN also was preoccupied with extending its presence and had located forward posts in areas frequently claimed by the communists."[35]

Pike and Don were correct. The advent of the Paris Peace Accords did not settle the all-important question of who controlled what in South Vietnam. The issue of which side held dominion in the countryside and wielded power in the villages and hamlets remained as much a matter of dispute after the ceasefire had gone into effect as before. "Any agreement that failed to answer the basic question of the war, namely who was to hold political power," one observer declared, "could scarcely have been expected to lead to anything but a continuation, and possibly a gradual re-escalation of the war."[36] The irony of the Paris Peace Accords is that the agreement had perpetuated rather than ended the conflict in Vietnam—something very few observers failed to notice. "The ceasefire exists," a Vietnamese villager declared less than ten days after the armistice had gone into effect, "but the fighting must go on."[37] Also in February, a COSVN directive made the following observation: "Although there is an Agreement to End the War and Restore Peace, in truth in [South Vietnam] there are many places where the shooting and bombing and shelling continue, and in some places more than prior to the Agreement."[38] Ten months later, Kissinger underscored this contradiction. In high-level talks with delegates from Hanoi and Saigon, Kissinger lamented what he considered the "paradoxical consequence" of the armistice. "Each time we have proclaimed a ceasefire," Kissinger stated, "each side has tried to grab as much territory as it could just before the ceasefire. And then whichever side lost territory spent the next weeks trying to get it back. So the ceasefire led first of all to an upsurge of military activity, and then to a period of more intense activity to try to undo the consequences of the ceasefire."[39]

GRABBING LAND, GRABBING PEOPLE: THE STRUGGLE FOR CONTROL, 1973–1974

The struggle to control the operational environment, regardless of a formal peace treaty and the U.S. exit from Southeast Asia, continued unabated during 1973–74. Much as they had done for nearly two decades prior to the agreement in Paris, the communists and the RVN jockeyed for advantage in the control war and struggled to expand and/or consolidate their hold on South Vietnam's operational environment. The objective, as before, was to seize and hold territory and population. "As expected, the ceasefire did not end the fighting in South Vietnam," the *Washington Post* stated in the spring of 1973. "Military activity continued in the form of a land rush as both armies tried to grab each other's territory. The land rush was followed by a less obvious but no less violent political struggle."[40] Immediately after the Paris Peace Accords had gone into effect, both sides began maneuvering for advantage while accusing one other of ceasefire violations and "land-grabbing" operations. By 31 March, barely two months after an armistice had been signed, the Office of the Assistant Secretary of Defense (OASD) estimated that more than 1,300 major ceasefire violations had already occurred in South Vietnam. Approximately half of these took place in MR 1, primarily in the northernmost province of Quang Tri. Another quarter took place in and around the Mekong Delta. The remaining 25 percent of ceasefire violations were divided equally between MRs 2 and 3. Casualties during this period also ran high despite the fact that main-force units on both sides continued to avoid contact.

It was determined that 18,000 military personnel were killed during the first two months of the ceasefire, including 12,000 VC/NVA and 6,000 RVNAF. In spite of such high casualties, the conflict was shaped more by efforts to expand political presence than to destroy one another. "In essence, the military activity which has continued since January 28 has been for political purposes and designed by both sides to expand the base of the population they control," the OASD determined.[41] Despite openly denouncing the actions of their adversaries while pledging fealty to the Paris Peace Accords, the communists and the RVN both continued efforts to tighten their grip on rural areas and expand influence and control over the civilian environment. When asked if RVNAF military operations against communist areas were a violation of the Paris agreement, one ARVN officer brazenly replied, "Sure it is, but the communists are violating it too."[42] The U.S. Senate Committee on Foreign Relations recognized the unfortunate reality behind this comment. In a report issued a year after the Paris agreement had gone into effect, the committee made the following observation: "Although both sides continue to charge the other with ceasefire violations, lack of respect for the agreement is so widespread that it is impossible to apportion responsibility for the continued fighting."[43]

For Hanoi and the PRG, the primary objective was still to attack pacification, inflict damage on the RVNAF, prevent government encroachment into liberated areas, and, when possible, advance the revolution and expand the insurgent state. On 3 February, members of the Politburo held a plenary conference to discuss the RVN's pacification and "land-grabbing" operations. Convinced that the RVNAF would launch a series of counterattacks in an effort to regain populated areas that were lost to the communists in the run-up to the ceasefire, they recommended to COSVN that revolutionary forces within the region "rigorously maintain" their offensive posture in order to counter Saigon's attempt to recover lost population and territory. The group also urged a continuation of the political struggle, particularly proselytizing activities, in order to "maintain the gains made by the revolution and continue to advance the movement."[44]

Eight months later, these same orders were still in effect. In October, communist leaders convened the Twenty-first Conference of the Party Central Committee to discuss efforts "toward completing the liberation of the South." As it had done in previous years, the committee defined military activity largely as efforts to inflict losses on RVNAF, conserve and expand liberated areas, disrupt the pacification program, and blunt Saigon's efforts to expand its control over population and territory. "We must resolutely counterattack and attack the enemy and maintain and develop our initiative in all regards, in order to defeat the enemy's pacification and encroachment plans, especially in the lowland and contiguous areas," it declared. "The application of that motto must be tied in with the requirement of winning control of the population and winning the right of mastery, in order to gain the strength with which to defeat the enemy."[45] Similar directives called for stepped-up resistance against RVN "land-grabbing" and "nibbling" operations, attacks against pacification, and the expansion of liberated areas. "We should expand liberated hamlets and villages and gain control on many levels, from weak hamlets and villages to strong, from strong hamlets and villages to liberated ones," read a political reorientation and training manual issued to infrastructure cadres. The manual also urged social, political, and economic development in villages and hamlets under communist control to support military activities and serve as "a banner appealing to the people living in the enemy-held areas to support the revolution."[46]

Saigon was equally committed to gaining the upper hand in the control war. Although South Vietnamese officials frequently claimed the government would observe the rules of the Paris treaty, RVNAF officers openly planned pacification drives to shrink and/or eliminate communist enclaves after the ceasefire had gone into effect.[47] In July, the DAO acknowledged that RVN activity in the first half of 1973 was intended to "eliminate spheres of communist influence."[48] Saigon's decision to continue pacification operations and erase communist enclaves despite the Paris agreement comported with Thieu's rigid "four no's" policy and his conviction that Hanoi's most probable course of action would be to attempt to seize political power through guerrilla warfare and political subversion. According to one Saigon official, Thieu maintained that Hanoi and the PRG would pursue a protracted grassroots struggle in order to expand communist enclaves, grow the insurgent state, and eventually seize control of the country. Although Thieu suspected Hanoi could launch another major military offensive to capture territory, negotiate another in-place ceasefire, and consolidate its position in rural areas to prepare for future attacks, he believed the more likely scenario would be a more protracted approach. As in the past, the aim would be to isolate and encircle South Vietnam's cities and district capitals by gradually disintegrating Saigon's grip on rural areas. Thus, the Saigon regime would be relegated to the cities and surrounded by a hostile countryside.[49]

By September 1973, all pretenses to following the Paris agreement were gone, as communist units and the RVNAF launched small but intensive combat operations in select areas to shrink enemy enclaves and strengthen friendly control over territory and population. "South Vietnamese government forces, using air strikes, naval gunfire, artillery, and elite troops, are waging a sustained campaign to wrest land and people from communist control," read one newspaper account of the intense fighting in northern Binh Dinh Province. "Caught by the January ceasefire agreement with little more than the shoulders of the main north-south highway under their control, the South Vietnamese are now pushing the communists back and are beginning to repopulate villages that were lost in the North Vietnamese Easter offensive of 1972."[50] Heavy fighting was also reported that month in Kontum and Tay Ninh Provinces as well as near the ancient city of Hue, the capital of Thua Thien Province, located near the coast.[51] Also in September, communist units began an extensive campaign to capture remote RVN bases and outposts throughout South Vietnam's central highlands. According to ARVN general Cao Van Vien, a number of bases in these areas were overrun by communist forces that fall.[52] Less than ten months after it had been implemented, observers and analysts recognized that the Paris Peace Accords was a dead letter. "The current situation in Vietnam is one of 'less fire' rather than 'cease fire,' and there is no real peace," declared an October national intelligence estimate. "Both sides have initiated or provoked some of the fighting that has occurred since the January accords, primarily by attempting to seize or retake territory deemed strategic."[53]

The war in South Vietnam settled into a discernible pattern in 1973 and 1974. For its part, the DRV maintained a high level of infiltration of men and supplies into South Vietnam using the well-worn routes of the Ho Chi Minh Trail as it snaked through Laos and Cambodia. Hanoi also improved NVA combat capabilities by strengthening the firepower and mobility of main-force units, reequipping and retraining local forces, and expanding guerrilla organizations. In the northern and western potions of South Vietnam under their control, the communists expanded their network of supply bases and macadamized roads along the Ho Chi Minh Trail complex and inside the country. They also constructed or reconditioned a number of airfields and built a system of fuel storage areas and pipelines in these areas. In the densely populated RVN-controlled areas along the coast and in the Mekong Delta, NLF/NVA units continued to wage protracted warfare using a combination of main-force and guerrilla operations. The objective here was threefold: (1) to expand communist control over the operational environment and grow the insurgent state where possible; (2) to maintain pressure on the RVNAF wherever and whenever the tactical situation permitted; and (3) to stretch out and ultimately exhaust government forces. The RVN, meanwhile, worked to strengthen its grip on rural areas and deny the enemy control of the population by expanding pacification activities and attacking communist base areas and infiltration routes in harassment and interdiction campaigns. "Each adversary has attempted to consolidate his area of predominant control," declared the Senate Committee on Foreign Relations, "the North Vietnamese and Viet Cong in the mountains and the South Vietnamese on the coastal plain. Most of the significant military activity has been concentrated where these two geographies meet or where pockets of resistance oppose the predominant force."[54]

Throughout 1973 and into 1974, the RVNAF performed relatively well in the control war, maintaining and in some cases reestablishing Saigon's hold over territory and population that had been lost to communist forces. According to U.S. congressional estimates, Saigon added 770 hamlets to the "relatively secure" category and reduced the number of contested hamlets by well over a third between the time the ceasefire went into effect and May 1974. The communists, conversely, had during the same time frame lost 90 hamlets that were under their firm control at the time of the armistice as well as an additional 300 hamlets they had seized immediately after the Paris treaty was implemented. Analysts also reported that 93 to 94 percent of the population were living in government-controlled areas, a 5-point increase since November 1972.[55] Overall, however, the situation on the ground remained static; there was no change in control over the operational environment despite fourteen months of fighting and more than 80,000 Vietnamese deaths. "After more than a year's experience with the 'ceasefire,'" the Senate Committee on Foreign Relations declared in May 1974, "the consensus of official U.S. opinion is that there has been no major countrywide shift in control of territory or population between the government and the communists."[56] The U.S. Senate was not alone in this assessment. Other observers, both inside and outside of South Vietnam, realized that 1973 was not a year of decision. "Though bloodletting has been as extensive as it was before," the *New York Times* observed on the one-year anniversary of the Paris treaty, "neither side could claim stunning victories or significant territorial acquisitions. Nor, despite continual accusations, could either side demonstrate convincingly that it was more sinned against than sinning."[57] This indecisiveness, however, along with the slight edge enjoyed by the RVN in the control war, was about to change.

Unlike its calendrical predecessor, 1974 was a decisive period. The "Year of the Tiger" was a turning point in the Second Indochina War because it brought powerful changes both in the strategic balance of forces and in the conflict's trajectory. Saigon began the year in a strong position militarily. With more than a million men under arms, improved logistics, and more tanks, artillery, and aircraft in its possession than ever before, it appeared the RVN had a strong chance of surviving. "The Republic of Vietnam Armed Forces . . . today constitutes the largest, most well-equipped force in Southeast Asia," an April CIA intelligence estimate declared. "In the year since the ceasefire and withdrawal of U.S. forces, the RVNAF has proven capable of defending GVN-controlled territory. Indeed, in the fighting over the past year, the South Vietnamese have generally been able to prevail over the communists, and the GVN now controls slightly more people and territory than it did at the time of the Paris agreement."[58] This dominant position in the control war did not last, however. Within a matter of months, the RVN lost its edge and was forced into an increasingly defensive posture as the result of cumulative war fatigue among the RVNAF and the country at large, a sharp decline in U.S. military aid, and a resurgent communist attack.[59] By the end of 1974, the "correlation of forces" in Southeast Asia had shifted in favor of the communists, who, thanks to Soviet largesse, employed an increasingly mobile and technologically advanced war machine at a time when their adversary was forced to fight with diminishing military resources. In sharp contrast to the popular image of an indigent revolutionary force armed with primitive weapons—David versus Goliath—the NVA had been transformed as early as 1972 into a thoroughly modern military, one that was fully equipped with an array of highly sophisticated weapons in great numbers, including artillery, armor, and surface-to-air missiles. One year after the Paris Peace Accords had gone into effect, Hanoi began to flex its increasingly powerful military muscle.

Despite its growing strategic advantage, Hanoi did not launch a large, conventional offensive in 1974. In fact, most communist military activity was modest and geared toward isolated outposts and other low-hanging fruit in remote areas. Nonetheless, these efforts were effective. "Territorial security of the rural countryside has worsened [as] enemy forces continued to intensify their efforts at applying moderate to heavy pressure against territorial force units, outposts, [lines of communication], static defense lines, and subsector headquarters," the DAO concluded late in the year. "The enemy continued to . . . subvert GVN control at the village/hamlet level."[60] North Vietnamese general Van Tien Dung confirmed this assessment after the war. "In every place where the Resolution of the Twenty-first Conference of the Central Committee and the Resolution of the Central Military Committee were understood clearly, where the standpoint of revolutionary violence and the concept of taking the offensive were grasped firmly," Dung recalled, "not only was the area able to destroy the enemy's 'pacification' plans and preserve our liberated zones and our bases scattered through the enemy-controlled zones, but they were even able to expand many liberated zones and back the enemy into a position of confused passive resistance."[61] The losses in government control described by Dung and the DAO had pushed the RVN to the precipice by the end of 1974. According to former CORDS officer Stuart Herrington, who was working for the DAO at the time, the communists had severely damaged Saigon's strategic position that year and brought about an "impending communist victory." This, he recalled, raised a profoundly important existential question as the Year of the Tiger gave way to the Year of the Cat: how much longer could the RVN stave off the inevitable?[62]

THE FAILURES OF PACIFICATION-VIETNAMIZATION AND THE END OF THE WAR IN SOUTH VIETNAM

The shift in the strategic balance of forces in South Vietnam during 1974 came about due to a confluence of complex factors and developments. Some were the product of outside circumstances, particularly the diminishment of American aid and an increasingly muscular and assertive North Vietnamese military offensive. Others, however, were rooted in the control war and the overall failure of the pacification and Vietnamization programs. As discussed in previous chapters, the pacification-Vietnamization strategy put in place by the RVN and the United States in 1968 was intended not only to establish a high degree of government control over the national territory and develop an effective South Vietnamese military; it also was meant to construct a national political community and build a resilient economy. Despite achieving unprecedented levels of security in the countryside and government control over the civilian environment, the strategy failed to meet these objectives. The social disruption and despondence brought on by years of sustained conflict and a permanent state of war fatally undermined comprehensive nation building in South Vietnam. Despite years of strenuous effort, Saigon was unable to construct either an economically prosperous or sociopolitically cohesive society. Nor was it able to fully exorcise the communist presence from the national territory or develop armed forces that were capable of defending the national territory against the communists and their unique and ubiquitous brand of hybrid warfare. Each of these deficiencies was significant and contributed in some way to the RVN's diminishing fortunes in the final two years of the war. Thus, the failure to eliminate the communist insurgent state or establish control over the operational environment proved decisive, facilitating both the strategic collapse and the eventual overthrow of the RVN in the spring of 1975.

In large part, the RVN's deteriorating strategic position in 1974–75 stemmed from the cumulative effect of adverse economic conditions. The national economy, which was never strong to begin with, became increasingly dysfunctional in the years following the Paris Peace Accords and American withdrawal. Although to a certain extent the agricultural sector had continued to prosper after January 1973, the overall economic situation deteriorated largely because of the costs of sustained military activity involving forces in excess of a million men. The communists' ability to maintain the insurgent state, sustain military pressure throughout South Vietnam, and station large numbers of combat forces in the northern and western regions of the country forced Saigon to preserve a robust defense posture and allocate a significant portion of the nation's resources to war, resources that would otherwise have been available for economic development. In May 1974, the U.S. Congress estimated that the RVNAF consumed approximately one-fifth of the country's manpower and more than one-half of the national budget.[63] Overall, the money used to maintain the government and its armed forces ($870 million in 1974–75) far exceeded what the country generated in revenues. As a result, the South Vietnamese economy was not self-sustaining; it was wholly dependent on large-scale U.S. assistance.[64] Besides high levels of deficit spending, the RVN was beset by rampant inflation and high unemployment. According to a congressional accounting, the inflation rate was 65 percent in 1973 and was expected to exceed 110 percent in 1974. Unemployment figures were equally grim, running as high as 40–50 percent in some areas. Declining standards of living and even food shortages in select provinces were additional byproducts.[65] "The South Vietnamese economy after ten years of implacable war," recalled Cao Van Vien and Dong Van Khuyen, "was in a desperate situation."[66] Making matters worse, U.S. economic aid was drastically cut during this period, falling from $2.3 billion in 1973 to about $1 billion in 1974. In September 1974, Congress reduced aid even further, approving a mere $700 million out of a $1.4 billion proposal for what would be the United States' last aid package to the RVN. Approximately half of this final installment went to shipping costs.[67] These reductions brought about a dramatic disparity in combat proficiency. At a time when the NVA was becoming more modernized and lavishly supplied owing to generous assistance from the Soviet Union, the RVNAF was increasingly forced to fight a "poor man's war" as American aid diminished and the qualitative edge of its military equipment and weaponry degraded. "I am apprehensive over the GVN future," Major General John E. Murray, defense attaché to Saigon, wrote in 1974, "as the enemy builds up and our support of the RVNAF lessens."[68]

After Paris, the people of South Vietnam as a whole suffered from rampant inflation, chronic unemployment, and reduced American aid; the military was among those groups hardest hit. To begin with, the majority of RVNAF troops were poorly paid—below subsistence levels in many cases. Severe unemployment also took its toll. While ARVN soldiers and territorial forces were obviously employed, their family members increasingly were not. This was highly disconcerting for military personnel because their families simply could not survive without some type of supplemental income other than military pay. Polls taken in 1974 indicated that low pay and increasing concern for their families' well-being as a result of high unemployment were taking a massive toll on the morale and esprit de corps of the

RVNAF. According to one survey, more than 90 percent of those interviewed indicated that their pay and allowances were insufficient to meet the everyday needs of their families. More than 88 percent claimed that their standard of living had deteriorated since 1973. Such circumstances, needless to say, were highly detrimental to motivation and morale. "It is quite clear," noted one report, "that South Vietnamese military personnel are forced to live at less than reasonable subsistence levels, and that performance and mission accomplishment are seriously affected. Day-to-day survival in the face of worsening economic conditions has caused a deterioration of performance which cannot be permitted to continue, if the South Vietnamese military is to be considered a viable force."[69] When combined with severe inflation, low pay and high unemployment crippled the fighting spirit of the RVNAF. According to the DAO, the country's poor economy not only "drained the soldier's pay and his morale"; it "contributed to corruption, desertion, and lessened his will to fight."[70]

The RVNAF's dwindling morale, high desertion rates, and diminishing combat proficiency also reflected Saigon's inability to forge a strong, lasting sense of national political community. While it is true that the pacification program had made progress in developing democratic institutions, particularly universal suffrage and rural land tenancy, and that the population had increasingly turned away from the NLF/PRG and toward the RVN in the years after *Tet Mau Than*, the loss of American backing along with the jarring impact of perpetual war and unending civil strife undermined societal cohesion and prevented Saigon from generating genuine enthusiasm or support among the people. As throughout most of its history, the RVN remained a fractured society after the Paris Peace Accords. "In the midst of these difficult times, South Vietnamese society, which had never been united in the anti-communist struggle, was sinking deeper into division and selfishness," ARVN officers Nguyen Duy Hinh and Tran Dinh Tho later recalled of 1974–75. "Internal strife and the threat of collapse were also aggravating. The authority of government and national leaders as well as popular faith sank to an all-time low. It was no longer possible to mobilize the masses, and the survival strength of South Vietnam was visibly on the decline with each new communist attack."[71]

A lack of social and political cohesion in the RVN was in many ways the product of an unrelenting war, economic decline, and a sense of abandonment by its former ally. It was also attributable, however, to the failures of pacification, particularly the Saigon regime's inability to build a truly national political community. Despite more than five years of dominant government control in the countryside, the rural population remained war weary, numb, and unsure of what the RVN brand of social revolution meant other than what Hinh and Tho described as "an uncompromising anti-communist stance embedded in a vague nationalistic ideal."[72] While much of the public's apathy was rooted in the pragmatic rationality of the peasantry and a conscious decision to withhold (for reasons of survival) support until the balance of forces decisively favored one side or the other, the overall disaffection among the population also stemmed from the fact that Saigon lacked a positive national ideology—one that would inspire the people to support the government. Despite Tu Tuong and government lip service to creating an "authentically Vietnamese" national ideology, the reality is that the RVN failed to develop a unifying principle or political philosophy that inspired the population or rallied them behind the nationalist banner. In March 1970, just prior to his resignation as assistant deputy for CORDS, Willard E. Chambers succinctly captured the frustration felt by many over Thieu and the RVN's failure to garner genuine public enthusiasm for the national cause. "The most significant of our shortcomings has been that we have failed to provide a dream," Chambers wrote in a letter to his boss, William Colby.

> We have not succeeded, nor even seriously attempted to involve the people in the war in an emotional level. We have asked the people to support a government and yet we should know that people will not fight and die for a government. What they will fight for is a great dream, an impossible ideal, a great leader, or a great hate. None of these factors are present nor are they being developed. At no time have we attempted to tell the people what is at the end of the rainbow. What are the great aspirations which can make their suffering tolerable and their sacrifices worthwhile? We have been opposed to communism and we have somehow assumed that because we think communism is bad that others would find resistance to it to be worth fighting for and dying for. People will not fight and die for negative values. And yet no positive value has been suggested.

To some degree, Chambers's statement was an accurate indictment of Saigon's lack of positive messaging. After the 1966 Honolulu Conference, Thieu and the Saigon regime dropped most pretenses to "social revolution" and the verbiage of positive values and replaced them with a language that spoke predominantly in terms of controlling the population—something that while necessary, expedient, and successful in the short term did not generate the mass support necessary for long-term nation building. "All of human history is ample evidence that people *will not be controlled*," Chambers declared emphatically. "They can be led, they can be motivated, they can be inspired to support or oppose; but a nation whose fundamental principle is control of its people, is a nation which is doomed to fail."[73]

Chambers's argument was accurate in some regards; yet it overstated the case concerning population control and popular support. History, contrary to his assertion, is rife with instances where people can and have been controlled by authoritarian regimes. Moreover, his statements overlooked the fact that building political community in a time of civil war and social turmoil was a near impossible task. As noted by Tran Dinh Tho, the RVN faced a particularly acute problem over the span of its short lifetime. "The regime of the Republic of Vietnam," he declared after the war, "was . . . faced with a double challenge, that of nation building and the threat of insurgency and outright invasion from the north."[74] A number of American policymakers agreed. In 1969, Melvin Laird informed Nixon that nation building was "a difficult task" under peaceful circumstances. "It is herculean," he told the president, "while hostilities continue at the present level."[75] According to Henry Kissinger, "Thieu was expected to accomplish within months and amidst a civil war what no other Southeast Asian leader had achieved in decades of peace. He was being asked simultaneously to win a war, adjust his own defense structure to the withdrawal of a large American military establishment, and build democratic institutions in a country that had not known peace in a generation or democracy in its history."[76] Nonetheless, Chambers's overall argument concerning the lack of a unifying political ideology and the detrimental impact this had on South Vietnamese society cannot be dismissed. It would be incorrect and overly simplistic to attribute the RVN's defeat predominantly to a lack of popular support and the absence of a unifying nationalistic ideology. Politico-military control, after all, trumped politico-ideological supremacy and was the more desirable and determining factor in the control war. It would be equally problematic, however, to underestimate the importance that popular motivation and morale played in the nation's demise.

Much like the poor economy, Saigon's failure to articulate a vibrant national vision or arouse its population impacted negatively on the morale and effectiveness of government armed forces. As John Paul Vann pointed out in 1971, Saigon's soldiers and civil servants alike fought for a salary—not for an inspiring ideology. This grossly undermined both the military and the civil dimensions of pacification. "The morale of South Vietnamese soldiers is low," Vann told reporters. "In the political field the situation is not much better. The government does not have dedicated cadres, but only civil servants who work for salary, not for ideology. . . . Government programs which look good on paper do not get carried out well because civil servants, like soldiers, do not have motivation."[77] Speaking before the House of Representatives Committee on Appropriations a year earlier, William Westmoreland expressed similar sentiments. When asked if the people and the armed forces of South Vietnam lacked the "discipline and determination" of their adversaries, Westmoreland replied: "I do not think it is an unfair assessment. . . . Democracy and the doctrine of freedom and independence, of course, had not been well codified in the minds of the South Vietnamese. The South Vietnamese were fighting based on a negative motivation: that they wanted no part of communist domination and subjugation. The political turmoil in the south has been of such magnitude that it has created problems coalescing the people."[78]

The combination of diminishing military aid, the attenuating impact of sustained combat, a poor economy, and low morale had exacerbated what was already a crushing shortage of manpower by the fall of 1974. When the Paris Peace Accords went into effect in January 1973, the RVNAF appeared to be in a relatively strong position vis-à-vis the communists. Not only were they well supplied and in possession of vast material stockpiles, but they had more than a million men under arms, about half of which were in the regular army, navy, and air force. The other half served in the territorial forces, police, and paramilitary formations. Despite these numbers, however, many questioned whether or not the RVN could sustain these forces over time given the country's demographics and available manpower base.[79] According to former ARVN generals Vien and Khuyen, the country's overall population of 18 million people simply could not provide the manpower necessary to sustain a protracted war now that the U.S. was gone. Although Saigon had been able to fulfill RVNAF personnel requirements after the general mobilization had been implemented in 1968, it became increasingly difficult to fill the ranks as the American withdrawal neared completion. "In 1972 manpower became a major problem again despite all the positive actions taken by the GVN to keep the RVNAF up to strength and replace losses," they recalled. "It was obvious that South Vietnam's draft-age resources were drying up and even the yearly emerging generation of 18-year olds was not adequate to replace losses."[80]

By 1975, the answer to whether or not the RVN could sustain a million-man force in a long war had become clear. To begin with, the ARVN and the RF/PF lost more than 353,000 personnel to desertion alone between January 1973 and November 1974, an average of approximately 15,000 per month.[81] These losses were predominantly desertions by underpaid and despondent soldiers who simply walked off the job and went home; they were not defections to the communists. Nonetheless, a considerable number of regular army and territorial troops dropped out of the ranks. Combined with combat and administrative losses, the RVNAF was hemorrhaging manpower. According to Hinh and Tran, the RVN only recruited around 150,000 draft-age men per year during this period, while its annual losses, including those attributable to desertion, combat casualties, and administrative turnover, amounted to around

250,000 per annum. As these numbers indicate, the RVN's military, economic, and spiritual malaise had produced a manpower crisis by early 1975. "The country was virtually on the point of total exhaustion," the two generals recalled. "The ranks of combat units were thinning out. The material and manpower depletion occurred at an accelerated pace while the erosion of morale and the will to fight, which had been smoldering for years, was reaching its climax."[82] Historian William Le Gro underscores the cyclical nature of RVNAF losses: "Quite understandably, as unit strengths declined, so did combat ability and morale, while desertions increased."[83]

High manpower losses were an added stressor on a poorly designed force structure that was already overtaxed. Of all the weaknesses that plagued the RVNAF, perhaps the most serious, particularly in relation to the control war, was South Vietnam's inflexible system of territorial defense and lack of general reserve forces. By the fall of 1974, there were more NVA troops in South Vietnam than at any time in its history; more than half of these belonged to regular infantry divisions. Although each side enjoyed relative parity in artillery and armor assets, Hanoi enjoyed a qualitative advantage. Intelligence sources estimated that 10 percent of all communist artillery pieces could be fired beyond the range of ARVN field guns. Moreover, NVA units had stockpiled enough ordnance to support an invasion similar to the 1972 offensive for up to twelve months. Hanoi's logistical infrastructure both inside and outside the RVN had also been vastly improved. North Vietnamese engineers built twelve supply airfields and surfaced or rebuilt nearly a thousand miles of road. Together, these logistical improvements provided a capability to support large conventional offensives. Arrayed against this burgeoning military juggernaut were 1.1 million South Vietnamese regular and paramilitary defense forces. A full one-half to two-thirds of RVNAF units, however, were deployed in static area defense missions. This stood in sharp contrast to less than 10 percent of NVA forces that were similarly deployed. Tasked with defending 90 percent of the population against a much more mobile NVA, the RVNAF was seriously disadvantaged. The balance of forces clearly favored the communists. "Basically I have one division and I have to provide security for six provinces with two million people," declared one exasperated ARVN officer. "I have to provide security everywhere."[84]

Providing security "everywhere" and defending all of the national territory entailed an immense geographical obligation, one that Vietnamization and the Area Security concept were unable to satisfy. Under these programs, the RVN's Joint General Staff planned to upgrade and expand the RF into larger tactical formations so that they could assume the burden of territorial defense and fill the gap left by the departure of ten U.S. and FWMAF combat divisions. In theory, ARVN was to be relieved of responsibility for territorial security and pacification support, replaced by RF/PF, and reallocated to movable defense tasks. The objective was to construct mobile reserves that could deploy anywhere they were needed. Under the combined campaign plans devised by MACV and the JGS, territorial forces would progressively relieve ARVN divisions from pacification and local security missions so that the latter could take over mobile large-unit operations from the departing American and allied forces. Ideally, this reassignment was to take place once territorial forces were fully trained and organized, and after RF/PF units were upgraded into larger tactical formations. In reality, however, this never happened for a number of practical as well as political reasons.

According to Vien and Khuyen, ARVN divisions were never extracted from their territorial responsibilities because many officers and civilian policymakers believed relocation would create a physical and psychological vacuum in the countryside, one that could not be adequately filled by the RF/PF. Although they constituted a formidable force, territorial units proved incapable of assuming the immense tasks of providing regional security *and* pacification support. The decision to retain the ARVN in a territorial defense posture, therefore, not only overextended RVNAF ground forces and prohibited the development of adequate tactical and strategic reserves, it seriously limited the courses of action open to Saigon by preventing regular forces from maintaining the initiative for long periods of time. "As a result," the former generals stated, "all the ARVN could do was to protect the populated 'national priority' areas, which left the enemy almost entirely free to use his base areas."[85] The highly political nature of the South Vietnamese army also lent to this problem. As Henry Kissinger discussed in his memoirs, ARVN generals put in charge of military regions refused to relinquish troops under their command because they were the "indispensable props" that sustained political standing in Saigon. "They judged—correctly," Kissinger recollected, "that their political influence at the capital bore some relationship to the number of troops they commanded."[86]

Another reason Saigon refused to free the RVNAF from defending specific areas was because adverse economic circumstances demanded that soldiers serve in close proximity to familial dependents. Practical considerations, particularly maintaining morale and limiting desertion among military personnel who were concerned for the health and welfare of their families, made it necessary for Saigon to impose a strict and inflexible territorial security structure. Past experience had shown that ground forces deployed to regions far removed from parents, wives, and children suffered from high desertion rates and low morale. As previously discussed, desertion was a major problem for the RVNAF. It was not, however, equally spread among regular army units and territorial forces. Low morale and desertion had always been a particular problem for ARVN because its members served at much greater distances from their home villages than did territorial forces. ARVN operated at the military region level, whereas the RF/PF remained within the province or district in which they were recruited. In 1971, Defense Secretary Laird informed Congress of this situation. The major impetus behind ARVN's high rates of desertion, he noted, was the troops' concern for their families. "A man leaves to take some money home on payday to see that his family is provided with rice and food," the secretary stated. "Usually you find these people coming back into the regional forces, or popular forces, but they are listed as deserters from the regular forces when they take off and go home to see how the family is and to see that they are being fed."[87] In order to alleviate this problem, Saigon stationed regular army divisions in the regions where they drew recruits. Although the geographical scope of ARVN's operational activity was much larger than that of the territorial forces, its divisions became rooted to the defense of particular areas, which hamstrung strategic flexibility and operational mobility.

A third reason the RVNAF clung to a rigid and inflexible territorial defense posture was Thieu's stubborn refusal to yield ground held by the RVN and to defend every inch of national territory already under government control. Although Saigon had succeeded in advancing pacification and frustrating Hanoi's land-grabbing attempts during the first year of the "ceasefire," the South Vietnamese president's decision to maintain an unyielding grip on isolated areas and remote outposts proved to be much more difficult the following year. "Thieu's instructions forbade their evacuation," Cao Van Vien later wrote. "At all costs, they were to be defended and held." While this policy may have been politically beneficial, Vien recalled, it was not militarily sound. "It was fairly easy for the enemy to concentrate a force five or six times greater than ours at any remote place and with abundant firepower overwhelm an outpost at will," he wrote. "To attempt to hold all remote outposts, therefore, amounted to sacrificing a substantial number of troops who could be employed effectively elsewhere. . . . [M]aintenance of these outposts, although costly, was thought of as visible proof of the extent of our control. Besides, abandoning these outposts would be tantamount to turning over to the enemy a sizable part of the national territory."[88] According to another former South Vietnamese officer, Thieu rejected all proposals to abandon remote outposts because he believed them "necessary to assure South Vietnamese control of the countryside."[89] Thieu's decision to commit the RVNAF to defending the entire national territory reflected a mind-set that fused politico-military control of the operational environment to issues of national sovereignty and political legitimacy—two cornerstones that buttressed pacification and the control war. "President Thieu's policy of holding on to every inch of national territory at all costs was consistent with South Vietnam's anti-communist policy and entirely responsive to the situation created by a standstill ceasefire," Vien and Khuyen wrote after the war. Thus, because the communists could lay claim to any area lacking a government presence, Thieu considered it "imperative to hold on to all outposts, hamlets, and district towns, especially those located in sparsely populated, outlying corners or even in the heart of enemy base areas. All of these symbolized the RVN national authority, and they concretely demonstrated . . . that our control extended to these areas."[90]

Throughout 1974, Saigon maintained a defensive strategy founded on a "hold all and at all costs" policy. Moreover, it did so despite the increasingly powerful and mobile posture of communist main-forces and the concomitant degradation of RVNAF defense capabilities. When tied to the RVN's personnel shortages and a ubiquitous insurgent threat that incessantly harried and harassed ARVN and territorial forces in nearly every district and every province in the country, Thieu's fourth "no"—no surrender of territory—proved disastrous. The fixed system of territorial defense imposed on ARVN precluded any action aside from staying alert and qualitatively reinforcing the defense forces already in waiting. Moreover, it rendered the RVNAF incapable of meeting a concentrated enemy attack by overextending its ground forces and consuming manpower that could have been held in abeyance and used as mobile reserves to meet the increasingly potent communist threat.[91]

Postwar interviews with South Vietnamese military and civilian leaders reveal a general consensus that the "no surrender of territory" strategy employed by Thieu was a gross miscalculation in nearly all respects. Three years after the fall of Saigon, former ARVN officer Tran Van Don admitted that the RVN did not adopt a "correct military strategy" to deal with what he called the "inexorable communist steamroller." According to Don, "We spread our forces too thin, trying to maintain a presence in and defend each province town, an ambition clearly beyond our capability. Although by this time we had an armed force of over one million men, such a method of defense did not have a chance for success."[92] More than three decades later, another former ARVN officer, Nguyen Cong Luan, confirmed Don's analysis that Saigon's preoccupation with maintaining an iron grip on the national territory had seriously weakened the RVN's strategic posture. "As far as territorial security was concerned," Luan wrote in 2012, "the ARVN high commands deployed regular combat units to the fullest extent to control the population with little reserve force. Those units were stretched thin and had to conduct operations to safeguard the area when local regional force units failed to fight off the enemy. The regular forces had little time to rest, and combat soldiers seldom got leave. That seriously affected their spirits."[93] According to Luan, even the RVNAF's elite units—airborne, marine, and ranger—were tied to territorial defense and worn out through incessant fighting. The result, he concluded, was a complete loss of tactical and strategic mobility.

The communist strategy that forced the RVNAF to disperse and stretch itself thin in defense of territory and population was an intentional tactic as well as a fundamental tenet of protracted warfare. As can be seen in any number of directives, the objective was to incessantly attack Saigon's military and administrative forces in as many places as possible and force them to defend everywhere. "The enemy's troops are limited," COSVN declared in December 1968. "If they disperse troops and use a major element to support their pacification teams according to their large-scale plan, the contradictions between concentration and dispersion will become serious. Therefore, we will have many advantages to wear down and annihilate [the] enemy's potential."[94] The following year, COSVN Resolution 10 indicated that the advent of Vietnamization and American withdrawal meant that the RVNAF would have to increasingly assume "all combat burdens" at a time they were subject to "continuous attacks from different directions" by communist guerrillas and political cadres. It was imperative, therefore, that revolutionary forces sustain the attack against government forces in rural areas. "The destruction of puppet forces, especially mobile forces and tyrannical puppet units in local areas, will be of particular significance in turning the tide of war in South Vietnam . . . in our favor," the resolution declared.[95] Three years later, the objective of incessantly attacking the RVNAF to stretch it out was still in effect. COSVN Directive 43, issued in March 1972 in support of the Nguyen Hue Offensive, contained the following passage: "If cadres of the revolution can successfully force the government to defend large areas, it will spread so thin that our forces will seize major victories." The objective, the directive declared, was to "create a state of perennial unrest in Saigon, province capitals, and district towns [to] pin down government forces while our forces liberate the bulk of the countryside."[96]

Saigon's inflexible system of territorial defense not only robbed the RVNAF of the mobility necessary to meet NVA main-force attacks; it spent all the available manpower in a manner that left no strategic reserve. "The government has a few resources to respond to the communist thrusts," the *New York Times* noted in January 1975. "But with no real strategic reserve, it is preoccupied with holding what it has; the communists, as the attackers, have their pick of targets."[97] Postwar analyses have confirmed that the RVN's strategic position suffered greatly from the lack of a reserve force. According to interviews with high-ranking South Vietnamese policymakers conducted after the war, the RVNAF's rigid territorial posture weakened its armed forces and produced "other adverse military consequences." Such an arrangement, a RAND Corporation analysis argued, led to "a firm commitment of all available forces to the defense of their respective areas so that these forces had no strategic mobility and ARVN had no strategic reserves."[98] Vien and Khuyen agreed with this assessment. "Our strategic reserves were not sufficient for South Vietnam's extended defense posture in the face of an enemy who held the offensive initiative," they wrote after the war. "The RVNAF force structure expansion and development programs failed to solve the problem of strategic reserves which became even more critical after the redeployment of . . . U.S. combat divisions. Our plans to release ARVN infantry divisions from territorial responsibilities were not practically feasible."[99]

Historian Graham Cosmas believes much of the blame for the RVNAF's inadequate force structure and lack of reserves lies with American policymakers and the failure of Vietnamization. The continued use of ARVN for territorial defense, he asserts, was a grave miscalculation that produced dire consequences. Saigon's ground forces lacked adequate mobile reserves because most of its infantry divisions were tied to particular localities due to territorial defense obligations. Thus, they lacked the type of forces necessary to counter concentrated attacks like those launched by Hanoi in 1972 and 1975. During the Nguyen Hue Offensive, U.S. airpower was the deciding factor that enabled the RVNAF to repel the communist invasion and emerge victorious. The days of lavish tactical air support supplied by the Americans, however, were over by January 1973. According to Cosmas, Nixon and Kissinger had urged the Defense Department to take action on this issue during the summer of 1972. Secretary Laird, however, backed by the Joint Chiefs, General Abrams, and General Fred C. Weyand (Abrams's successor as head of MACV), balked. They thought it impossible to drastically reorganize or expand RVNAF ground forces at that time. The only course of action available, Laird maintained, was a "sequential and evolutionary approach" that emphasized the continued improvement of territorial forces and gradual steps toward freeing regular divisions from static security tasks. Cosmas argues that in the end South Vietnamese ground forces could not cope with the large-scale conventional invasion that toppled the RVN in 1975. Vietnamization, therefore, had failed because it did not help RVNAF ground forces develop the defense capacity necessary to defeat—alone and without the assistance of plentiful American airpower—a major conventional NVA ground assault aided by heavy mechanized and artillery formations.[100]

By 1974, the combination of an overextended RVNAF and the lack of military reserves had a deleterious impact on the relative state of control over territory and population. In October, a Party Committee conference in Binh Dinh Province declared that ARVN had been forced to "deploy thinly to hold people and rural areas." This, the committee members argued, had created "many gaps in many rural areas" and left Saigon's territorial forces and local militia vulnerable to attack by communist guerrillas and local forces.[101] North Vietnamese general Hoang Van Thai confirmed this situation after the war. "The enemy forces had been stretched out, were resisting passively, and no longer had mobile troops to support pacification and encroachment," he wrote. "Our liberated area was expanding rapidly and was being further consolidated in all theaters, from [Tri-Thien-Hue] and Military Region 5 to eastern Nam Bo and the Mekong Delta. Our network of strategic and campaign roads was becoming increasingly complete and had created an interdicting status which directly threatened the enemy-controlled areas."[102] Thai's assessment was correct. By autumn 1974, Hanoi's "strategic raids campaign" in MR 3, the central highlands, and throughout remote areas in MR 1 had exploited the RVNAF's overextended posture, expanded the insurgent state by capturing a number of strategic outposts and districts, and established territorial platforms for launching military offensives against key strategic areas. Thus, the point of contact between the opposing forces was now directly adjacent to many of the country's most densely populated regions.[103] The gravity of the situation was not lost on observers in the United States. As indicated by the *Los Angeles Times*, communist units were occupying large portions of ground that thousands of GIs had been sent to Southeast Asia to liberate. It was estimated that communist forces controlled more territory inside of the RVN than when the U.S. Marines first landed at Da Nang in March 1965. "American troops fought and died during seven years of war in South Vietnam to win land from the communists," the paper stated despondently. "Now, two years after the last American attack, the major battlefields . . . are almost exclusively occupied by Viet Cong and North Vietnamese soldiers." Among the battlefields that had come under communist occupation were Khe Sanh, the Ia Drang Valley, and Hamburger Hill in the western portion of MR 1, and War Zone C and the Iron Triangle just north of Saigon in MR 3.[104]

On 14 December, communist forces launched an attack on Dong Xoai, a district capital located in Phuoc Long Province approximately fifty miles northeast of Saigon. After a short respite in which COSVN and the Politburo waited for an RVNAF counteroffensive, revolutionary forces resumed the attack and quickly captured the province.[105] Although not foreseen at the time, the fall of Phuoc Long marked the beginning of the end for the RVN. By January 1975, the RVN's military position had deteriorated to an alarming degree. It lost significant ground in 1974 as a result of Hanoi's strategic raids campaign along the northern and western frontier, and now it lost an entire province within easy striking distance of Saigon. Faced with this dire strategic situation, Thieu had no choice but to go on defense. "The RVNAF has been forced into an increasingly defensive and reactive posture," a December national intelligence estimate stated. "It is likely that heavier combat would force the GVN to dilute further the strategy and tactics that have stood it in good stead since the ceasefire."[106] The following month, the exhaustion and weakened strategic position of America's former ally were becoming more obvious. "The strains on the Saigon government are beginning to show," one U.S. newspaper reported. "The South Vietnamese army, according to most analysts, is dangerously overextended and is almost bound to lose some of its more vulnerable defensive positions in the months ahead."[107] According to the director of the Defense Intelligence Agency, Lieutenant General Daniel O. Graham, Saigon was now in a position where it would likely have to abandon a number of positions in contested territory in order to retain any hope of defending the country's vital population centers and rice-growing regions. "The strategic and tactical advantage," he declared, "has passed to the communists in South Vietnam."[108] Six weeks later, Congressman Donald M. Fraser of Minnesota informed President Gerald Ford that in addition to inflicting losses on the RVNAF in the main-force war, communist units had made gains in the local control war as well because of a revived insurgent political apparatus in the countryside. "I received reliable reports that the VC infrastructure is coming back," Fraser reported to a high-level cabinet meeting. "It is a smoldering fire." He went on to report that 600,000 South Vietnamese citizens had recently come under communist control in one month—a 10 percent overall shift in population control that doubled the total number of people living under communist dominion.[109]

The final offensive of the Second Indochina War began on 5 March, when communist units launched an attack on Ban Me Thuot, the largest city in the central highlands located in Darlac Province. After Washington had failed to respond to the capture of Phuoc Long in December, Le Duan and others in the Politburo determined it was unlikely, though not impossible, that the United States would reintervene in the conflict should Hanoi continue to press the attack. With this in mind, communist units cautiously initiated the attack on Ban Me Thuot and set in motion a string of events that culminated in the fall of Saigon on April 30. As with other large-scale communist offensives, planning for the winter–spring campaign began months earlier. In July and August 1974, COSVN and the Central Military Committee drew up separate plans for a two-stage offensive that would culminate in the liberation of South Vietnam in 1976 or, at the latest, early 1977. Although the time frame and scope of each plan differed, both shared two primary objectives for the first phase: to inflict as much damage as possible on the RVNAF; and to expand and consolidate communist control over the operational environment in preparation for the final offensive that would begin in the fall of 1975.[110] The planned campaign also resembled earlier offensives with its emphasis on destroying the pacification-Vietnamization strategy and growing liberated areas and the insurgent state. One NLF resolution contained the following excerpt:

> The enemy continues to implement the "Vietnamization" of the war and use "pacification" as its primary strategic measure to hold on to and expand its area of control, increase its strength, and weaken ours. We must defeat the strategy of the enemy in order to win over the people, put the people in control, expand our areas, increase our strength in all respects, and fundamentally change the balance of forces between the enemy and ourselves, in order to overthrow the enemy. The defeat of the enemy's primary strategic measure will lead to overthrow of the enemy. That is the key objective, and it is the axis of the entire apparatus of the life-and-death struggle between the enemy and ourselves in the new phase.

The means to these ends, the resolution continued, were to initiate a "high tide" of attacks and uprisings in rural areas, to coordinate local actions with main-force NVA operations, and to intensify the political struggle in urban areas. Together, they were meant to achieve a single objective: to "annihilate the enemy and liberate each and every area" in South Vietnam.[111]

In March and April, the communist offensive rapidly captured large swaths of South Vietnamese territory, even if it did not "annihilate" the RVNAF. Although much of the success was attributable to careful planning and the skill of the communist war machine, the offensive was helped along by a string of poor strategic decisions by Thieu and a disorganized withdrawal from Military Region 1 and the central highlands, a retreat that quickly degenerated into a panic-stricken rout. On 14 March, one day after the fall of Ban Me Thuot, Thieu ordered the evacuation of all remaining South Vietnamese forces from the northern provinces and central highlands. According to the U.S. embassy in Vietnam, Thieu recognized he could no longer defend every inch of the national territory in light of the deteriorating situation. "The president said he could no longer fight what from a military standpoint had for the last two years been a 'stupid' way to fight a war," read a memorandum issued to the State Department on 20 March. "He had no choice but to give up territory or face the prospect of having his own forces destroyed piece by piece, eventually producing a situation where because of effort to defend every part of the country, he could no longer defend any of it." Rather than blame the situation on himself and the last of his "four no's" (no surrender of territory), however, Thieu instead denounced the Paris Peace Accords and the provision that had allowed Hanoi to leave nearly 200,000 NVA soldiers inside South Vietnam. According to a telegram sent to the U.S. State Department, "The president returned to the theme that until now the RVN had fought a war based on the premises of the Paris Agreement. This premise had made for a stupid war militarily, with scattered units on the government's side trying to defend every piece of territory."[112]

By 24 March, communist forces had "liberated" Kontum, Pleiku, and Darlac Provinces—nearly the entire central highlands. The ancient imperial city of Hue fell on 25 March, followed four days later by Da Nang. Once Hue had been taken, Le Duan and other high-ranking leaders determined that the time had come to rapidly accelerate the timetable for total victory. Over the next several weeks, communist forces continued to advance, mopping up pockets of enemy resistance in South Vietnam's northern and central provinces and inching closer toward Saigon. It was then that Thieu made a desperate last-minute appeal to the United States for assistance. Congress rebuffed his plea. Rather than supply additional funds or provide air support, American policymakers instead made preparations to evacuate U.S. personnel still in South Vietnam along with select Vietnamese who were closely identified with the Saigon regime.[113] On 30 March, communist forces captured Qui Nhon, the capital of Binh Dinh Province and the fourth-largest city in South Vietnam. By then more than half of the country's territory had come under communist control. Hanoi and the PRG stood on the verge of a complete victory. The final act of what the Vietnamese communists would later call "our great spring victory" came on 30 April, when NVA forces rolled into Saigon and took possession of the capital. By then, President Thieu had resigned and the last remaining Americans along with hundreds of desperate South Vietnamese climbed frantically onto the rooftop of the U.S. Embassy in Saigon for evacuation by helicopter. After providing more than twenty years of support—material, financial, and spiritual—to the fledgling South Vietnamese republic, the United States had thrown in the towel. Rather than reintervene and attempt to save the nation it had shed so much blood and treasure to preserve, America instead rowed away in the last lifeboat from a sinking ship of state, one that was rapidly descending below the waterline and would soon disappear forever. The war in South Vietnam was over.

CONCLUSION

The Second Indochina War was many things: a war of foreign aggression waged by one Vietnamese state to subjugate another; a "people's war" where an indigenous insurgency rose up to throw off the yoke of neocolonial despotism; a war of reunification waged to fuse together two halves of a nation artificially divided by Cold War conflict; a village war that involved two competitors vying for the support of the rural peasantry; a war of "big battalions" where large conventional military forces conducted a grueling "war of attrition" against a wily and elusive foe; a guerrilla war in which a poor, ill-equipped, yet tenacious insurgency defied its much wealthier and more powerful adversaries; a "hybrid" or "mosaic" war that involved a complex mixture of conventional and guerrilla warfare. It was also a competition to control contested political spaces—a "control war" in which Hanoi and the National Liberation Front on one side and the Republic of Vietnam and its allies on the other vied for politico-military supremacy and waged a violent struggle for power and dominion over South Vietnam's physical and ideational landscape. At the heart of this revolutionary civil war lay two competing but incompatible visions for modern Vietnam, dueling nationalisms that struggled to actualize their conflicting political and social blueprints for a brand-new society based on independence, social justice, and economic prosperity. This highly partisan and ideological bid for political supremacy centered on a bitter zero-sum contest to control the operational environment—that is, the people, the territory, and the resources of South Vietnam.

The operational environment was not only central, but a decisive factor in the war for South Vietnam. The struggle for control of territory and population shaped and ultimately determined the outcome of the war for Indochina. Like other revolutionary civil wars, the key to winning political power in South Vietnam was to establish dominion over the physical world (territory, population, resources) and the ideational world (the political organization of occupied territory). The means to do so were communist revolutionary warfare and pacification—two approaches that pursued the same goal (population and territory control) and used the same methods (a blend of military force, political violence, and socioeconomic policy) despite their countervailing purposes. Spatial contestation, territoriality, and efforts to control geography and the human environment, then, were salient features in the fight to establish political hegemony and determine the future social and political order of South Vietnam. As this study argues, the operational environment, including both the natural and the human-made world, were more than just a setting or backdrop to the politico-military contest that consumed the Republic of Vietnam for two decades; they were the locus, medium, and determinant of this historical conflict.

Contrary to popular conceptions, the crux of the war was not winning the hearts and minds or gaining the love and affection of the rural peasantry. It was instead, as one American military adviser put it, to "get control of the people first and teach them to love us later."[1] This perception was widely shared by observers as well as participants. "The Vietnam war has never been a popular contest," declared an American journalist. "Control has always been a much more important factor than popularity."[2] The reality behind these statements explains why communist revolutionaries and the Saigon regime focused on establishing political and/or military control over the operational environment. Communist revolutionary warfare and pacification were both hybrid systems specifically designed to establish and grow political authority over land and people via an assortment of military and political means. Control was central to this process. This explains why—despite happy talk about constructing a new society backed by the people and built on peace, prosperity, and social justice—both sides regularly resorted to violence and coercion to advance their brand of revolution. Communists and anticommunists alike earnestly desired to improve Vietnamese society and construct a vibrant, independent nation. Yet they understood that political domination could not be achieved through peaceful political competition. Given the ideological dimension of the war and the incompatible visions advanced by Hanoi and Saigon, issues of legitimacy, authority, and the right to rule simply could not be settled through nonviolent means. Military force and other forms of coercion were the surest methods for laying the foundation for political power.

Establishing and maintaining control over the operational environment had been a primary consideration in South Vietnam throughout the Second Indochina War. They assumed unprecedented emphasis, however, between 1968 and 1973, a period when Vietnamization, American withdrawal, and the Paris peace talks made maximizing politico-military control over territory and population the defining issue of the war. The control war shaped events on the ground and profoundly affected diplomatic efforts during this period. Negotiations, in turn, influenced events inside South Vietnam. In this way, the control war and the Paris peace talks shared a symbiotic relationship. On the one hand, the level of control each side held over the operational environment directly influenced the progress of the ongoing negotiations in Paris. The willingness or unwillingness of both sides to make concessions and/or agree to a settlement depended on the amount of territory and population they each controlled. On the other hand, the issues that dominated the negotiations—a brokered settlement, an in-place ceasefire, and postwar territorial accommodation based on the disposition of forces and areas of occupation—spurred efforts to maximize control over territory and population before a deal was reached, one that that would have ended hostilities, frozen opposing political and military forces in place, and apportioned political and administrative rights over occupied territory once the armistice went into effect. Conventional wisdom has largely credited the communists for using the technique of fighting while negotiating during this phase of the war. The reality, however, is that both sides employed a "talk-fight" strategy that wed force and diplomacy in the years after *Tet Mau Than*.

The struggle to control the people and territory of South Vietnam continued to define the conflict even after the Paris Peace Accords were signed and the Americans went home. The ability of the RVNAF to adequately defend the national territory and maintain politico-military supremacy in the countryside had not only become increasingly difficult after 1968, but impossible in the years that followed the 1973 treaty and America's exit from Southeast Asia. Ultimately, it was within the realm of spatial contestation, territorial defense, and the civilian environment that the stage was set for the final climactic military showdown between Saigon and Hanoi. The outcome of the war was largely determined by the battle for control. Although there were a host of issues that shaped the conflict and influenced the balance of forces after 1968, particularly Vietnamization and the level of aid Hanoi and Saigon received from their superpower sponsors, the ability of each side to effectively control territory and population decided the contest. The control war did not take place in a vacuum; it was highly contingent on other military and political factors, including diplomacy and outside military and economic aid. Yet the war in South Vietnam was ultimately won/lost within the realm of territorial conflict and spatial contestation. The trajectory and outcome of the conflict were highly contingent on the relative strengths and weaknesses of each side in the struggle to expand and consolidate control over the countryside.

The RVN and the United States lost this struggle. Although Hanoi and the National Liberation Front never gained control over a majority of the population or territory of South Vietnam prior to April 1975, they succeeded in building and maintaining an insurgent antistate within the national boundaries of South Vietnam, an achievement that not only preserved the Vietnamese revolution but helped the communists topple the Saigon regime in the years after American withdrawal.

Saigon's loss in the control war and the RVN's eventual collapse rest on the failure of pacification, particularly the program's inability to crush the communist insurgency or destroy the insurgent antistate inside South Vietnam. Despite the demonstrable success of pacification after 1968, particularly in expanding security and control over territory and population, the program ultimately failed. The revisionist argument that pacification had won the war by 1970 is simply wrong. Over the years, some scholars and participants have argued that the RVN and the United States achieved "victory" by 1971. This, they contend, was largely due to pacification—a program that after years of failure and setbacks achieved spectacular success following the Tet Offensive. "The fighting wasn't over . . . but the war was won," historian Lewis Sorley asserts. "This achievement can probably best be dated in late 1970. . . . [B]y then the South Vietnamese countryside had been widely pacified. . . . The successful pacification program, one repeatedly cited in enemy communications as a threat that had to be countered, was extending not only security but also elected government, trained hamlet and village officials, and economic gains to most of the population."[3] Sorley is not alone in his assessment. Others, including key participants in the program, agree that pacification had achieved its purpose during the final four years of American intervention. William Colby, the director of CORDS from 1968 to 1971, also maintained that pacification had achieved victory in Vietnam. By late 1970, he argued, the war for control of the countryside "essentially had been won."[4] Other commentators, including General Westmoreland's onetime deputy,

General Bruce Palmer, made similar claims. Writing in 1984, Palmer asserted: "By the end of 1970 and certainly by mid-1971 the pacification campaign was well on the way to success. Ambassador Bunker to this day insists that this essential and integral part of the war had been won in 1971."[5]

Pacification had indeed made unparalleled gains in security and government control over territory and population between 1969 and 1971. However, claims that the program achieved "victory" in South Vietnam are specious given the highly fragmented politico-military situation and the persistence of a communist insurgent antistate inside the geographical boundaries of South Vietnam. The RVN's national territorial integrity and area security were fatally compromised by a landscape honeycombed with enemy territory and spaces that were either influenced or outright controlled by communist military forces and political cadres. "As long as there are areas under VC control," a South Vietnamese district chief lamented in 1968, "no hamlet in Vietnam is completely safe."[6] MIT professor Ithiel de Sola Pool agreed. "In the checkerboard pattern of control that has existed," he explicated the same year, "no government hamlet could be fully secured while there were VC hamlets three or four miles away. Thus while we concentrate effort on areas with a substantial block of government population, still their neighboring VC centers have to be pacified too."[7] The "leopard skin spots" left by multitudinous pockets of communist control, South Vietnamese general Nguyen Duy Hinh noted bitterly, "were just so many knives thrust at the nation's back."[8]

Whether or not pacification could have won the war in South Vietnam, as Sorley and others have suggested, had it been implemented more vigorously prior to 1968 or had the United States sustained economic and military aid to the RVN in 1974 and 1975, will never be known. One thing is certain, however: the program did not achieve its purpose. Although pacification attained great success in security levels between 1968 and 1972 (the portion of the overall population living in "relatively secure" areas during this period was close to 90 percent or better), the program failed to exorcise the communist presence from South Vietnam, destroy the insurgent antistate, or blunt the communists' capacity to metastasize and expand influence and control over land and population. Had the allies implemented a vigorous pacification program earlier or had the United States sustained robust military and economic aid to the RVN after 1974, history may have proven Sorley, Colby, and Palmer correct. History as it happened, however, produced a reality much different from the counterfactual revisionist narrative. The RVN and its American ally lost the war in South Vietnam because pacification failed to provide meaningful and sustainable national territorial security. In the end, the program did not fashion a "lost victory," as some have claimed. Rather, it suffered defeat in the control war and loss in the revolutionary struggle to govern the people and territory of South Vietnam.

When coupled with the withdrawal of American combat forces, diminished economic and military aid from Washington, and the need to maintain a high level of military preparedness vis-à-vis conventional military invasion by an increasingly muscular and modern North Vietnamese military, the RVN's inability to destroy the communist insurgency after 1968, along with its decision to apply a rapid but thin method of pacification, proved disastrous. To begin with, the RVNAF was compelled to defend of the entire national territory against an enemy that threatened the government throughout the length and breadth of South Vietnam. The insurgency, while decimated in the 1968 and 1969 general offensives, managed to survive the crushing blows of allied military efforts owing to the durability of its political infrastructure, the infusion of NVA personnel into dangerously depleted units, and its skill in maintaining "liberated" areas throughout South Vietnam. The insurgency endured in large part because of the tenacity of its guerrillas and political cadres and the presence of conventional communist forces—divisions, battalions, and regiments—both inside and outside the country. Without its infrastructure, NLF/NVA main-force units, and the constellation of base areas and other liberated terrain throughout South Vietnam, Laos, and Cambodia, the communist insurgency would likely not have survived or triumphed. The hybrid conventional-guerrilla approach employed by Hanoi and the NLF along with the strategic use of geography and the environment allowed revolutionary forces to maintain a presence throughout Indochina after 1968. This was the key to victory.

The permanent addition of nearly 125,000 NVA regulars as a result of the 1972 Nguyen Hue Offensive placed further strains on the RVNAF. Faced with a large conventional military threat residing in the northern and western portions of the country, and a guerrilla presence nearly everywhere else, the RVN had no choice but to maintain an expansive and ultimately exhausting garrison state. The immense size and scope of the RVN's strategic posture consumed all of the nation's available manpower and forced Saigon to maintain a rigid system of territorial security based on static defense of specific regions. This combination of limited human capital and vast geographical obligations not only left the RVN with no strategic reserve; it stretched its armed forces dangerously thin. The pacification program's objective of near-total control of territory and population, along with President Nguyen Van Thieu's insistence that every inch of South Vietnam be held at all costs, proved impossible to achieve. According to North Vietnamese general Tran

Van Tra, Thieu's orders to defend "everywhere," along with his stubborn refusal to surrender a single village or hamlet to insurgent forces, was a "stupid strategic line" that brought ruin to Saigon and its military forces in 1975. "The tendency of defeat for the U.S.-puppets and victory for us in the war, which began with the 1973–1974 dry season and ended with the total defeat of the enemy," Tra declared, "had its origins in an erroneous strategy that was pursued from the beginning by the Americans and puppets: the strategy of defending all parts of the country [and] spreading troops thinly to prevent the 'Viet Cong' from taking a single inch of the land, in accordance with Thieu's 'four no's' policy."[9]

The collapse of the Saigon government in April 1975 stemmed from a confluence of diverse factors, including low morale and high rates of desertion among the RVNAF, widespread public despondency and war-weariness both in the RVN and in the United States, an economic crisis that crippled the RVN, and the United States' refusal to provide adequate military aid in 1974 and 1975. Nonetheless, the fall of the Saigon government resulted largely from impossible territorial responsibilities rooted in the failure of pacification, particularly the program's inability to suppress the communist insurgency and remove the insurgent antistate from the national territory. "Had our army been able to devote itself alone to combating enemy soldiers we would have been able to continue the fight for a long time," Tran Van Don, a former South Vietnamese officer, wrote in 1978. "But the army also had to perform a territorial mission. . . . The fighting strength of the army was frittered away in trying to secure the countryside, and thus it became too weak to engage successfully with regular enemy troop formations."[10] Others attributed the RVN's defeat to an unmanageable system of territorial defense rooted in the allies' overall defensive strategy. General Westmoreland, for instance, argued that Hanoi and the NLF had won the war because their forces possessed the advantages of strategic flexibility and initiative—two critical elements the RVN lacked. Saigon, meanwhile, had lost because its armed forces were stretched thin and bound to the defense of specific areas. "The North and South Vietnamese armies could not be viewed as two boxers in a ring with no initial advantage to either," Westmoreland declared. "North Vietnam, as the invader, was capable of massing its forces at points of its own choosing without fear of an attack on its home territory. . . . South Vietnam's sole aim was to hold on to its territory and avoid encroachment along the ceasefire line. Its troops were spread, making them vulnerable to defeat at any point that the enemy chose to mass his forces for attack."[11] Historian Dale Andrade agrees. The fact that the RVN fought on the strategic defensive, he argues, put the young republic at a serious disadvantage throughout the war. "This defensive strategy had been the crux of South Vietnam's disadvantageous position throughout the war," Andrade writes. "While North Vietnam could prepare interminably for an offensive with little fear of a pre-emptive strike, and then pick the time and place of its attack, Saigon was forced to remain forever vigilant, maintaining a thinly spread defensive network all over South Vietnam."[12]

The argument advanced by Westmoreland and Andrade is correct; the strategic situation facing each side strongly determined their combat options. The RVN fought exclusively to protect the national territory. It did not conduct persistent cross-border ground attacks against communist base areas, cordon off Laos and Cambodia, or use military force other than airpower to preemptively strike against infiltration. These inactions severely handicapped Saigon's range of action. The bulk of RVN ground forces were tied down in static security missions to defend large swaths of territory and population, including lines of communication, government and military installations, and provincial and district capitals. Hanoi and the National Liberation Front, for their part, were not immobilized by crushing spatial defensive obligations. The communists had very little territory to defend, and their base areas and lines of communication were not held or defended by large numbers of troops. Moreover, they had little to fear outside of the geographical confines of South Vietnam. As previously noted, Saigon and the United States never conducted sustained military operations outside of South Vietnam. The majority of communist forces, therefore, were given free rein to engage in battle at a time and place of their own choosing.

The fall of Saigon in April 1975 resulted from more than the RVN's failures, the inherent advantages/disadvantages of insurgent-counterinsurgent conflict, and the varied territorial defense obligations of the control war. Although there were a number of complex causal factors behind the collapse of the South Vietnamese government in 1975, the triumph of the communist revolution was in large part a product of the strategic use of geography and the environment. Hanoi and the National Liberation Front emerged victorious not only because they developed a unique method of protracted warfare—one that meshed political organization, guerrilla tactics, and conventional operations—but because they achieved several key spatial objectives using this hybrid politico-military approach. First, they expanded areas under communist control and grew an insurgent antistate within South Vietnam. Second, they disintegrated Saigon's hold over the country's territory and population. Third, they overextended and ultimately exhausted government forces through constant military and paramilitary pressure along a wide geographical front. Each of these strategies played a key role in communist efforts in the control war. In tandem, they helped the communists achieve victory in Vietnam.

Hanoi and the NLF succeeded in the end because they expanded liberated areas after *Tet Mau Than* at the same time they reduced Saigon's hold over territory and population. This was particularly true during the first half of 1972 and, to a lesser degree, the latter half of 1974. Although the antipacification program that dominated communist strategy failed to achieve its larger objectives (the RVN continued to exercise authority over most of the population), the communists successfully expanded the insurgent state in a number of key strategic areas, particularly in the central highlands and along the DMZ. Moreover, they consolidated and transformed areas under their control into springboards for future offensives. In the northern and western portions of the country, they macadamized roads, built airfields, and constructed a network of supply bases, fuel storage areas, and pipelines. As noted by American officials in 1974, the development of these facilities greatly enhanced Hanoi's ability to launch conventional attacks. "Because they are located in inaccessible areas of the interior and are well defended by anti-aircraft units," read a report generated for the Senate Committee on Foreign Relations, "the new North Vietnamese installations are relatively secure from government attack. Their overall effect is to diminish significantly the logistical advantage which the South enjoyed in the past and to give the North Vietnamese a capability to move and mass troops in a manner hitherto impossible for them."[13] This diagnosis was prescient; in line with communist strategy, these base areas later served as both a marshaling area and a springboard for main-force units as they launched the final offensive of the war in the spring of 1975.

The enlargement of the communist "rear area" and the development of a large and unimpeded supply corridor along South Vietnam's western border allowed Hanoi to funnel manpower and matériel directly to NLF/NVA units without interference. This was instrumental to the success of the 1975 offensive. The Vietnamese communists' ability to maintain and expand a military and political foothold in South Vietnam, as it turned out, was a decisive factor in the communist victory in Southeast Asia. According to the PAVN official history, the vast network of liberated zones and local base areas largely untouched by pacification played a significant role in cementing the communists' road to power in Vietnam:

> The intermingling of our areas within enemy-controlled areas throughout South Vietnam gave our troops solid footholds for their operations. Without operations conducted by local armed forces, without civilians supporting our forces and fighting beside them there would have been no rear base areas, and without those bases our military units would

> not have been able to maintain firm footholds, conduct protracted operations, and continually grow in strength. Thanks to those local bases, the vast liberated zones in South Vietnam, and the immediate rear areas, our army was able to build and consolidate itself after every battle, after every campaign, and was able to continuously prepare for new battles and new campaigns.[14]

Thus, over time, the balance of forces not only shifted in the communists' favor, but Hanoi was enabled to achieve victory through a conventional coup de main launched from both inside and outside the RVN. Although the communists failed to destroy pacification and were unable to overturn predominant RVN control over populated areas, NVA main-forces had, by 1975, "eliminated step by step isolated government outposts, most of which interfered in some degree with the communist plan for developing sparsely populated regions and securing the expanded and modernized logistical system supporting the rapidly growing expeditionary force in the South," according to historian William Le Gro. Hanoi and the NLF's success in expanding the insurgent state was significant because it gradually pushed the RVNAF out of select rural areas and transformed the defenses immediately surrounding South Vietnam's major population centers into "the forward line of contact" between the opposing forces.[15]

The communists were also victorious because their spatial strategy tied down large numbers of RVN forces through widespread incessant attack. According to Tran Van Tra, "It was the all-encompassing strategy of revolutionary warfare which caused the enemy troops to suffocate, to worry apprehensively day and night, and think that all places had to be defended and they could be safe only with large forces."[16] Graham Cosmas agrees that the protracted war strategy implemented by the communists allowed them to capitalize on the RVN's territorial weaknesses and gradually expand their hold on the countryside. According to Cosmas, "The communists, with an army that was smaller overall than the South Vietnamese army . . . could concentrate superior strength at any point of attack, and Saigon lacked mobile reserves to counter them. . . . Exploiting this advantage, the North Vietnamese step by step eliminated the outer layers of South Vietnam's defenses, securing their expanding logistical system and compressing the South Vietnamese back into the major population centers."[17]

One of the remarkable aspects of communist revolutionary warfare was that it recognized and manipulated the asymmetrical geostrategic commitments that are inherent within irregular wars. The spatial strategy put in place by Hanoi and the

National liberation Front forced Saigon to defend every inch of national territory and provide adequate military forces to protect a population dispersed throughout an incredibly diverse and complex operational environment. This stretched the South Vietnamese forces well beyond their capacity to effectively defend the nation against pervasive attack. Moreover, it robbed them of strategic mobility—something that proved fatal when Hanoi launched its massive conventional ground offensive in the spring of 1975. The communists, unlike the government of South Vietnam, had very little territory to defend. This stemmed, in part, from the evasive and rootless nature of guerrilla warfare. It was also a product of the narrow political objectives put in place by the United States and the Republic of Vietnam. As highlighted by Westmoreland and Andrade, the allies fought exclusively on the strategic defense during the Second Indochina War—they never invaded North Vietnam, nor did they ever plan to. The overall goal in Southeast Asia was carefully proscribed: to preserve an independent, noncommunist South Vietnam. It was not to destroy North Vietnam, liquidate Hanoi's military power, or overthrow the communist regime.

The self-limiting geostrategic commitments put in place by the Republic of Vietnam and the United States not only allowed communist forces to maintain a high level of mobility throughout the war; they permitted Hanoi and the National Liberation Front to husband their resources, choose the time and place of attack, and concentrate their forces at key areas. While these advantages failed to break the political and military stalemate during the years of American intervention, they proved decisive in the years following the withdrawal of U.S. combat forces and the termination of American military aid. Exhausted by years of war, overextended in its defense of territory and population, and forced to fight with diminishing material resources, the RVN and its military simply could not defend the national territory. The ability of Hanoi and the National Liberation Front to claim space and retain control over territory south of the seventeenth parallel proved to be a Trojan horse. In the end, the Saigon regime was simply overwhelmed by insurmountable strategic obligations and an enemy who successfully applied a brilliant geospatial strategy, one that succeeded in assembling and maintaining an insurgent antistate within the host nation and that overstretched and ultimately exhausted the enemy through a hybrid approach attacking incessantly across a broad multidirectional geographical front. "Few people ever really believed that the South Vietnamese alone would be able to do what the Americans and South Vietnamese together had never completely managed," one observer noted in 1975, "namely to defend the entire geographical border of South Vietnam."[18]

Spatial contestation and efforts to control politico-military space and through the strategic use of geography and the operational environment remain a fundamental tenet of irregular conflicts and wars of insurgency-counterinsurgency. The ongoing conflicts in Syria, Iraq, Gaza, Nigeria, Afghanistan, and the Philippines are local and regional conflicts where incumbent regimes are pitted against insurgents in a bitter struggle to control territory, population, and resources. Most, although not all, are also "hybrid" wars that involve a complex mixture of conventional combat operations, guerrilla warfare, terrorism, and grassroots political action. As such, they have developed a spatial composition very similar to the chaotic patchworks and vague territorial boundaries during the Second Indochina War. There are other similarities as well.

To begin with, the insurgencies in these regions, mostly Islamist terrorist organizations, have conquered large swaths of territory and established networks of terrorist base areas and safe havens across portions of the host nations. Insurgent zones not only provide these movements with the resources they need to survive, but serve as springboards for launching armed attacks against the indigenous governments, their armed forces, and local populations. The seizure of key cities and other territory has also allowed insurgent groups such as the Islamic State, Boko Haram, the Taliban, and Hamas to establish "state" functions and implement a religiopolitical weltanschauung (in this case a puritanical form of Sharia law) in occupied areas. Such measures are intended to serve the intermediate and long-range objectives of these organizations: first, to establish an Islamist anti-state within the national borders of host nations; and second, to create an Islamic caliphate that spans the greater Middle East and portions of western Europe, Asia, and northern Africa. Next, rather than contests to win the "hearts and minds" of indigenous populations in the Middle East, Africa, and Asia, these conflicts represent efforts to establish and expand politico-military control over population and territory through violence and raw coercive power. Although contemporary irregular wars are highly ideological contests between competing social, political, and religious systems, and the support of the people remains critical to long-term success, these conflicts are dominated by efforts to seize control of the civilian environment through military force, terrorism, and localized violence. Finally, the social grievances and antipathies that produce intrastate civil conflicts have been intensified by transnational ideologies and political movements that are grafted on to what are often local and/or regional clashes over power, identity, and resources. Overall, the similarities between the Second Indochina War and the wars in Syria, Iraq, and Afghanistan are outweighed by their dissimilarities (as is almost always the case with historical comparison). Nonetheless, the parallels between the struggle for South Vietnam and today's irregular wars are strong enough to warrant attention.

Protracted war strategies that fuse conventional combat, guerrilla warfare, and terrorism, and that rely on taking, consolidating, and growing insurgent territory as a means to overwhelm the state and seize power are as relevant to today as they were during the Cold War. Warfare in the twenty-first century has been dominated by small- to medium-sized conflicts that involve a complex mixture of conventional and unconventional means; intense competition to control population, territory, and resources within a single geographical area; and conflict zones where opposing forces are interspersed rather than separated behind traditional front lines. This is likely to continue in the years and decades to come. The relative decline of Western power and influence in recent years, coupled with a dramatic increase in identity politics, fundamentalism, millennialism, violent extremism, and economic disparity, and a rising sense of nationalism throughout the third world, will continue to generate violent clashes, including revolutions, independence movements, religious and ethnic sectarianism, political instability, and social and political competition, and military interventions by larger powers. Thus, it is vital we look to the past, particularly the Second Indochina War, for insight into the nature of hybrid wars of innumerable fronts.

NOTES

ABBREVIATIONS

CDEC	Combined Documents Exploitation Center
CMH	U.S. Army Center of Military History. Fort McNair, Washington, D.C.
CORDS	Civil Operations and Rural Development Support
CPDC	Central Pacification and Development Council
FRUS	*Foreign Relations of the United States*. U.S. Department of State.
MACV	U.S. Military Assistance Command, Vietnam
MHI	U.S. Army Military History Institute. Carlisle Barracks, Pa.
NARA II	National Archives and Records Administration. College Park, Md.
TTUVA	The Vietnam Center and Archive, Texas Tech University. Lubbock.
VNIT	Vietnam Interview Tapes, U.S. Army Center of Military History
VNDRN	*Vietnam Documents and Research Notes*. U.S. Mission in Vietnam.

INTRODUCTION

1. Elliott, *Vietnamese War*, 164.
2. McAlister, *Vietnam*, 4.

3. Jamieson, *Understanding Vietnam*, 233, x.
4. Smith, *Utility of Force*, 5.
5. Mao, *On Guerrilla Warfare*, 93.
6. Galula, *Pacification in Algeria*, 246.
7. Kissinger, "Vietnam Negotiations," 212; Dunn, "New Look at Pacification," 85.
8. Fall, "South Vietnam at the Crossroads," 142.
9. "Ambassador Komer's Final Press Conference," 3 November 1968, DepCORDS Files, U.S. Army Center for Military History (hereafter cited as CMH).
10. VSSG, "The Situation in the Countryside," 13 May 1970, 3–5; RG 472, A1 462, Box 17, Folder: 1601–04 VSSG Final Study #55, Folder 1, 1970, National Archives and Records Administration (hereafter cited as NARA II).
11. Bullington and Rosenthal, "South Vietnamese Countryside," 651.
12. Tran Dinh Tho, *Pacification*, 3.
13. Document No. 70, "The Party's Military Line is the Ever-Victorious Banner of People's War in Our Country," 32, U.S. Mission in Vietnam, *Vietnam Documents and Research Notes* (hereafter cited as VNDRN).
14. See Nguyen, *Hanoi's War.*
15. "Le Duan: Letters to the South, February 1965," in *Encyclopedia of the Vietnam War: A Political, Social, and Military History*, edited by Spencer C. Tucker (ABC-CLIO, 2011), http://search.credoreference.com/content/entry/abcvw/le_duan_letters_to_the_south_february_1965_excerpts/0.
16. A Resolution passed at the Ninth Plenum of the Central Committee of the Vietnam Workers' Party contains the following passage: "The general guideline for our people's revolutionary war in [South Vietnam] is to conduct a protracted war, relying mainly on our own forces, and to combine political struggle and armed struggle in accordance with each area and time." Document No. 96, "The Vietnam Workers' Party's 1963 Decision to Escalate the War in the South," 13, VNDRN.
17. Ibid., 14.
18. Quoted in Elliott, "Hanoi's Strategy in the Second Indochina War," 73.
19. "Speech Given by Party First Secretary Le Duan to the 12th Plenum of the Party Central Committee," 1 December 1965, History and Public Policy Program Digital Archive, ed. Pham Thi Vinh, Van Kien Dang Toan Tap, Tap 26, 1965 [Collected Party Documents, vol. 26, 1965] (Hanoi: National Political Publishing House, 2003), 593–96, trans. for CWIHP by Merle Pribbenow, http://digitalarchive.wilsoncenter.org/document/113970.
20. Document No. 102, "The People's Revolutionary Party of South Vietnam, Part I," 14, VNDRN.
21. Ibid., 26.
22. "Le Duan: Letter to the Central Office for South Vietnam, November 1965," in *Encyclopedia of the Vietnam War: A Political, Social, and Military History*, edited by Spencer C. Tucker (ABC-CLIO, 2011).

23. The official Department of Defense definition for operational environment is "A composite of the conditions, circumstances, and influences that affect the employment of capabilities and bear on the decision of the commander." According to the DoD, a holistic view of the operational environment included the physical area and factors, the information environment, and other factors such as weather and climate, sociocultural factors, and time. DoD Joint Publication 1–02, *Department of Defense Dictionary of Military and Associated Terms*, 8 November 2010, 175; DoD Joint Publication 2–01.3, *Joint Intelligence Preparation of the Operational Environment*, 16 June 2009, 1.2–1.4.
24. Harmon, Dillon, and Garver, "Perspectives on Military Geography," 8.
25. "Le Duan: Letter to the Central Office for South Vietnam, November 1965."
26. Lefebvre, *Production of Space*.
27. Lefebvre, "Reflections on Politics of Space," 32.
28. Ibid., 31.
29. Elden, *Understanding Henri Lefebvre*, 185.
30. Ibid., 190–91.
31. Lefebvre, "Reflections on Politics of Space," 33.
32. Foucault, *Discipline and Punish*, 231–56.
33. Foucault, "Space, Knowledge, and Power," 252.
34. Lefebvre, *Production of Space*, 26.
35. Pred, *Making Histories*, 5.
36. Sack, *Human Territoriality*, 19.
37. Cowen and Gilbert, "Politics of War, Citizenship, Territory," 16.
38. Mitchell, "End of Public Space?," 115.
39. Harmon, Dillon, and Garver, "Perspectives on Military Geography," 8.
40. Flint, *Geography of War and Peace*, 4–5.
41. Kalyvas, *Logic of Violence*.
42. The battlefield analogy used here was inspired by Stuart Elden's *Understanding Henri Lefebvre: Theory and the Possible*, 151.
43. Clarke, *Advice and Support*, 8–9.
44. Ibid.; Rottman, *Viet Cong and NVA Tunnels*, 7–8.
45. Mao, *Selected Military Writings*, 269.
46. Document No. 101, "The Provisional Revolutionary Government of the Republic of Vietnam, Part II: The Founding of the PRG," 73, VNDRN.
47. Hosmer, *Viet Cong Repression*, 1.
48. Liberation Press Agency, "Revolution Power Gains in Liberated Zone," 9 December 1970, Folder 02, Box 16, Douglas Pike Collection: Unit 05—National Liberation Front, The Vietnam Center and Archive, Texas Tech University (hereafter cited as TTUVA).
49. Mao, "On Protracted Warfare," in *Selected Military Writings*, 167–86.
50. McColl, "Insurgent State," 613–31.
51. "Le Duan: Letter to the Central Office for South Vietnam, November 1965."

52. Hoang Van Thai, *Decisive Years*, 18.
53. Westmoreland, *Soldier Reports*, 68.
54. House Committee on Appropriations, *Supplemental Appropriation Bill, 1971*, 1249.
55. Sorley, *Vietnam Chronicles*, 456.
56. MACV, "The Concept of Pacification and Certain Definitions and Procedures," 2 March 1965, 2, U.S. Army Military History Institute (hereafter cited as MHI).
57. Ibid., 1.
58. Vietnam Special Studies Group, "The VSSG Control Indicator," 41, NARA II; Hunt, "Strategies at War," 29–31.
59. Pike, *War, Peace, and the Viet Cong*, 168.
60. Huntington, "Bases of Accommodation," 643–44.
61. Gil Carl Alroy, "Winning People's Hearts and Minds Isn't Crucial to Winning a War," *Los Angeles Times*, 31 March 1968.
62. See Kalyvas and Kocher, "Dynamics of Violence in Vietnam," 335–55; Kocher, Pepinsky, and Kalyvas, "Aerial Bombing and Counterinsurgency," 201–18.
63. Kalyvas and Kocher, "Dynamics of Violence in Vietnam," 339.
64. Willbanks, *Abandoning Vietnam*, 2.
65. House Committee on Appropriations, *Department of Defense Appropriations for 1971*, 6.
66. Senate Committee on Foreign Relations, *Vietnam Policy and Prospects, 1970*, 348.
67. Cooper et. al., *American Experience with Pacification*, 3:278.
68. Goodman, "End of the War," 343.
69. McAlister and Mus, *Vietnamese and Their Revolution*; Fitzgerald, *Fire in the Lake*; Kolko, *Anatomy of a War*; Wiegersma, *Vietnam*; Jamieson, *Understanding Vietnam*; Hunt, *Vietnam's Southern Revolution*.
70. Hess, *Vietnam*, 122.
71. Palmer, *25-Year War*; Colby, *Lost Victory*; Sorley, *Better War*; Collins, "Vietnam Postmortem," 32–39.
72. See Catton, *Diem's Final Failure*; Ang Cheng Guan, *Vietnam War from the Other Side*; Elliott, *Vietnamese War*; Ang Cheng Guan, *Ending the Vietnam War*; Wiest, *Rolling Thunder in a Gentle Land*; Anderson and Lexington, *War That Never Ends*; Andrade, "Westmoreland Was Right," 145–81; Wiest, *Vietnam's Forgotten Army*; Hunt, *Vietnam's Southern Revolution*; Pribbenow, "General Vo Nguyen Giap," 1–33; Milam, *Not a Gentleman's War*; Asselin, "Revisionism Triumphant," 101–37. Daddis, *No Sure Victory*; Nguyen, *Hanoi's War*; Asselin, *Hanoi's Road to the Vietnam War*; and Daddis, *Westmoreland's War*.
73. *Foreign Relations of the United States* (hereafter cited as *FRUS*), *1964–1968*, 4:659.
74. U.S. Department of the Army, *Program for Pacification and Long-Term Development*, 1.
75. Central Intelligence Agency memorandum, "The Situation in Vietnam: Overview and Outlook," 24 January 1969, A-1, CIA-RDP80R01720R00600020056–9, CIA Records Search Tool (CREST) (accessed 14 December 2013), NARA II.

CHAPTER 1

1. Mao, *On Guerrilla Warfare*, 42.
2. Ibid.
3. Katzenbach, "Time, Space, and Will," 36–40.
4. Mao, "Strategy in Guerrilla War," 167.
5. Mao, *On Guerrilla Warfare*, 68.
6. Ibid.
7. Taber, *War of the Flea*, 20.
8. Thompson, "Squaring the Error," 445.
9. Document No. 81, "COSVN Resolution 14," 8, VNDRN.
10. Thayer, *War without Fronts*, 90–91.
11. Elliott, "Hanoi's Strategy in the Second Indochina War," 70.
12. Duncanson, "Pacification and Democracy," 414.
13. Lin, *Long Live the Victory*, 24–26.
14. Ibid., 48.
15. Vo Nguyen Giap, "The Political and Military Line of Our Party," in *Military Art of People's War*, 180.
16. Pike, *Viet Cong*, 270.
17. COSVN, Resolution 9, 75, MHI.
18. Southeast Asia (SEA) Analysis Report April–May 1970, 33, RG 472, A1 472, Box 6, Folder: Southeast Asia Analysis Reports, March–December 1970, NARA II.
19. Report, "VC/NVA Base Area Study," 1 January 1972, 2, RG 472, A1 531/Box 1, Folder: Base Area Studies 1970–73, NARA II.
20. Document No. 96, 35, VNDRN.
21. Vo Nguyen Giap, *Banner of People's War*, 49.
22. Ibid.
23. U.S. Department of the Army, *Program for Pacification and Long-Term Development*, 38–39; Pribbenow, *Victory in Vietnam*, 82.
24. Davison, *Some Observations*, 77.
25. Elliott, *Pacification*, 33.
26. Nguyen Hoai, "Expansion of Liberated Zones Reemphasized," 9 December 1969, Folder 16, Box 15, Pike Collection: Unit 05—National Liberation Front, TTUVA.
27. McColl, "Insurgent State," 614.
28. Lin, *Long Live the Victory*, 25.
29. Kernig, *Marxism, Communism, and Western Society*, 279.
30. Document No. 93, "Viet-Cong Political Geography of South Vietnam," 7–8, VNDRN.
31. Anderson, Arnsten, and Averch, *Insurgent Organization and Operations*, 6.
32. Document No. 93, 11–14, VNDRN.
33. Ibid., 26.

34. RVNAF/MACV, Strategic Research and Analysis Branch, Combined Intelligence Center, Vietnam, "VC Political Infrastructure," 5, RG 472, A1 474, Box 7, Folder: #101080—VC Political Infrastructure, NARA II.
35. "Le Duan: Letter to the Central Office for South Vietnam, November 1965."
36. Document No. 70, 44, VNDRN.
37. Document No. 96, 18, VNDRN.
38. Ibid.
39. Le Duan, "Political and Military Forces," 223.
40. Duiker, *Communist Road to Power*, 226–27.
41. Document No. 70, 31–32, VNDRN.
42. Document No. 96, 18, VNDRN.
43. Pribbenow, *Victory in Vietnam*, 83.
44. Huntington, "Bases of Accommodation," 648.
45. Joiner, "South Vietnam," 153–54.
46. Document No. 45, "The Process of Revolution and the General Uprising," 11, VNDRN.
47. Ibid., 23.
48. Elliott, "Hanoi's Strategy in the Second Indochina War," 71.
49. Pike, *PAVN*, 213.
50. Davidson, *Secrets of the Vietnam War*, 18.
51. Pike, *PAVN*, 215–30.
52. Ibid., 222.
53. Elliott, *Pacification*, 100.
54. Pike, *PAVN*, 233.
55. MACV, Combined Intelligence Center, Vietnam, "The Enemy System, Ben Tre Province," MHI.
56. RVNAF/MACV, Strategic Research and Analysis Branch, Combined Intelligence Center, Vietnam, "VC Political Infrastructure," 1, NARA II.
57. Document No. 40, "The Central Office of South Vietnam," 1, VNDRN.
58. RVNAF/MACV, Strategic Research and Analysis Branch, Combined Intelligence Center, Vietnam, "VC Political Infrastructure," NARA II.
59. VSSG, "Situation in the Countryside," 22, Final Study #55, NARA II.
60. Anderson, Arnsten, and Averch, *Insurgent Organization and Operations*, 8–9; Allen, Carver, and Elmore, *Estimative Products on Vietnam*, 440–44, 480.
61. MACV, Combined Intelligence Center, Vietnam, "The Enemy System, Ben Tre Province," 3, MHI.
62. Elliott, *Pacification*, 35–36.
63. Liberation Radio Directs Establishment of Revolutionary Administrations to Enlarge PLAF Base Areas, 3 May 1972, Folder 05, Box 05, Pike Collection: Unit 05—National Liberation Front, TTUVA.
64. Galula, *Counterinsurgency Theory and Practice*, 56.

65. Mao, *Selected Military Writings*, 170–71.
66. Vo Nguyen Giap, "The War of Liberation, 1945–1954," in *Military Art of People's War*, 87–88.
67. Galula, *Counterinsurgency Theory and Practice*, 85–86.
68. Mao, "Strategy in China's Revolutionary War," in *Selected Military Writings*, 138.
69. Truong Chinh, *Primer for Revolt*, 141.
70. Vo Nguyen Giap, *Military Art of People's War*, 88.
71. Thayer, "Patterns of French and American Experience in Vietnam," 20.
72. Taylor, *Swords and Plowshares*, 405.
73. For examples of this line of argument see Fall, *Two Vietnams*, 133; Pike, "Guerrilla Warfare in Vietnam," 49; and Race, *War Comes to Long An*.
74. Document No. 81, 21, VNDRN.
75. Hoang Van Thai, *Some Aspects of Guerrilla Warfare*, 17.
76. Document No. 77, "Under the Glorious Party Banner, for Independence, Freedom, and Socialism, Let Us Advance and Achieve New Victories," 5, VNDRN.
77. Ibid., 42.

CHAPTER 2

1. Tran Dinh Tho, *Pacification*, 5.
2. Vietnam Information Notes, "Pacification in Vietnam," July 1969, 1, Folder 16, Box 02, Glenn Helm Collection, TTUVA.
3. Robert W. Komer, "The Pacification Program in Perspective," n.d., 4, DepCORDS Files, CMH.
4. Pamphlet, "The Declaration of Honolulu," March 1966, Folder 15, Box 05, Pike Collection: Unit 01—Assessment and Strategy, TTUVA.
5. Civil Operations and Rural Development Support (hereafter cited as CORDS), Report, "Revolutionary Development Concept—Record of MACV Part 1," n.d., 1, Folder 0804, Box 0029, Vietnam Archive Collection, TTUVA.
6. MACV, *Handbook for Military Support of Pacification*, February 1968, 4–5, Folder 08, Box 04, Pike Collection: Unit 11—Monographs, TTUVA.
7. Tran Dinh Tho, *Pacification*, v.
8. Cao Van Vien and Dong Van Khuyen, *Reflections on the Vietnam War*, 156.
9. Memo, "Revolutionary development in Vietnam," 1966, 16, Folder 09, Box 05, Pike Collection: Unit 01—Assessment and Strategy, TTUVA.
10. Senate Committee on Foreign Relations, *Vietnam Policy and Prospects, 1970*, 49.
11. Pamphlet, "Pacification," 4, Robert Montague Papers, Vietnam Pacification 1968–1971, Box 7, Folder: 1969, MHI.
12. "Revolutionary development in Vietnam," 1966, 16, TTUVA.
13. Memo, "Draft Plan for People's Self Defense," RG 472, A1 462, Box 4, Folder: PSG Studies, 1968, Book I, Folder IV, 1968, NARA II.

14. Sharp and Westmoreland, *Report on the War in Vietnam*, 235.
15. Republic of Vietnam, CPDC, *1969 Pacification and Development Plan*, "Annex XII: Geographic Areas of Precedence," 1–2, MHI.
16. Memo, Train to Brown, 24 June 1975, *FRUS, 1969–1976,* 10:966; Cao Van Vien and Dong Van Khuyen, *Reflections on the Vietnam War*, 114, 134.
17. Truong Ngo Quang, *Territorial Forces*, 16–17.
18. MACV, *Handbook for US Forces in Vietnam*, December 1968, 2, Folder 11, Box 04, United States Armed Forces Manuals Collection, TTUVA.
19. CORDS, *1970 Vietnamese Village Handbook for Advisors*, 2 May 1970, 6–8, Folder 03, Box 01, Pike Collection: Unit 03—Civil Operations, Revolutionary Development Support, TTUVA.
20. Richard C. Kriegel, "Vietnamese Attitudes and Behavior Related to Management Problems of the Revolutionary Development Program," Research Report #88, March 1969, 8, RG 472, A1 462, Box 12, Folder: 1601–10A, Comments & Res. Rept., VN Attitudes & Behavior Related to Mgmt. Problems, RD Program, 1969, NARA II.
21. CORDS Reports and Analysis Directorate, *Hamlet Evaluation System Handbook*, June 1969, 5–6, MHI.
22. Kriegel, "Vietnamese Attitudes and Behavior," 14, NARA II.
23. Hickey, *Village in Vietnam*, 19.
24. MACV, *The 1971 Vietnam Village Handbook for Advisors*, 24 June 1971, 6–7, Folder 04, Box 01, Pike Collection: Unit 03—Civil Operations, Revolutionary Development Support, TTUVA.
25. CORDS, "Civil Operations and Rural Development Support—Revolutionary Development Concept—Record of MACV Part 1," n.d., 1, TTUVA.
26. Pacification Studies Group, "Binh Dinh Province—The Challenge—1971," 12 June 1971, 1, Province Reports, II CTZ, Kontum & Binh Dinh/Folder: 07—Binh Dinh, CMH.
27. ODASD Regional Programs Report, "The Vietnam Conflict—1972 and Beyond, February 1973, 17, RG 472, A1 210, MACV, Box 5, Folder: Booklet, The Vietnam Conflict—1972 and Beyond, ODASD (SA), February 1973, NARA II.
28. VSSG, "Situation in the Countryside," 1, Annex A, Kien Hoa Province, Final Study #55, NARA II.
29. Pacification Study Group, Study, "Evaluation Report: Hamlet Regression," 23 August 1968, 1–2, RG 472, A1 462, Box 5, Folder: PSG Studies, 1968, Book II, Folder II, NARA II.
30. Ellsberg, *Day Loc Tien Was Pacified.*
31. William Westmoreland, "Final Report on Pacification," 1969, Pike Collection, Box 10, Folder 5, TTUVA, http://www.vietnam.ttu.edu/virtualarchive/items.php?item=2131005006.
32. Thompson and Frizell, *Lessons of Vietnam*, 232.

33. Thayer, *War without Fronts*, 138–41.
34. Brigham, "Pacification Measurement," 49.
35. CORDS, *User's Guide to the Hamlet Evaluation System 1970 (HES/70)*, February 1970, 2, MHI; Komer, *Impact of Pacification on Insurgency*, 9.
36. VSSG, "The VSSG Control Indicator," 41, RG 472, A1 473, Box 6, Folder: 1601–11A, VSSG, 1969–1972, NARA II.
37. VSSG, "Situation in the Countryside," 2, Final Study #55, NARA II.
38. Ibid., 10.
39. Memo, "National Security Study Memorandum (NSSM) One," March 1969, 15–5, Folder 07, Box 01, Pike Collection: Unit 01—Assessment and Strategy, TTUVA.
40. VSSG, "Situation in the Countryside," 4–5, Final Study #55, NARA II.
41. House Committee on Foreign Affairs, *Measuring Hamlet Security in Vietnam*, 5.
42. Schultz, "Pacification-Vietnamization Strategy," 62.
43. Kolko, *Anatomy of a War*, 241.
44. House Committee on Foreign Affairs, *Measuring Hamlet Security in Vietnam*, 9.
45. Daddis, *No Sure Victory*, 121.
46. Komer, *Impact of Pacification on Insurgency*, 10.
47. House Committee on Foreign Affairs, *Measuring Hamlet Security in Vietnam*, 8.
48. Bole and Kobata, *Hamlet Evaluation System*, xi.
49. Corson, *Betrayal*, 237.
50. Daddis, *No Sure Victory*, 120.
51. Bole and Kobata, *Hamlet Evaluation System*, 34.
52. Colby, *Lost Victory*, 263.
53. Sorley, *Vietnam Chronicles*, 123.
54. Although literally hundreds of documents containing such language can be found, some examples include Document No. 113, *COSVN's Directive 02/73 "On Policies Related to the Political Settlement and Ceasefire,"* 14, 16, VNDRN; and "The New Stage of Development of the Revolution in South Vietnam," *Hoc Tap*, January 1975, Folder 04, Box 31, Pike Collection: Unit 01—Assessment and Strategy, TTUVA.
55. Vietnam Information Notes, "Pacification in Vietnam," July 1969, 7, TTUVA.
56. Thayer, "On Pacification," 234–35.
57. Bole and Kobata, *Hamlet Evaluation System*, 109.
58. Kinnard, *War Managers*, 108.
59. Bole and Kobata, *Hamlet Evaluation System*, xi–xii.
60. Komer, *Current Pacification Program in Vietnam*, 15.
61. Komer, *Impact of Pacification on Insurgency*, 9–10.
62. CORDS, "Revolutionary Development Concept," 1, TTUVA.
63. Report, "The Principles and Application of Area Security," 10, RG 472, A1 455, Box 4, Folder: Area Security, NARA II.
64. Hickey, *Accommodation and Coalition*, 37.

65. MACV, "Commander's Summary of the MACV Objectives Plan," 22–23, RG 472, A1 455, Box 4, Folder: Readings Files, Misc., Folder II, NARA II.
66. For a historical overview of how the U.S. military incorporated these elements into its counterinsurgency operations in Vietnam and elsewhere, see Birtle, "Persuasion and Coercion," 45–53.
67. MACV, Memo, "Tactics and Techniques for Employment of U.S. Forces in the Republic Of Vietnam," 17 September 1965, 2–3, Folder 03, Box 03, Larry Berman Collection, TTUVA.
68. RVNAF Joint General Staff and MACV, *1969 Combined Campaign Plan*, 23 October 1968, 5, RG 472, A1 120, Box 2, Folder: Combined Campaign Plan, 1969 (pt. 1 of 2), NARA II.
69. MACV, *Guide for Province and District Advisors*, 1 February 1968, 2.3, Folder 32, Box 01, Robert M. Hall Collection, TTUVA.
70. RVNAF Joint General Staff and MACV, *1969 Combined Campaign Plan*, 5, NARA II.
71. Republic of Vietnam, CPDC, *1969 Pacification and Development Plan*, "Annex I: Military Support 1969 Pacification and Development Plan," 1, MHI.
72. MACV, *Handbook for Military Support of Pacification*, 1, TTUVA.
73. Memo, W. W. Rostow to President Lyndon B. Johnson, Re: Cy Vance's Response to Senator Edward Kennedy, 1 February 1968, Folder 16, Box 09, Larry Berman Collection, TTUVA.
74. Senate Committee on Foreign Relations, *Vietnam Policy and Prospects, 1970*, 4.
75. Quoted in Cosmas, *MACV*, 245.
76. MACV, *Handbook for Military Support of Pacification*, 43, TTUVA.
77. Ibid., 43; CORDS, "Revolutionary Development Concept," 2–3, TTUVA.
78. Thayer, *War without Fronts*, 155.
79. Ibid., 156.
80. CORDS, "Revolutionary Development Concept," 4, TTUVA.
81. Memo, "Revolutionary Development/Pacification," n.d., 22, Folder 01, Box 03, John Donnell Collection, TTUVA.
82. CORDS, "Revolutionary Development Concept," 4, TTUVA.
83. CORDS, *1970 Vietnamese Village Handbook for Advisors*, 61–63, TTUVA.
84. Republic of Vietnam, CPDC, *1969 Pacification and Development Plan*, "Annex V: People's Self Defense Forces," 1, MHI.
85. Memo, "Utilization of People's Self-Defense Forces in the Counteroffensive," October 1968, 6, DepCORDS Files, CMH.
86. CORDS, *1970 Vietnamese Village Handbook for Advisors*, 63–65, TTUVA.
87. Pacification Studies Group, Evaluation Report, "National Police Field Force, Thua Thien Province, I CTZ," 13 September 1968, 1, RG 472, A1 462, Box 4, Folder: PSG Studies, 1968, Book II, Folder II, NARA II.
88. Vietnam Information Notes, "Pacification in Vietnam," July 1969, 3–4, TTUVA.

89. Republic of Vietnam, CPDC, *1970 Plan for Pacification and Development*, "Annex X: Revolutionary Development Cadre," 1, MHI.
90. Republic of Vietnam, CPDC, *1969 Pacification and Development Plan*, "Annex VII: RD Cadre," 2, MHI.
91. Donnell, "Pacification Reassessed," 567–69.
92. Earl J. Young, "Stability in Rural Vietnam," December 1968, 16, Folder 41, Box 01, John B. O'Donnell Collection, TTUVA.
93. Vietnam Information Notes, "Pacification in Vietnam," July 1969, 3, TTUVA.
94. Donnell, "Pacification Reassessed," 570.
95. CORDS, *1970 Vietnamese Village Handbook for Advisors*, 65, TTUVA.
96. Thayer, *War without Fronts*, 169.
97. RVNAF Joint General Staff and MACV, *1968 Combined Campaign Plan*, 23 March 1968, B-5, RG 472, A1 120, Box 2, Folder: Combined Campaign Plan, 1968 (pt. 1 of 2), NARA II.
98. Study: *South Vietnam's Internal Security Capabilities* [resulting from National Security Study Memorandum 19], May 1970, i, 48, Folder 16, Box 15, Pike Collection: Unit 01—Assessment and Strategy, TTUVA.
99. Office of the Assistant Secretary of Defense, Southeast Asia Programs (OASD [SA]), "Application of the Area Security Concept," January 1970, 22; OASD (SA), "Application of the Area Security Concept: Interim Report #2," February 1970, 14, RG 472, A1 472, Box 6, Folder: Southeast Asia Analysis Reports, October 1969—February 1970, NARA II.
100. Ibid.
101. Report, "Principles and Application of Area Security," 5–7, NARA II.
102. MACV, "Commander's Summary," 25, RG 472, A1 455, Box 4, Folder: Readings Files, Misc., Folder II, NARA II.
103. Ibid., 12–19.
104. Report, "Operational Guidance for Implementation of Territorial Security Concept," 30 November 1969, 2, RG 472, A1 733, Box 1641, Folder: 201–07, Misc. Reports, 1969, NARA II.
105. Cosmas, *MACV*, 138.
106. Thayer, *War without Fronts*, 144.
107. Ibid., 149.
108. Peter Osnos, "The War: Uneasy Lull, Fading Optimism," *Washington Post*, 1 August 1971.
109. Vietnam Information Notes, "Pacification in Vietnam," July 1969, 3, TTUVA.
110. For Diem's efforts at developing local self-government, see Catton, "Counterinsurgency and Nation Building," 918–40.
111. Republic of Vietnam, CPDC, *1970 Plan for Pacification and Development*, "Annex IV: Local Administration," 2, MHI.

112. Philip Fradkin, "South Vietnam Hamlets Hold Local Elections," *Los Angeles Times*, 23 March 1969.
113. Republic of Vietnam, CPDC, *1969 Pacification and Development Plan*, "Annex III: Strengthening Village and Hamlet Government," 6, MHI.
114. Republic of Vietnam, CPDC, *1971 Community Defense and Local Development Plan*, 4–5, RG 472, A1 473, Box 6, Folder: 1601–11A, CDLD Plan, 1971, NARA II.
115. Republic of Vietnam, CPDC, *1970 Plan for Pacification and Development*, "Annex VIII: Prosperity for All," 1, MHI.
116. Republic of Vietnam, CPDC, *1971 Community Defense and Local Development Plan*, 5–9, NARA II; Republic of Vietnam, CPDC, *1970 Plan for Pacification and Development*, "Annex VIII: Prosperity for All," 1–8, MHI.

CHAPTER 3

1. Scott, *Seeing Like a State*, 87–102. Scott argues that massive national development projects and state-managed social engineering projects such as Mao's Great Leap Forward, collectivization in Russia, and compulsory villagization in Tanzania, Mozambique, and Ethiopia were all products of high modernist ideology.
2. Jamieson, *Understanding Vietnam*, 233, x.
3. McAlister and Mus, *Vietnamese and Their Revolution*, 155.
4. "Liberation Is Double Speak," April 1971, 23, Folder 09, Box 04, Pike Collection: Unit 05—National Liberation Front, TTUVA.
5. McAlister, *Vietnam*, 10.
6. Lefebvre, *Production of Space*, 53.
7. Ibid., 53–59.
8. Document No. 110, "PRP Plans for Controlling 'Liberated Areas' in South Vietnam," 1, VNDRN.
9. Document No. 77, 89, VNDRN.
10. Document No. 102, 74, VNDRN.
11. For a treatise on the differences between rebellion, guerrilla warfare, and revolution, see Johnson, "Third Generation of Guerrilla Warfare," 435–47; Johnson, *Autopsy on People's War*; and Pike, *Viet Cong*, 32, 92–93.
12. Conley, "Communist Thought," 206.
13. Document No. 77, 104, VNDRN.
14. Vo Nguyen Giap, *South Vietnamese People Will Win*, 58–59.
15. "Consolidation, Protection of Southern Liberated Areas Urged," 16 August 1972, Folder 09, Box 24, Pike Collection: Unit 01—Assessment and Strategy, TTUVA.
16. Davison, *Some Observations*, 77.
17. "Expansion of Liberated Zones Reemphasized," December 1969, Folder 16, Box 15, Pike Collection: Unit 05—National Liberation Front, TTUVA.
18. Document No. 91, "The 19th Plenary Session of the Central Committee of the Vietnam Workers' Party and Its Reference Documents," 16, VNDRN.

19. Vo Nguyen Giap, *South Vietnamese People Will Win*, 59.
20. Document No. 91, 20, VNDRN.
21. "The Liberated Zone: Image of Future South Vietnam," May 1967, Folder 13, Box 15, Pike Collection: Unit 05—National Liberation Front, TTUVA.
22. Document No. 91, 36, VNDRN; "Consolidation, Protection, of Southern Liberated Areas Urged," 16 August 1972, Folder 09, Box 24, Pike Collection: Unit 01—Assessment and Strategy, TTUVA.
23. Vo Nguyen Giap, *South Vietnamese People Will Win*, 59–60.
24. "The Liberated Zone: Image of Future South Vietnam," Pike Collection, TTUVA.
25. "VNA Describes Life, Progress in 'Liberated Zones,'" 7 June 1973, Folder 07, Box 16, Pike Collection: Unit 05—National Liberation Front, TTUVA.
26. "Consolidation, Protection of Southern Liberated Areas Urged," Pike Collection, TTUVA.
27. "Political Program of the South Vietnam National Liberation Front" (unnumbered), *Text of the NLF Program*, 1967, VNDRN.
28. Ibid., 7.
29. Ibid., 7–14.
30. Pike, *Viet Cong*, 287.
31. Ibid., 271.
32. "The Binh Thuan Document," February 1974, Folder 10, Box 22, Pike Collection: Unit 04—Political Settlement, TTUVA.
33. Conley, "Communist Thought," 207–8.
34. Davison, *Some Observations*, 45–46.
35. Huntington, "Bases of Accommodation," 644–45.
36. Combined Document Exploitation Center (CDEC), Captured Enemy Document: "Attitudes in Viet Cong Controlled Areas," 22 May 1967, Folder 1271, Box 0154, Vietnam Archive Collection, TTUVA.
37. Report, "The Viet Cong Infrastructure Modus Operandi of Selected Political Cadres," July 1968, Folder 02, Box 03, Pike Collection: Unit 05—National Liberation Front, TTUVA.
38. Elliott, *Pacification and the Viet Cong System*, 32.
39. Kellen, *View of the VC*, 11.
40. Davison and Zasloff, *Profile of Viet Cong Cadres*, 41.
41. Pacification Studies Group, Evaluation Report, "VC/NVA Governments in 'Liberated' Areas, Quang Tri, Binh Long, Kontum, and Binh Dinh Provinces," 22 September 1972, 6, RG 472, A1 462, Box 41, Folder: 1601–10A, Evaluation: VC/NVA Governments in Liberated Areas, NARA II.
42. Davison, *Some Observations*, 77–94; Tom Buckley, "What Life's Like in Viet Cong Territory," *New York Times*, 23 November 1969.
43. Davison, *Some Observations*, 103.
44. Davison and Zasloff, *Profile of Viet Cong Cadres*, 41.

45. Davison, *Some Observations*, 103–22; "Viet Cong Laid a Heavy Hand on Village," *Washington Post*, 31 October 1969.
46. Davison, *Some Observations*, 122–30.
47. Ibid., 130–33.
48. Hosmer, *Viet Cong Repression*, 86–109; Davison, *Some Observations*, 139–49; Document No. 5, "Repression of Counter-Revolutionaries: The Viet Cong System of Punishment," 6, VNDRN.
49. Davison, *Some Observations*, 142.
50. Ibid., 148–49.
51. For statistical data and discussion of this issue, see Thayer, *War without Fronts*, 132–33; Kalyvas and Kocher, "Dynamics of Violence in Vietnam," 335–55; and Kocher, Pepinsky, and Kalyvas, "Aerial Bombing and Counterinsurgency," 201–18.
52. Kellen, *Conversations with Enemy Soldiers*, 85.
53. Ibid., 87.
54. Davison, *Some Observations*, 123–24.
55. Arthur J. Dommen, "Our Great Illusion about Peace in Vietnam," *Los Angeles Times*, 26 May 1968.
56. Kellen, *Conversations with Enemy Soldiers*, 82.
57. "Revolutionary Development in South Vietnam," 1966, 3, Folder 09, Box 05, Pike Collection: Unit 01—Assessment and Strategy, TTUVA.
58. Ibid., 4.
59. "Johnson Welcome to Saigon Leaders at Honolulu," *New York Times*, 7 February 1966.
60. "Honolulu Declaration and Communique," *New York Times*, 9 February 1966.
61. "The People in the Countryside Firmly Stand Up to Abolish Their Old Life and Build Up a New One!," 27 December 1966, Folder 02, Box 03, John Donnell Collection, TTUVA.
62. "Honolulu Declaration and Communique."
63. Military Cadre Administration: National RD Congress—1966, 27 December 1966, "Destroy Old Life to Build New Life," Folder 08, Box 04, Pike Collection: Unit 11—Monographs, TTUVA.
64. "Honolulu Declaration and Communique."
65. "Johnson Welcome to Saigon Leaders at Honolulu."
66. "Honolulu Declaration and Communique."
67. "The Great Society in Asia," *Wall Street Journal*, 22 April 1966.
68. "Declaration of Honolulu," March 1966, 7, Pike Collection, TTUVA.
69. "Johnson Statement on Honolulu Talks," *New York Times*, 9 February 1966.
70. "Revolutionary Development in South Vietnam," 4, TTUVA.
71. Tran Dinh Tho, *Pacification*, 160–61.
72. Ibid., 8.
73. Kriegel, "Vietnamese Attitudes and Behavior," 64–68, NARA II.

74. Republic of Vietnam, CPDC, *1970 Pacification and Development Plan*, 1, MHI.
75. Ibid., 9.
76. US MACV/RVNAF JGS, "Combined Strategic Objectives Plan, 1971," 11.2, Folder 01, Box 19, Pike Collection: Unit 01—Assessment and Strategy, TTUVA.
77. Ibid., 11.2–11.3.
78. For sociopolitical divisions in South Vietnam, see Dodd, "Faction and Failure," 173–78.
79. Southeast Asia (SEA) Analysis Report November–December 1970, 1; RG 472, A1 472, Box 6, Folder: Southeast Asia Analysis Reports, March–December 1970, NARA II.
80. Memo, Chambers to Colby, "A Personal Mid-Tour Report," 10 March 1970, 2, RG 472, A1 456, Box 11, Folder: 1601–11A, End of Tour Reports/1970, NARA II.
81. MACV, "Bi-Weekly Summary of VCI Activities," 4–17 January 1970, RG 472, A1 210, Box 1, Folder: Kien Hoa Province—Reports—1970–1972, NARA II.
82. Popkin, "Pacification," 664.
83. MACV, "Commander's Summary of the MACV Objectives Plan," 18, RG 472, A1 455, Box 4, Folder: Readings Files, Misc., Folder II, NARA II.
84. Memo, "Civilian Victims of Communist Terrorism, Republic of Vietnam: 1957–1974," RG 472, A1 531, Box 8, Folder: Terrorism Summary, 1974, NARA II.
85. For a full discussion on communist terror, including its purposes and modus operandi, see Hosmer, *Viet Cong Repression*.
86. Senate Committee on Foreign Relations, *Vietnam: May 1972*, 14; Pacification Studies Group, Evaluation Report, "Impact of the Enemy Offensive on Pacification," 5 October 1972, 1, RG 472, A1 462, Box 42, Folder: 1601–10A, Evaluation–Impact of Enemy Offensive on Pacification, 1972, NARA II.
87. Schultz, "Pacification-Vietnamization Strategy," 69.
88. Peter R. Kann, "Death Marches On," *Wall Street Journal*, 4 October 1972.
89. Pacification Studies Group, Evaluation Report, "RVNAF Depredation," 10 March 1972, RG 472, A1 462, Box 42, Folder: 1601–10A, Memos, Spot Reports, 1972; Pacification Research Report, "The ARVN Units' Conduct and the Popular Opinion towards Them," 7 November 1972, RG 472, A1 462, Box 42, Folder: Kien Hoa Province, 1971–1973, NARA II.
90. Pacification Studies Group, Evaluation Report, "Anti-American Demonstrations in Qui Nhon," 18 April 1971, 3–5, RG 472, A1 462, Box 117, Folder: PSG Studies 1971, Book I, Folder I, 1971, NARA II.
91. Report, "Opinions on the Reduction of U.S. Troops in Binh Duong," 17 November 1969, 3, RG 472, A1 462, Box 112, Folder: Binh Duong, 1969, NARA II.
92. Hickey, *U.S. Strategy in South Vietnam*, 2.
93. VSSG, "Situation in the Countryside," 29, Final Study #55, NARA II.
94. MACV, Strategic Research and Analysis Division, Directorate of Intelligence Production, "VC Efforts to Gain Popular Support," 3, MHI.

95. VSSG, Study, "The Revolutionary Development Program in Thua Thien Province," n.d., 23, RG 472, A1 462, Box 4, Folder: PSG Studies, 1968, Book I, Folder I, 1968, NARA II.
96. Donnell, "Expanding Political Participation," 690.
97. Cooper et. al., *American Experience with Pacification*, 1:7.
98. Steve Young, "Political Attitudes and Organization," n.d., 3, RG 472, A1 474, Box 70, Folder: 101984—The Situation in the Countryside: Binh Duong Province, NARA II.
99. Don Oberdorfer, "Doubt Clings to Saigon Efforts at Pacification," *Los Angeles Times*, 2 December 1969.
100. Kriegel, "Vietnamese Attitudes and Behavior," 39, NARA II.
101. Elliott, *Vietnamese War*, 1382.

CHAPTER 4

1. Kalyvas, *Logic of Violence*, 114.
2. Quoted in Document No. 51, "Let Us Be Grateful to Karl Marx," February 1969, 16, VNDRN.
3. Hoang Van Thai, *Some Aspects of Guerrilla Warfare*, 10.
4. Vo Nguyen Giap, *South Vietnamese People Will Win*, 53.
5. Document No. 102, 2, VNDRN.
6. Document No. 97, "Communist Repression, Part I," and Document No. 2, "Directive No. 167/AN, Security Agency, COSVN," September 1968, 2, VNDRN.
7. Document No. 96, 96, VNDRN.
8. COSVN, Resolution 9, 27–28, MHI.
9. Document No. 77, 29, VNDRN.
10. Hosmer, *Viet Cong Repression*, 16.
11. Mallin, *Terror in Vietnam*, 2.
12. Document No. 109, "Communist Aborted Plans for Seizure of Danang," December 1972, 30, VNDRN.
13. Hosmer, *Viet Cong Repression*, 31–36, 121–42; Republic of Vietnam Ministry of Foreign Affairs, Report, "The Communist Policy of Terror," December 1972, 16, Folder 13, Box 14, Pike Collection: Unit 05—National Liberation Front, TTUVA.
14. COSVN, Resolution 9, 45–47, MHI; "VC Terrorism, Kidnapping, Etc.," RG 472, A1 455, Box 9, Folder: Misc., Folder III, NARA II; Hosmer, *Viet Cong Repression*, 63–120.
15. Mao, *On Guerrilla Warfare*, 46.
16. Report, "COSVN Resolution 14 and Its Implications," 29 August 1970, RG 472, A1 456, Box 6, Folder: Viet Cong Documents and Resolutions 1968–1970, NARA II. See also Duiker, *Communist Road to Power*, 305–7; and Hunt, *Pacification*, 218.
17. CDEC, Captured Enemy Document: "Guerrilla Warfare Setback," RG 472, A1 456, Box 7, Folder: VC Documents 1969–1970, NARA II.

18. CDEC, Captured Enemy Document: "Tan Uyen District PRP Anti Pacification Program," 12 December 1968, Folder 07, Box 12, Pike Collection: Unit 01—Assessment and Strategy, TTUVA.
19. Document No. 81, 8–9, VNDRN.
20. Report, "Indicators of Enemy Activity Levels in Vietnam," *Southeast Asia Analysis Report Nov/Dec 1969*, 5, RG 472, A1 472, Box 6, Folder: General Records, Southeast Asia Analysis Reports, October 1969–November 1970, NARA II.
21. HQ Third Regional Assistance Command, Report, "The Military Region Overview for Period Ending 29 February 1972," 1–2, RG 472, P106, Box 6, Folder: 1601–11A, Field Overviews and Province Reports, February 1972, NARA II.
22. COSVN, Resolution 9, 19–20, MHI.
23. Report, "Campaign X and the Mini-War," 9, RG 472, A1 456, Box 6, Folder: Viet Cong, 1969–1970, NARA II.
24. Report, "Terrorism in South Vietnam," *Southeast Asia Analysis Report Sep/Oct 1970*, 8, RG 472, A1 472, Box 6, Folder: General Records, Southeast Asia Analysis Reports, October 1969–November 1970, NARA II.
25. Document No. 97, "Communist 'Repression' in South Vietnam—Part II," August 1971, 7–7, VNDRN.
26. Hosmer, *Viet Cong Repression*, 16.
27. CDEC, Captured Enemy Document: "Agency of the VC Quang Ngai Province Party Committee, VC Region 5," 23 September 1971, 19, Folder 01, Box 21, Pike Collection: Unit 01—Assessment and Strategy, TTUVA.
28. Hosmer, *Viet Cong Repression*, 18–19.
29. COSVN, Resolution 9, 44, MHI.
30. MACV Office of the Assistant Chief of Staff, Report, "Analysis of Viet Cong Infrastructure in VC Subregion 1," 28 November 1969, 9, MHI.
31. "Viet Cong Activity Plan—Vietnam," 1 July 1971, Folder 04, Box 18, Pike Collection: Unit 02—Military Operations, TTUVA.
32. "Communist 'Repression' in South Vietnam—Part I," 10, VNDRN.
33. Ibid., 4–5.
34. Report, "Terrorism in South Vietnam," *Southeast Asia Analysis Report June/July 1970*, 10, RG 472, A1 472, Box 6, Folder: General Records, Southeast Asia Analysis Reports, March 1970–December 1970, NARA II. According to the Systems Analysis Division of the Office of the Assistant Secretary of Defense, more than three-fourths (76–88%) of all terrorist victims were ordinary civilians.
35. Memo, "Terrorist Incidents in SVN," 7 June 1972, MACV Command Historian's Collection, Series II: Staff Sections, MACV J-3: CORDS, Reports, Briefings, 1967–1973, MHI.
36. CORDS, *1970 Vietnamese Village Handbook for Advisors*, 3, TTUVA.

37. Senate Committee on Foreign Relations, *Impact of the Vietnam War*, 18.
38. Thayer, *Systems Analysis View*, 3:185–90.
39. Mallin, *Terror in Vietnam*, 57.
40. Douglas Pike, "The Viet-Cong Strategy of Terror," February 1970, 28, Folder 06, Box 15, Pike Collection: Unit 01—Assessment and Strategy, TTUVA; Hunt, *Pacification*, 218–20.
41. John T. Wheeler, "Reds Bring Death, Terror to Hamlets," *Los Angeles Times*, 6 October 1968.
42. Ibid.
43. Hosmer, *Viet Cong Repression*, 131.
44. Davison, *Some Observations*, 26.
45. Shultz, "Pacification-Vietnamization Strategy," 53.
46. Report, "The Viet Cong Political Strategy," 25 December 1969, 4, MHI.
47. Report, "People's Impression of the Present Security Situation in Two Villages, Binh Duong Province," 8 August 1972, 1, RG 472, A1 462, Box 112, Folder: Binh Duong 1972–1973, NARA II.
48. Michael Parks, "War of Terror Goes Unabated," *Baltimore Sun*, 7 December 1970.
49. Republic of Vietnam Ministry of Foreign Affairs, "Communist Policy of Terror," 12–13.
50. West, *Area Security*, 6.
51. Ibid.
52. Herrington, *Silence Was a Weapon*, 39.
53. "Terror Becomes the Norm in Once-Model Viet Village," *Baltimore Sun*, 7 December 1970.
54. Kellen, *Conversations with Enemy Soldiers*, 54.
55. Davison, *Some Observations*, 16–20.
56. US MACV/RVNAF JGS, "Combined Strategic Objectives Plan, 1971," 5.3–5.4, Pike Collection, TTUVA; MACV, "The Viet Cong Guerrilla," 20 June 1969, 6, RG 472, A1 474, Box 43, Folder: 101557-The Viet Cong Guerrilla, NARA II.
57. Memo, "Civilian Victims of Communist Terrorism, Republic of Vietnam: 1957–1974," RG 472, A1 531/Box 8/Folder: Terrorism Summary 1974; and Memo, "VC Kidnapping, Assassination, and Terrorism," October 1970, RG 472, A1 456, Box 6, Folder: VC Terrorism 1968–1970, NARA II.
58. Thayer, *Systems Analysis View*, 10:112.
59. Report, "Binh Dinh Province Refugee Overview," 30 June 1969, 3, RG 472, A1 505, Box 7, Folder: 1618–05A, Refugee & Returnee Files, 1969, #4 of 17, NARA II.
60. CDEC, Captured Enemy Document: "Agency of the VC Quang Ngai Province Party Committee, VC Region 5," 23 September 1971, 23, Folder 01, Box 21, Pike Collection: Unit 01—Assessment and Strategy, TTUVA.
61. Report, "VC Efforts to Gain Popular Support," 12 November 1969, 3, MHI.
62. MACV, Report, "Terrorist Incidents in SVN," MHI.

63. Christopher Luella, *South Vietnamese Civilian Casualties Inflicted by Enemy through Terrorism*—Foreign Affairs Division Report, 19 May 1972, Folder 03, Box 44, Pike Collection: Unit 03—Statistical Data, TTUVA.
64. Senate Committee on Foreign Relations, *Vietnam Policy and Prospects, 1970*, 12.
65. MACV, "Terrorist Incidents in SVN," MHI.
66. Pacification Studies Group, Report, "Completed Ap Doi Moi Khanh Loc, Binh Duong Province," 4 February 1968, 5, RG 472, A1 462, Box 5, Folder: 1601–10A Completed Ap Doi Moi 1968, NARA II.
67. Advanced Research Projects Agency, *Village Defense Study—Vietnam*, 1:11.
68. "National Security Study Memorandum (NSSM) One," March 1969, 17–1, Pike Collection, TTUVA.
69. Thayer, *Systems Analysis View*, 1:24.
70. MACV, Report, "Viet Cong Political Strategy," 4, MHI.
71. MACV, Report, "VC Terrorism, Kidnapping, Assassination," RG 472, A1 456, Box 6, Folder: VC Terrorism 1968–1970, NARA II.
72. Nguyen Duy Hinh and Tran Dinh Tho, *South Vietnamese Society*, 105–6.
73. Republic of Vietnam, CPDC, *1970 Plan for Pacification and Development*, I-1–1, MHI.
74. Deputy for CORDS, II FFV/III CTZ, Pacification Research Report, "Popular Anxiety about the War in the 32nd DTA," 17 April 1970, 2, RG 472, A1 462, Box 112, Folder: Binh Duong 1970, NARA II. See also Elliott, *Vietnamese War*, 258–65; and Race, *War Comes to Long An*, 188. Although the discussion of insurgent terrorism and its effect on the rural population found within these two historical studies takes place prior to 1968, it is safe to assume that assassination, torture, and other forms of individualized violence had an equally strong impact on the attitudes and behaviors of the peasantry during the post-Tet period.
75. Anderson, *Columbia Guide to the Vietnam War*, 98–99. In February 1968, 2,810 bodies were discovered in mass graves after ARVN/U.S. forces had regained control of the city. An additional 3,000 residents were reported missing.
76. Pacification Studies Group, Report, "The Situation in the Countryside, Thua Thien Province," 10 March 1970, RG 472, A1 474, Box 70, Folder: 101981, The Situation in the Countryside: Thua Thien Province, NARA II.
77. Pacification Studies Group, Report, "Hamlet Regression," 23 August 1968, 10–11, RG 472, A1 462, Box 5, Folder: PSG Studies, Book II, Folder II, 1968, NARA II.
78. Elliott, *Vietnamese War*, 1380.
79. Arthur J. Dommen, "Viet Cong: Now Fish in a Hostile Sea," *Los Angeles Times*, 2 March 1969.
80. Thayer, *Systems Analysis View*, 3:201.
81. HQ Third Regional Assistance Command, "Military Region Overview for Period Ending 31 October 1971," 12 November 1972, 3, RG 472, P 106, Box 7, Folder: 1601–11A, Field Overview and Province Reports, October 1972, NARA II.

82. The argument that Abrams rejected Westmoreland's heavy-handed strategy for a more enlightened approach can be seen in Sorley, *Better War*; Nagl, *Learning to Eat Soup*; Corum, *Bad Strategies*; Colby, *Lost Victory*.
83. The number of combat operations battalion size and above launched in 1971 (715) was roughly half the number of similar-sized operations conducted in 1967 (1,451). Source: Office of the Secretary of Defense, Comptroller, "Unclassified Statistics on Southeast Asia," 267–72.
84. Birtle, *Army Counterinsurgency and Contingency Operations Doctrine*, 367.
85. The exact number of Vietnamese civilians killed during the Second Indochina War has been widely debated. According to Thayer, *War without Fronts* (125–29), the overall number of civilian casualties in South Vietnam during the period of direct American intervention was estimated to be between 1.2 million (195,000 killed and 1,030,000 wounded) and 1.3 million (415,000 killed and 935,000 wounded).
86. Although the numbers of refugees fluctuated over time throughout the war, as crises generated new refugees and U.S./RVN forces worked to resettle displaced persons, it was estimated that the war displaced more than 3 million people (almost 20 percent of South Vietnam's total population) between 1964 and 1970. According to CORDS, a refugee was defined as "people who have had to leave their houses and their means of livelihood to escape from communist pressures, from artillery or bombardment, or from the crossfire of war." Aside from refugees, well over a million South Vietnamese had seen their homes destroyed or had their lives disrupted by the war during this period. Source: Senate Committee on Foreign Relations, *Vietnam Policy and Prospects, 1970*, 215–16.
87. Special National Intelligence Estimate, "Capabilities of the Vietnamese Communists for Fighting in South Vietnam," 17 July 1969, in Allen, Carver, and Elmore, *Estimative Products on Vietnam*, 481.
88. Ibid., 475.
89. See MACV, *Handbook for Military Support of Pacification*, 4–5, TTUVA; Headquarters Department of the Army, *Field Manual (FM) 31-16 Counterguerrilla Operations*, March 1967, MHI; and RVNAF Joint General Staff and MACV, *1969 Combined Campaign Plan*, NARA II.
90. Headquarters 9th Infantry Division, Report, "LOI 43–69 (1969 Pacification and Development Plan)," 12 May 1969, 5, RG 472, A1 472, Box 3, Folder: 9th Infantry Division, Old Reliables, 1969, NARA II.
91. Report of the Department of the Army Review of the Preliminary Investigations into the My Lai Incident: Volume I of The Report of the Investigation, 14 March 1970, 9–1, Folder 12, Box 38, Pike Collection: Unit 03—War Atrocities, TTUVA.
92. Julian J. Ewell, Senior Officer Debriefing Report, 17 September 69, 12, available online from the Defense Technical Information Center (DTIC) at http://handle.dtic.mil/100.2/AD505846.

93. Memo, LTC Clifford Crosmun to J. P. Vann, "Interim Report, DMAC Fire Support Committee," 22 October 70, Folder: 1970, John Paul Vann Papers, MHI.
94. Sorley, *Vietnam Chronicles*, 12.
95. Ibid., 66–67.
96. Lewy, *America in Vietnam*, 97.
97. Thompson, *No Exit from Vietnam*, 129.
98. Quoted in Krepinevich, *The Army and Vietnam*, 222.
99. Julian J. Ewell, Senior Officer Debriefing Report, 17 September 69, 12. For Ewell's strategic and operational weltanschauung, see Ewell and Hunt, *Sharpening the Combat Edge*.
100. Kinnard, *War Managers*, 47.
101. "Report of the Chairman, Joint Chiefs of Staff, Gen. Earle G. Wheeler, on the Situation in Vietnam, February 27, 1968 (Extracts)," in Porter, *Vietnam*, 2:501–4.
102. "Colby Talking Paper," 1 May 1968, DepCORDS Files, CMH.
103. MACV, *Command History, 1968*, 1:521–22.
104. See Republic of Vietnam, CPDC, *1969 Pacification Campaign Guidelines*, MACV Command Historian's Collection, Series II: MACV Staff Sections, MACV J-3, CORDS, MHI; Republic of Vietnam, CPDC, *1970 Plan for Pacification and Development*; and Republic of Vietnam, CPDC, *1971 Community Defense and Local Development Plan*, Republic of Vietnam, CPDC, MHI.
105. 9th Infantry Division, Combat After Action Report, Operation Speedy Express, 14 June 1969, RG 472, Box 209, Folder: Speedy Express Report, 14 June 1969, NARA II.
106. Ibid., 43.
107. Ibid., 1.
108. Ibid., 39.
109. Operational Report Lessons Learned, Department of the Army Headquarters 1st Brigade 9th Infantry Division, sub: Operational Report for Quarterly Period Ending 31 January 1969, 9th Infantry Division Operational Reports Lessons Learned, 19–20, RG 472, Headquarters United States Army Vietnam, Command Historian, Box 43, Folder: ORLL Ending 31 January 1969, NARA II.
110. After Action Report, Speedy Express, 14 June 69, 39, 41.
111. Hunt, *9th Infantry Division in Vietnam*, 141–42.
112. After Action Report, Speedy Express, 14 June 69, 50.
113. Pacification Studies Group, Report, "Effects of 9th Division Redeployment on Dinh Tuong and Kien Hoa Provinces," 9 October 1969, 2, RG 472, A1 462, Box 8, Folder: PSG Studies, 1969, Book II (folder 2), NARA II.
114. Pacification Studies Group, Report, "Redeployment Effects of the 9th U.S. Division from Dinh Tuong and Kien Hoa Provinces," 15 March 1970, 2, RG 472, A1 462, Box 16, Folder: PSG Studies, 1970, Book I (folder 2), NARA II.
115. Kevin P. Buckley, "Pacification's Deadly Price," *Newsweek*, 19 June 1972.

116. For additional heavy-handed pacification operations, see Hunt, *Pacification*, 227–33; Lewy, *America in Vietnam*, 139–47; and Krepinevich, *The Army and Vietnam*, 256–57. For the correlation between the airstrikes/artillery bombardment and HES security ratings, see Kalyvas and Kocher, "Dynamics of Violence in Vietnam." Using quantitative analysis of HES, Kalyvas and Kocher illustrate that villages and hamlets with low HES scores (denoting predominant or total control by the NLF) were subject to artillery bombardment and airstrikes at a much higher rate than those that were under the predominant or total control of the RVN. Thayer also illustrates this phenomenon in *War without Fronts*, 129–34.
117. Pacification Studies Group, Evaluation Report, "Territorial Security in Huong Tra, Buong Thuy, and Phu Vang Districts, Thua Thein Province, I Corps Tactical Zone," 25 June 1968, 8, RG 472, A1 462, Box 4, Folder: PSG Studies, 1968, Book I (folder 2), NARA II.
118. Advisory Team 86, Province Report, "Long An Province, Period Ending 31 May 1969," RG 472, A1 495, Box 7, Folder: Province Reports, III CTZ, May 1969, NARA II; Peter Arnett, "A Decision in Vietnam: Death of a Hamlet," *New York Times*, 1 June 1969, 3.
119. MACV, Report, "CORDS Field Overview for the Month Ending 31 May 1969," 14 June 1969, and Report, "CORDS Field Overview for the Month Ending 30 June 1969," 13 July 1969, RG 472, A1 493, Box 1, Folder: 501–08, CORDS Overviews, April–June 1969 (folder 2 of 2), NARA II; Advisory Team 88, Province Report, Kien Hoa Province, [Period Ending 31 May 1969] 2 June 1969, RG 472, A1 495, Box 7, Folder: Province Reports, IV CTZ (Folder 1)/May 1969, NARA II; Assistant Director, CORDS/Refugee, IV CTZ to Deputy for CORDS, IV CTZ, Memo, "Apparent Increase in the Number of Refugees in IV CTZ," 25 June 1969, RG 472, A1 505, Box 9, Folder: 1618–05A, Refugee Questions, 1969, NARA II.
120. Scales, *Firepower in Limited War*, 148.
121. Hoang Ngoc Lung, *Strategy and Tactics*, 134.
122. "Military Region Overview for the Period Ending 31 October 1972," and "Military Region Overview for the Period Ending 30 November 1972," RG 472, P 106, Box 7, Folder: 1601–11A, Field Overviews and Province Reports, November 1972, NARA II; Advisory Team 91, "Province Report, Binh Duong Province, Period Ending 31 October 1972," and "Province Report, Binh Duong Province, Period Ending 30 November 1972," RG 472, A1 495, Box 27, Folder: Province Reports, III Corps, October–November 1972, NARA II.
123. "Minutes, National Security Council Meeting," 12 September 1969, *FRUS, 1969–1976*, 6:400.
124. House Committee on Appropriations, *Department of Defense Appropriations for 1971*, 95.
125. VSSG, "Situation in the Countryside," 13, RG 472, A1 462, Box 17, Folder: 1601–04A Situation in the Countryside, Draft #39, 1970, NARA II.

126. Denton, *Consequences of Bombardment of Civilian Areas*, 11–12.
127. Senate Committee on Foreign Relations, *Vietnam Policy and Prospects, 1970*, 637.
128. Tom Buckley, "What Life's Like in Viet Cong Territory," *New York Times*, 23 November 1969, SM48.
129. Memo, "Senior Advisor's Policy for Combat in Populated and/or Built-up Areas," April 1972, 3, Folder: 1972, John Paul Vann Papers, MHI.
130. Memo, "Use of Firepower," 11 November 1970, 1, RG 472, A1 700, Box 525, Folder: 1605–04A, Regional and Popular Force Advisory Files, 1970, NARA II.
131. Denton, *Consequences of Bombardment of Civilian Areas*, 7–10.
132. Memo, "John Paul Vann to Lt. Gen. Fred C. Weyand," 29 February 1968, Folder: 1968, John Paul Vann Papers, MHI.
133. DepCORDS Policy Statement, "Attacks on Pagodas, Churches or Populated Areas," 23 May 1969, Folder: DepCORDS Policy Statements, 1 February 1969–7 May 1971, John Paul Vann Papers, MHI.
134. Ibid.
135. "Senior Advisor's Policy for Combat in Populated and/or Built-up Areas," 2, Vann Papers, MHI.
136. Ibid., 7.
137. Nguyen Duy Hinh and Tran Dinh Tho, *South Vietnamese Society*, 109–10.
138. Ian Wright, "Hamlet under Siege," *Guardian*, 20 December 1968.

CHAPTER 5

1. Duiker, *Sacred War*, 208–10.
2. Tran Van Tra, "Tet," 40.
3. Ibid.
4. CORDS Field Overview for March 1968, 13 April 1968, 3, VNIT, Box 482 7–8, CMH.
5. Bradley, *Vietnam at War*, 152.
6. Tran Van Tra, "Tet," 60.
7. Joiner, "South Vietnam," 152.
8. Ibid.
9. "After the Tet Offensive," *New York Times*, 8 February 1968.
10. CBS News Special Report on Vietnam, 7 February 1968, quoted in Hammond, *Reporting Vietnam*, 121.
11. Herring, *America's Longest War*, 192, 201.
12. Max Frankel, "Westmoreland to Leave Vietnam, Head Army," *New York Times*, 23 March 1968.
13. Herring, *America's Longest War*, 196.
14. "Text of President Johnson's Speech to the Nation," *Baltimore Sun*, 1 April 1968.
15. A number of top officials, particularly Ambassador Robert Komer, were highly optimistic in the wake of the Tet Offensive. As such, they urged Washington and

allied commanders to take advantage of the weakened NLF and go on the offensive. See Memo, Komer to Bunker and Abrams, 16 July 1968, DepCORDS Files, CMH; "Westmoreland Requests 206,000 More Men, Stirring Debate in Administration," *New York Times*, 10 March 1968.

16. Davidson, *Vietnam at War*, 529.
17. Johnson's Wise Men included an impressive host of American political and military luminaries such as Dean Acheson, George Ball, McGeorge Bundy, Cyrus Vance, General Omar Bradley, General Maxwell Taylor, General Matthew Ridgway, Henry Cabot Lodge, and Abe Fortas.
18. Davidson, *Vietnam at War*, 524–25.
19. Clifford, "Vietnam Reappraisal," 613.
20. "Transcript of President Nixon's Address to the Nation on the War in Vietnam," *New York Times*, 15 May 1969.
21. Kimball, *Vietnam War Files*, 24.
22. Nguyen, *Hanoi's War*, 123.
23. Thompson and Frizell, *Lessons of Vietnam*, 84.
24. Nguyen, *Hanoi's War*, 130.
25. Document No. 39, "Indoctrination Notes on Peace Talks: A Call for 'Violent Revolution' to the End," 3, VNDRN.
26. Memo, Kissinger to Nixon, 15 April 1969, *FRUS, 1969–1976*, 6:204.
27. Nguyen, *Hanoi's War*, 130.
28. Duiker, *Communist Road to Power*, 307.
29. The essential elements of the communist negotiating platform can be seen in the National Liberation Front's "Principles and Main Content of an Overall Solution to the South Vietnam Problem—The Ten-Point Plan." See Murrey Marder, "VC Offer New Plan at Paris," *Washington Post*, 9 May 1969. See also Memo, Kissinger to Nixon, 3 July 1969, *FRUS, 1969–1976*, 6:1108–11.
30. Document No. 101, "The Provisional Revolutionary Government, Part IV: The PRG's Diplomatic Offensive," January 1972, 40, VNDRN.
31. Kissinger, *White House Years*, 979; emphasis in original.
32. Document No. 8, "The Position of North Vietnam on Negotiations," 5, VNDRN.
33. Nguyen, *Hanoi's War*, 194.
34. Memorandum of Conversation, 31 May 1971, *FRUS, 1969–1976*, 7:653.
35. Memo, Kissinger to Nixon, 20 July 1970, *FRUS, 1969–1976*, 6:1134.
36. Ibid.
37. Kissinger, *White House Years*, 974.
38. MACV, "Communist Plans for Coalition Government in the RVN," 16 May 1970, 1, RG 472, A1 455, Box 17, Folder: Reading Files, Folder I, NARA II.
39. Pike, *War, Peace, and the Viet Cong*, 131–32.

40. Document No. 101, "'The Provisional Revolutionary Government, Part I: Preparing to Form the PRG," January 1972, 20, and Document No. 101, "'The Provisional Revolutionary Government, Part IV: The PRG's Diplomatic Offensive," January 1972, 48, VNDRN.
41. MACV, "Communist Plans for a Coalition Government in the RVN," NARA II.
42. Document No. 20, "The New Situation and Mission: A Viet Cong Training Document," March 1968, 4, VNDRN.
43. Kissinger, *White House Years*, 269.
44. Memorandum, Kissinger to Nixon, 22 September 1970, *FRUS, 1969–1976*, 7:114.
45. Kissinger, "Vietnam Negotiations," 228.
46. Memorandum of Conversation, 26 June 1971, *FRUS, 1969–1976*, 7:762.
47. Memorandum of Conversation, 4 April 1970, *FRUS, 1969–1976*, 6:807, 821.
48. Editorial Note, Kissinger Summary of Conversation with Ambassador Dobrynin, 11 June 1969, *FRUS, 1969–1976*, 6:258.
49. Kissinger, *White House Years*, 282.
50. Nixon, *Memoirs*, 393.
51. "Address by Ambassador Ellsworth Bunker at the Council on Foreign Relations," 2 April 1969, DepCORDS Files, CMH.
52. "Nixon's Address to the Nation on the War in Vietnam."
53. William Tuohy, "Vietnam Cease Fire Could Cause Enormous Problems," *Los Angeles Times*, 1 November 1968.
54. Goodman, "South Vietnam," 110.
55. Minutes, National Security Meeting, 12 September 1968, *FRUS, 1969–1976*, 6:400.
56. Memo, Formation of the Office of Civil Operations, n.d., RG 472, A1 462, Box 89, Folder: 1601–11A, Ceasefire Planning, CD&LD, Folder I, 1973.
57. See Huntington, "Bases of Accommodation," 642–56; Kissinger, "Vietnam Negotiations," 211–34; and Samuel P. Huntington, "Getting Ready for Political Competition in South Vietnam," Southeast Asia Development Advisory Group Discussion Paper, 29 March 1969, Folder 06, Box 07, John Donnell Collection, TTUVA.
58. "Text of President's Vietnam Address," *Chicago Tribune*, 21 April 1970.
59. Memo, Sullivan to Kissinger, 27 June 1969, *FRUS, 1969–1976*, 6:278.
60. Memo, Kissinger to Nixon, 11 September 1969, *FRUS, 1969–1976*, 6:383.
61. Ibid., 385–86.
62. Kissinger, "Vietnam Negotiations," 226–27.
63. "Thieu Ends Rumors of Cease-Fire," *Baltimore Sun*, 1 August 1970.
64. Pauker, *Essay on Vietnamization*, 22.
65. Memorandum for the Record, DCI Meeting with the PFIAB, 6 August 1970, *FRUS, 1969–1976*, 7:17.
66. Ibid.; Alvin Shuster, "Thieu Says Peace Can't Be Obtained at Talks in Paris," *New York Times*, 1 November 1970.

67. Memo, "Cease-Fire Considerations," 9 November 1970, RG 72, A1 455, Box 5, Folder: Cease-Fire Considerations 1970, NARA II.
68. MACV, COMUSMACV OPLAN J-200 Ceasefire, 1 December 1971, B-2, RG 472, A1 210, Box 9, Folder: COMUSMACV OPLAN J 200 Ceasefire, December 1971, NARA II.
69. Ibid., B-3.
70. VSSG, "Situation in the Countryside," Annex B, Ceasefire, Final Study #55, NARA II.
71. The official designation for South Vietnam's four military regions changed during the war, from Corps Tactical Zone (CTZ) to Military Region (MR). Based on the earlier designation, they were also frequently referred to as Corps. Regardless of the name, the specific locations were the same and thus interchangeable: I CTZ = MR 1 = I Corps; etc.
72. Memo, Smyser to Kissinger, 18 June 1970, *FRUS, 1969–1976*, 6:1075–76.
73. "Text of President Nixon's Talk on Vietnam Proposals," *Boston Globe,* 8 October 1970.
74. Ralph Blumenthal, "Truce: A Jigsaw Puzzle," *New York Times*, 10 October 1970.
75. "The Ceasefire Proposal: Triumph of Ambiguity," *Washington Post*, 9 October 1970.
76. "Text of President Nixon's Talk," *Boston Globe.*
77. "Challenge to Hanoi," *New York Times*, 11 October 1970.
78. Murrey Marder, "Proposal Unlikely to Stir Foe," *Washington Post*, 8 October 1970.
79. "Nixon Seeks New Path to Indochina Peace," *New York Times*, 11 October 1970.
80. Alvin Shuster, "Saigon Backs Nixon's Proposals and Again Offers Free Election," *New York Times*, 9 October 1970.
81. Henry Giniger, "At Paris Talks, Nixon Plan Meets a 'Total' Rejection," *New York Times*, 16 October 1970.
82. "Three Interviews on Vietnam," 30 November 1970, 4, Folder 08, Box 13, Pike Collection: Unit 04—Political Settlement, TTUVA.
83. "In Order to Protect the People the PRG Cannot Extend Cease-Fire," 10 December 1970, Folder 12, Box 13, Pike Collection: Unit 04—Political Settlement, TTUVA.
84. Memorandum, Kissinger to Nixon, 23 November 1970, *FRUS, 1969–1976*, 7:186.
85. Harrison E. Salisbury, "Viet Cong Aide in Paris Is Adamant on Talks with Saigon," *New York Times*, 24 November 1969.
86. "Three Interviews on Vietnam," 19, Pike Collection, TTUVA.
87. "Issues and Current Status of the Paris Peace Talks," 7 October 1970, 49, Folder 11, Box 01, Pike Collection: Unit 04—Political Settlement, TTUVA.
88. "Nixon Seeks New Path."
89. Anatole Shub, "Reds Said to Fear Liquidation of VC during a Cease-Fire," *Washington Post*, 27 October 1970.
90. "To Protect the People," Pike Collection, TTUVA.
91. Shub, "Reds Said to Fear Liquidation."

92. "To Protect the People," Pike Collection, TTUVA; John Woodruff, "Viet Cong Already Recorded against Extended Ceasefire," *Baltimore Sun*, 8 December 1970.
93. "Hanoi Insisting upon September 17 Terms Again Rejects US Ceasefire Proposal," 30 November 1970, Folder 08, Box 13, Pike Collection: Unit 04—Political Settlement, TTUVA.
94. "Three Interviews on Vietnam," 4, Pike Collection, TTUVA.
95. "Nixon Seeks New Path."
96. Salisbury, "Viet Cong Aide in Paris."
97. Memo, Kissinger to Nixon, 22 August 1970, *FRUS, 1969–1976*, 7:53.
98. Document No. 109, 18, VNDRN.
99. Document No. 108, "PRP Abandons 1972 Plans for a Revolution in Saigon," November 1972, 28, VNDRN.
100. Craig Whitney, "His Answer Is Still Four No's: Thieu," *New York Times*, 22 October 1972.
101. Message, Haig to Kissinger, 16 October 1972, *FRUS, 1969–1976*, 9:163.
102. Message, Haig to Kennedy, 8 October 1972, *FRUS, 1969–1976*, 9:25–26.
103. Ibid., 119.
104. Memorandum of Conversation, 31 May 1971, *FRUS, 1969–1976*, 7:653.
105. Memorandum of Conversation, 26 June 1971, *FRUS, 1969–1976*, 7:762.
106. "Nixon's Report to Congress on Foreign Policy," *New York Times*, 19 February 1970.
107. House Committee on Appropriations, *Supplemental Appropriation Bill, 1971*, 1249.
108. "Nixon's Report to Congress."
109. Nixon, *Memoirs*, 392.
110. Larry A. Niksch, "Vietnamization: The Program and Its Problems," Congressional Research Service, 5 January 1972, 20, Folder 01, Box 19, Pike Collection, TTUVA.
111. Dong Van Khuyen, *RVNAF*, 18.
112. Thayer, *War without Fronts*, 34–35.
113. Ibid., 157.
114. Memo, Laird to Nixon, 13 March 1969, *FRUS, 1969–1976*, 6:115–16.
115. Sorley, *Vietnam Chronicles*, 184.
116. Memo, Laird to Nixon, 4 September 1969, *FRUS, 1969–1976*, 6:360–61.
117. House of Representatives, *Supplemental Appropriation Bill, 1971*, 1249.
118. House Select Committee on United States Involvement in Southeast Asia, *United States Involvement in Southeast Asia*, 71.
119. Hendrick Smith, "Nixon to Reduce Vietnam Force," *New York Times*, 9 June 1969.
120. Nguyen Duy Hinh, *Vietnamization and the Cease-Fire*, 22.
121. Robert B. Semple, "Nixon to Pull Out 150,000 from Vietnam in a Year," *New York Times*, 21 April 1970.
122. Nguyen Duy Hinh, *Vietnamization and the Cease-Fire*, 22.
123. Hunt, *Pacification*, 211.

124. Kissinger, *White House Years*, 272.
125. House Select Committee on United States Involvement in Southeast Asia, *United States Involvement in Southeast Asia*, 11.

CHAPTER 6

1. Ward Just, "Guerrillas Wreck Pacification Plan," *Washington Post*, 4 February 1968.
2. Memo, "Intelligence Note from the Director of the Bureau of Intelligence and Research (Hughes) to Secretary of State Rusk," 5 February 1968, *FRUS, 1964–1968*, 6:130.
3. "Report of the Chairman, Joint Chiefs of Staff, Gen. Earle G. Wheeler, on the Situation in Vietnam," 501–2.
4. Intelligence Memorandum, "Pacification in the Wake of the Tet Offensive in South Vietnam," 19 March 1968, 1, Folder 88, Box 06, Central Intelligence Agency Collection, TTUVA.
5. Advisory Team 3, "Monthly Report for Thuan Thien Province Period Ending 31 March 1968," 1, RG 472, A1 684, Box 96, Folder: Monthly Province Reports, March–June 1968, 221–08, NARA II.
6. Advisory Team 88, "Monthly Report for Kien Hoa Province," 5 March 1968, 1, RG 472, A1 495, Box 3, Folder: 1301–02, Province Reports, IV Corps/Folder II, February 1968, NARA II.
7. A 6 June 1968 special national security estimate reported that the RVN had exercised control over 1.1 million fewer rural inhabitants by the end of March than it had prior to Tet. The number of hamlets with a functioning government administration had also dropped, from 7,200 before 30 January to 4,400 by March. Special National Intelligence Estimate, "The Vietnam Situation," 6 June 1968, *FRUS, 1964–1968*, 6:763.
8. Pike, *Bunker Papers*, 385.
9. Memo, "Forsythe to Bunker," 4 April 1968, DepCORDS Files, CMH.
10. HQ II Field Force, CORDS Field Overview for April 1968, 13 May 1968, 1, RG 472, P 98, Box 4, Folder: 1601–09, Monthly Report, Folder I, April 1968, NARA II.
11. Advisory Team 96, CORDS Field Overview for Period Ending 30 April 1968, 13 May 1968, 1, RG 472, A1 583, Box 4, Folder: CORDS MR 4 Overviews ADV TM 96, 1968, NARA II.
12. Gene Roberts, "Komer Optimistic over Pacification Despite Foe's Drive," *New York Times*, 19 April 1968.
13. John Randolph, "U.S. Statistics Weigh Effect of Tet Offensive," *Los Angeles Times*, 19 April 1968.
14. *Bunker Papers*, 2:385–86.
15. Hendrick Smith, "Fear in Hamlets Is Still Hampering Pacification," *New York Times*, 29 March 1968.
16. Bernard Weinraub, "U.S. Admits Blow to Pacification," *New York Times*, 25 February 1968.

17. Nixon, *No More Vietnams*, 105.
18. *Bunker Papers*, 2:386–87.
19. Memo, "Forsythe to Bunker," 4 April 1968, DepCORDS Files, CMH.
20. "The War Starts Up Again," *Guardian*, 14 January 1969.
21. "Party First Secretary Le Duan to the 12th Plenum of the Party Central Committee," 1 December 1965.
22. Hoang Ngoc Lung, *General Offensives of 1968–69*, 126.
23. Document No. 8, 4, VNDRN.
24. John Randolph, "Hanoi Will Fight While Talking, Defector Says," *Los Angeles Times*, 21 May 1968.
25. Document Nos. 61–62, "'Decisive Victory': Step by Step, Bit by Bit," 5–6, VNDRN.
26. Terence Smith, "Strategy in Vietnam," *New York Times*, 23 May 1969.
27. Minutes of National Security Council Meeting, 25 January 1969, *FRUS, 1969–1976*, 6:40.
28. Enclosure, "Vietnam Policy Alternatives," July 1969, in Kimball, *Vietnam War Files*, 53.
29. Memo, Nixon to Kissinger, 1 February 1969, *FRUS, 1969–1976*, 6:52.
30. Nixon, *No More Vietnams*, 106.
31. Kissinger, *White House Years*, 261.
32. Nguyen, *Hanoi's War*, 2.
33. Ibid., 83.
34. According to CIA estimates, personnel losses in South Vietnam for communist forces were approximately 300,000 for 1968; a similarly high level was reported for the first half of 1969. Source: Special National Intelligence Estimate, "Capabilities of the Vietnamese Communists for Fighting in South Vietnam," 481.
35. *Bunker Papers*, 2:571–72.
36. CORDS Field Overview for November 1968, MR 4, 13 December 1968, 1, RG 472, A1 583, Box 4, Folder: CORDS MR 4 Overviews, ADV, Thu, 1968, NARA II.
37. CORDS Field Overview for November 1968, Military Region 2, 13 December 1968, 2, VNIT, Box 482 7–8, CMH.
38. CORDS Field Overview for November 1968, Military Region 3, 14 December 1968, 2, RG 472, P 98, Box 6, Folder: 1601–09 Monthly Report, Folder I, November 1968, NARA II.
39. CORDS Field Overview for November 1968, Military Region 1, 13 December 1968, 2, RG 472, P 63, Box 1, Folder: I CTZ CORDS Field Overviews January–December 1968, NARA II.
40. Sorley, *Vietnam Chronicles*, 79–80.
41. David Hoffman, "Troops in Delta Try Out Tactics for Small-Unit Warfare," *Washington Post*, 26 November 1968; Ian Wright, "NLF Keeps Finger on Trigger to Influence Paris Talks," *Guardian*, 3 December 1968.

42. MACV, "Campaign X and the Mini-War," n.d., 2–3, RG 472, A1 456, Box 6, Folder: Viet Cong 1969–1970, NARA II.
43. Testifying before Congress in February 1970, William Colby estimated that while the ratio of enemy forces was roughly 75 percent southerners and 25 percent NVA in 1965, these numbers had essentially reversed by 1970 with total enemy combat strength consisting of roughly 72 percent NVA and about 26 or 28 percent South Vietnamese. The accuracy of these numbers may be debatable; however, the overall trend in increasing numbers of NVA operating in the South and helping sustain the numerically depleted NLF insurgent organization after 1968 is not. Senate Committee on Foreign Relations, *Vietnam Policy and Prospects, 1970*, 198.
44. Pike, *PAVN*, 228.
45. Douglas Pike, "The 'Super Guerrilla' Concept: Oct. 1969 (Development of PLAF sapper teams)," October 1969, 1–4, Folder 10, Box 03, Pike Collection: Unit 05—National Liberation Front, TTUVA.
46. MACV, "COSVN Resolution 14 and Its Implications," 29 August 1970, 2, RG 472, A1 456, Box 6, Folder: Viet Cong Documents and Resolutions 1968–1970, NARA II.
47. Report, "Exploratory Examination of the Enemy's Recent Behavior," 24 November 1968, DepCORDS Files, CMH.
48. *Bunker Papers*, 2:626–27.
49. MACV, Strategic Research and Analysis Division, "Communist Strategy as Reflected in Lao Dong Party and COSVN Resolutions," 26 November 1969, 7–8, RG 472, A1 456, Box 6, Folder: Viet Cong Documents and Resolutions 1968–1970, NARA II.
50. CDEC, Captured Enemy Document: "Schemes to Implement the Pacification Plan and the Dissension within the U.S. Government," 23 July 1969, Folder 02, Box 14, Pike Collection: Unit 01—Assessment and Strategy, TTUVA.
51. MACV, Strategic Research and Analysis Division, "Communist Strategy as Reflected in Lao Dong Party and COSVN Resolutions," 8, NARA II.
52. Document No. 64, "Summer 1969: A Viet Cong Study of the Situation and Prospects," 8, VNDRN.
53. Document Nos. 61–62, 9, VNDRN.
54. Document No. 64, 8, VNDRN.
55. Hoang Ngoc Lung, *General Offensives of 1968–69*, 128.
56. Ang Cheng Guan, *Ending the Vietnam War*, 22.
57. "Campaign X and the Mini-War;" n.d., RG 472, A1 456, Box 6, Folder: Viet Cong 1969–70; Pacification Studies Group, "Study of Effects of COSVN 9," n.d., RG 472, A1 462, Box 16, Folder: PSG Studies 1970, Book II, Folder II, 1970, NARA II.
58. MACV, Briefing, n.d., RG 472, A1 474, Box 86, Folder: #102244, Significant Points from Captured COSVN "Absolute Secret" Document Winter Spring Campaign/1969–1970, NARA II.
59. COSVN, Resolution 9, 8, MHI.

60. MACV, "The Viet Cong Political Strategy," 25 December 1969, 2, MHI.
61. COSVN, Resolution 9, 12, MHI.
62. Ibid., 19–20.
63. Document No. 99, "The Study of COSVN Resolution 10," 13, VNDRN.
64. Document No. 81, 8, VNDRN.
65. Document No. 68, "Two Speeches by Giap," 35, VNDRN.
66. CDEC, Captured Enemy Document: "Directives at All Echelons to Disrupt the Pacification Program during Campaign X," 10 March 1970, Folder 04, Box 16, Pike Collection: Unit 01—Assessment and Strategy, TTUVA.
67. Guerrilla Warfare Movement in VC Military Region 5, 14 August 1970, 3–4, Folder 02, Box 04, Pike Collection: Unit 05—National Liberation Front, TTUVA.
68. CDEC, Captured Enemy Document: "Additional Activity Plan for Dry Season," 30 March 1971, 12, Folder 02, Box 20, Pike Collection: Unit 01—Assessment and Strategy, TTUVA.
69. COSVN, Resolution 9, 44–47, MHI; CDEC, Captured Enemy Document: Directive 33/CT70 on Counter-Pacification Activities, Current Affairs Committee, COSVN, 15 December 1970, 3, Folder 08, Box 18, Pike Collection: Unit 01—Assessment and Strategy, TTUVA.
70. Document No. 101, "The Provisional Revolutionary Government, Part I: Preparing to Form the PRG," 1, VNDRN.
71. Document No. 35, "'People's Revolutionary Councils' in Rural Areas," 3, VNDRN.
72. Memo, Helms to Kissinger, 24 January 1969, *FRUS, 1969–1976*, 6:12.
73. MACV, "Viet Cong Strategy and Tactics," MHI.
74. MACV, Combined Intelligence Center, "People's Liberation Committees: A Strategic Assessment," 25 November 1968, 2, RG 472, A1 474, Box 23, Folder: #101279—People's Liberation Committees: A Strategic Assessment, NARA II.
75. "'People's Revolutionary Councils' in Rural Areas," 3.
76. People's Admin Councils in Hamlets and Villages, August 1968, Folder 14, Box 08, Pike Collection: Unit 05—National Liberation Front, TTUVA.
77. "Viet Cong Strategy and Tactics."
78. "People's Liberation Committees," 15.
79. *Bunker Papers*, 2:603.
80. Republic of Vietnam Revolutionary Development Ministry National Training Center, "A Paper for Submission to the President of the Republic of Vietnam," 2 October 1968, Robert M. Montague Papers, Folder: 1968, MHI.
81. Arthur J. Dommen, "Viet Cong Reported Organizing Village 'Liberation Committees,'" *Los Angeles Times*, 28 October 1968.
82. Joseph B. Treaster, "U.S. Aides Say Viet Cong Held Elections Aimed at a Coalition," *New York Times*, 16 November 1968.
83. "Communist Strategy as Reflected in Lao Dong Party and COSVN Resolutions," 9.

84. Document No. 101, "The Provisional Revolutionary Government, Part II: The Founding Conference of the PRG," 65, VNDRN.
85. Ibid., 41.
86. Truong Nhu Tang, *Viet Cong Memoir*, 146–47.
87. MACV, Strategic Research and Analysis Division, "An Assessment of the Provisional Revolutionary Government," 26 October 1969, 1–2, MHI.
88. Nguyen, *Hanoi's War*, 140.
89. Drew Middleton, "Viet Cong and Their Allies Set Up Anti-Thieu Regime," *New York Times*, 11 June 1969.
90. "Saigon Dismisses Move," *New York Times*, 11 June 1969.
91. "The Provisional Revolutionary Government, Part II," 31–32.
92. Document No. 101, "The Provisional Revolutionary Government, Part III: The First Nine Months of the PRG," 41, VNDRN.
93. Ibid., 40.
94. Ibid., 50.
95. Ibid., 52.
96. CORDS estimates showed that the share of the population living in relatively secure areas had jumped from 47 percent (7.7 million people) in mid-1965 to 72 percent (11.9 million) by September 1967. Conversely, the share of people living in areas controlled by the NLF had dropped from 22 to 14 percent during the same period. Although Komer did state that MACV estimates for the "relatively secure" category were more modest (67 percent), these numbers still translated into an impressive gain in population security. Memo, Komer to Westmoreland, 11 November 1967, DepCORDS Files, CMH.
97. Memo, Komer to Abrams and Bunker, 16 July 1968, DepCORDS Files, CMH.
98. Colby, *Lost Victory*, 240.
99. Ibid., 253.
100. Memo, Colby to Khiem, 23 December 1968, DepCORDS Files, CMH.
101. Memo, "Ambassador Komer's Final Press Conference," 3 November 1968, DepCORDS Files, CMH.
102. *Bunker Papers*, 2:629.
103. Memo, "Remarks by President Thieu," 22 December 1968, DepCORDS Files, CMH.
104. Peter Braestrup, "Viet Pressure Keeps Up," *Washington Post*, 5 January 1969.
105. MACV, "Commander's Summary of the MACV Objectives Plan," 28–29, RG 472, A1 455, Box 4, Folder: Readings Files, Misc., Folder II, NARA II.
106. US MACV/RVNAF JGS, "Combined Strategic Objectives Plan, 1971," 5.2, Pike Collection, TTUVA.
107. Talking Paper, "Priorities for Pacification," 1 May 1968, DepCORDS Files, CMH.
108. *Bunker Papers*, 2:588.
109. Ibid., 616.

110. Memo, n.d., DepCORDS Files, CMH.
111. Memo, 23 November 1968, DepCORDS Files, CMH.
112. Memo, "Pacification," 16 December 1968, 2–3, DepCORDS Files, CMH.
113. CORDS Field Overview for November 1968, Military Region 2, 13 December 1968, 2, VNIT, Box 482 7–8, CMH.
114. CORDS Field Overview for November 1968, Military Region 3, 14 December 1968, 1, RG 472, P 98, Box 6, Folder: 1601–09 Monthly Report/Folder I, November 1968, NARA II.
115. CORDS Field Overview for November 1968, Military Region 4, 13 December 1968, 1, RG 472, A1 583, Box 4, Folder: CORDS MR 4 Overviews, ADV, Thu, 1968, NARA II.
116. CORDS Field Overview for November 1968, Military Region 1, 13 December 1968, 1, RG 472, P 63, Box 1, Folder: I CTZ CORDS Field Overviews January–December 1968, NARA II.
117. Memo, "The Accelerated Pacification Campaign," 4 February 1968, DepCORDS Files, CMH.
118. Memo, "Summary of APC 1 November—31 January," 28 February 1968, DepCORDS Files, CMH.
119. Republic of Vietnam, CPDC, *1969 Pacification and Development Plan*, 2, MHI.
120. Ibid., 2–3.
121. Southeast Asia (SEA) Analysis Report February 1970, 9, RG 472, A1 472, Box 6, Folder: Southeast Asia Analysis Reports, October 1969–February 1970; and VSSG, "Situation in the Countryside," 7–9, Final Study #55, NARA II.
122. Report, "Status of Pacification—31 December 1969—Final," RG 472, A1 474, Box 86, Folder: #102238 Status of Pacification Report for 1969, NARA II.
123. Minutes of National Security Council Meeting, 12 September 1969, *FRUS, January 1969—July 1970*, 6:400.
124. Quoted in Cosmas, *MACV*, 245.
125. House Committee on Appropriations, *Department of Defense Appropriations for 1971*, 95.
126. Komer, *"New Model" Pacification Program*, 90.
127. VSSG, "Situation in the Countryside," 13, Draft #39, NARA II.
128. Palmer, *Summons of the Trumpet*, 229.
129. Willbanks, *Abandoning Vietnam*, 70–76.
130. Ibid., 85.
131. Ibid., 115.
132. Cosmas, *MACV*, 343.
133. Southeast Asia (SEA) Analysis Report November/December 1970, 1, RG 472, A1 472, Box 6, Folder: Southeast Asia Analysis Reports/March–December 1970, NARA II.
134. Thayer, *War without Fronts*, 141–45.

135. Thayer, "On Pacification," 235.
136. Report, "Vietnam—Progress Despite the War," January 1969, 18–21. Folder 07, Box 01, Pike Collection: Unit 01—Assessment and Strategy, TTUVA.
137. House Committee on Government Operations, *U.S. Assistance Programs in Vietnam*, 3.
138. CORDS, Report, "Military Region Overview for the Month Ending 31 December 1971," 14 January 1972, CORDS Overviews 1971–1972, CMH.
139. Thayer, *War without Fronts*, 151–52; Joseph B. Treaster, "Delta Village Basks in New Prosperity," *New York Times*, 27 March 1972.
140. Robert S. Elegent, "South Vietnam Starts to Grow," *Los Angeles Times*, 16 April 1971.
141. Senate Committee on Foreign Relations, *Vietnam Policy and Prospects, 1970*, 224–27.
142. Ibid., 39.
143. Ibid., 218.
144. Subcommittee on Refugees and Escapees of the Senate Committee on the Judiciary, Statement of Ambassador William E. Colby, Deputy to COMUSMACV for Civil Operations and Rural Development Support, 21 April 1971, Folder 08, Box 31, Pike Collection: Unit 03—Refugees and Civilian Casualties, TTUVA.
145. House Select Committee on United States Involvement in Southeast Asia, *United States Involvement in Southeast Asia*, 75.
146. Senate Committee on Foreign Relations, *Vietnam Policy and Prospects, 1970*, 13.
147. A. S. Collins, Senior Officer Debriefing Report, 2, RG 472, A1 210, Box 10, Folder: Officer Debriefings 1971–1972, NARA II.
148. Spector, *After Tet*, 288.
149. Special National Intelligence Estimate, "Capabilities of the Vietnamese Communists for Fighting in South Vietnam," 499.
150. Brief, Intelligence: DIP—Re: Manpower Situation of Viet Cong Guerrilla Forces—Record of MACV, pt. 1, n.d., 1, Folder 0243, Box 0019, Vietnam Archive Collection, TTUVA.
151. Special National Intelligence Estimate, "The Outlook from Hanoi: Factors Affecting North Vietnam's Policies on the War in Vietnam," 5 February 1970, in Allen, Carver, and Elmore, *Estimative Products on Vietnam*, 509.
152. Ibid., 515.
153. CDEC, Captured Enemy Document: "Guerrilla Warfare Setback," 8 January 1970, 3–4, RG 472, A1 456, Box 7, Folder: VC Documents 1969–70, NARA II.
154. Interview DT (2)-71, 17, Reel 3, Interviews with Ralliers and Prisoners of War, David W. P. Elliott Collection of Vietnamese Communist Documents, Cornell University.
155. CDEC, Captured Enemy Document: "Weaknesses and Low Morale of Troops in VC Quang Da Special Zone, VC Military Region 5," 21 June 1971, 3–4, Reel 4, Interviews with Ralliers and Prisoners of War, Elliott Collection.
156. Pribbenow, *Victory in Vietnam*, 240, 247.

157. Parker, "Vietnam," 360–61.
158. Sorley, *Vietnam Chronicles*, 279.
159. Ibid., 280.
160. Memo, "Sir Robert Thompson's Report," 1, RG 472, A1 474, Box 71, Folder: #101991 Sir Robert Thompson's Views in Vietnam, January 1970, NARA II.
161. Senate Committee on Foreign Relations, *Vietnam Policy and Prospects, 1970*, 98.
162. Ibid., 100.
163. Nixon, *No More Vietnams*, 133, 140–41.
164. Thayer, *Systems Analysis View*, 9:265–70.

CHAPTER 7

1. House Committee on Armed Services, *Progress of the Pacification Program*, 5423.
2. Braestrup, "Viet Pressure Keeps Up."
3. Special National Security Estimate, "The Pacification Effort in Vietnam," 16 January 1969, 7, in Allen, Carver, and Elmore, *Estimative Products on Vietnam*.
4. "National Security Study Memorandum (NSSM) One," March 1969, 18–21, Pike Collection, TTUVA.
5. Memo, "Thoughts on GVN/VC Control," 2 April 1969, 2, John Paul Vann Papers, MHI.
6. Although he was discussing control vis-à-vis the ongoing negotiations in Paris, Kissinger made the following statement: "We would have been better off with 100 percent control over 60 percent of the country than with 60 percent control of 100 percent of the country." Kissinger, "Vietnam Negotiations," 214.
7. Memo, Kissinger to Nixon, 11 September 1969, *FRUS, 1969–1976*, 6:380.
8. Komer, *"New Model" Pacification Program*, 1242–44.
9. Special National Intelligence Estimate, "Capabilities of the Vietnamese Communists for Fighting in South Vietnam," 475.
10. "Army Activities Report: SE Asia," 20 December 1972, 40, MHI.
11. MACV, *Command History, 1971*, vol. 1, p. III-51.
12. Special National Intelligence Estimate, "Capabilities of the Vietnamese Communists for Fighting in South Vietnam," 501.
13. US MACV/RVNAF JGS, "Combined Strategic Objectives Plan, 1971," 4.6, Pike Collection, TTUVA.
14. Report, "VC/NVA Base Areas," *Southeast Asia Analysis Report April/May 1970*, 33, RG 472, A1 472, Box 6, Folder: General Records, Southeast Asia Analysis Reports, March 1970–December 1970, NARA II.
15. Pribbenow, *Victory in Vietnam*, 279.
16. CORDS, "DRAC Fact Book," 17 March 1972, 3, RG 472, A1 210, Box 4, Folder: Status Reports on P&D, CORDS, DRAC, Sent to Colonel D. E. Fanham, NARA II.
17. Peter Kann, "Death Marches On," *Wall Street Journal*, 4 October 1972.

18. Special National Intelligence Estimate, "Capabilities of the Vietnamese Communists for Fighting in South Vietnam," 475.
19. West, *Area Security*, 3.
20. Rowland Evans and Robert Novak, "Pacification Impossible in Vietnam while the Viet Cong's VCI Survives," *Washington Post*, 29 April 1970.
21. Document No. 99, 2–3, VNDRN.
22. VSSG, "Situation in the Countryside," 11–12, Final Study #55, NARA II.
23. Elliott, *Vietnamese War*, 1282–83.
24. US MACV/RVNAF JGS, "Combined Strategic Objectives Plan, 1971," 4.6, Pike Collection, TTUVA.
25. Advisory Team 18, "Monthly Province Report for Thua Thien Province Period Ending 31 May 1970," 4 June 1970, 2, RG 472, P 64, Box 5, Folder: I CTZ Province Reports, May 1970, NARA II.
26. "New Enemy Push Feared in Vietnam," *Baltimore Sun*, 7 May 1970.
27. Advisory Team 18, "Monthly Province Report for Thua Thien Province Period Ending 30 June 1970," 4 July 1970, 1–2, RG 472, P 64, Box 5, Folder: I CTZ Province Reports, June 1970, NARA II.
28. "Province Report, Binh Dinh Province, Period Ending 30 June 1970," 1–2, RG 472, A1 495, Box 14, Folder: Province Reports, II Corps, June 1970, NARA II.
29. Advisory Team 91, "Monthly Province Report for Binh Duong Province Period Ending 30 June 1970," 2, RG 472, A1 495, Box 14, Folder: Province Reports, III Corps, June 1970, NARA II.
30. "Province Report, Kien Hoa Province, Period Ending 31 May 1970," 2 June 1970, 1–2, RG 472, A1 495, Box 14, Folder: Province Reports, II Corps, June 1970, NARA II.
31. Pribbenow, *Victory in Vietnam*, 250.
32. Foreign Broadcast Information Service, Report, Vietnam: "The Anti-U.S. Resistance War for National Salvation 1954–1975: Military Events, 3 June 1982," 143–44, Folder 25, Box 17, Garnett Bell Collection, TTUVA.
33. Ang Cheng Guan, *Ending the Vietnam War*, 74.
34. Pribbenow, *Victory in Vietnam*, 283.
35. Truong Ngo Quang, *Easter Offensive*, 12–13.
36. Memo, Kissinger to Nixon, 26 April 1972, *FRUS, 1969–1976*, 8:329.
37. Memo, "Consolidation of GVN Control and Security of the Population," 25 July 1972, RG 472, A1 700, Box 528, Folder: 512–01A, PSA Counterpart Letters 1972, NARA II.
38. Pribbenow, *Victory in Vietnam*, 283–84.
39. Truong Ngo Quang, *Easter Offensive*, 157–58.
40. MACV, *Command History, 1972–1973*, 1:2.
41. Memo, Kissinger to Nixon, n.d., *FRUS*, 8:311.
42. Robert Keatley, "Current Hanoi Drive May Not Be as Vital as Thieu Has Declared," *Wall Street Journal*, 7 April 1972.

43. "Consolidation of GVN Control and Security of the Population," NARA II.
44. Andrade, *Trial by Fire*, 43.
45. Ang Cheng Guan, *Ending the Vietnam War*, 74.
46. Nguyen, *Hanoi's War*, 244.
47. Joseph Alsop, "Viet Year of Decision," *Washington Post*, 24 April 1972.
48. Memorandum of Conversation, 17 August 1972, *FRUS, 1969–1976*, 8:839.
49. Memo, Kissinger to Nixon, n.d., *FRUS, 1969–1976*, 8:308.
50. Memo, Kissinger to Nixon, 19 May 1972, *FRUS, 1969–1976*, 8:610.
51. CIA Memorandum, "Pacification in South Vietnam: A Preliminary Damage Assessment," 25 April 1972, 2, CIA-RDP80R01720R00600020056–9, CREST (accessed 14 December 2013), NARA II.
52. George McArthur, "With Saigon's Army Busy, Red Guerrillas Gain in Countryside," *Los Angeles Times*, 31 May 1972.
53. Pacification Studies Group, Evaluation, "Impact of the Enemy Offensive on Pacification," 16 September 1972, 2, RG 472, A1 462, Box 42, Folder: 1601–10A, Study/Evaluation, Enemy Offensive Impact on Pacification, 1972, NARA II.
54. CORDS Reports and Analysis Directorate, HES Summary Report(s) for 30 November 1971, 30 November 1972, and 31 January 1973, MHI.
55. "Impact of the Enemy Offensive on Pacification," 8.
56. MACV/CORDS, Brief, "Effects of the NVA Invasion on Pacification," n.d., 2, RG 472, A1 455, Box 16, Folder: HES Briefing Materials, NARA II.
57. "Effects of the NVA Invasion on Pacification," 10–11.
58. Memorandum, Carver to Kissinger, 30 October 1972, *FRUS, 1969–1976*, 9:334.
59. CIA Memorandum, "The Security Aspects of the Settlement Package," 16 October 1972, 1–3, CIA-RDP80R01720R00600020056–9, CREST (accessed 14 December 2013), NARA II.
60. Le Gro, *Vietnam from Ceasefire to Capitulation*, 28–30.
61. According to MACV, the number of communist forces reported inside South Vietnam in December of each year was as follows: 1965—230,000, 1966—280,000, 1967—228,000, 1968—259,000, 1969—237,000, 1970—230,000, and 1971—210,000. Source: MACV, Command Histories, 1965–71.
62. Nguyen Duy Hinh, *Vietnamization and the Cease-Fire*, 153.
63. Le Gro, *Vietnam from Ceasefire to Capitulation*, 28–30.
64. Mark Meredith, Viet Truce Could Bring 'Leopard Spot' Control," *Boston Globe*, 25 October 1972.
65. Isaacs, *Without Honor*, 23.
66. Rudy Abramson, "Despite Offensive, Reds Only Control about 2% of the Population," *Los Angeles Times*, 27 October 1972.
67. Joseph Treaster, "Foe Now Holds Large Part of 5 Northern Provinces," *New York Times*, 20 September 1972.

68. Abramson, "Reds Only Control about 2%."
69. CIA Memorandum, "The Security Aspects of the Settlement Package," 16 October 1972, 1–3, CIA-RDP80R01720R00600020056–9, CREST (accessed 14 December 2013), NARA II.
70. CIA Memorandum, "Perspectives on a Ceasefire in Vietnam," 8 September 1972, 2–3, CIA-RDP80R01720R00600020056–9, CREST (accessed 14 December 2013), NARA II.
71. Isaacs, *Without Honor*, 89.
72. Peter R. Kann, "Shape of Peace: Vietnam Accords Seem Likely but May Lead to a New Kind of War," *Wall Street Journal*, 14 November 1972.
73. Kissinger, "Vietnam Negotiations," 226.
74. "Thieu Ends Rumors of Ceasefire," *Baltimore Sun*, 1 August 1970.
75. Kann, "Shape of Peace."
76. Peter R. Kann, "Two Views in Vietnam," *Wall Street Journal*, 31 January 1975.

CHAPTER 8

1. Nguyen Duy Hinh, *Vietnamization and the Cease-Fire*, 162.
2. There were fewer than 70,000 U.S. troops in South Vietnam when the communists launched the Nguyen Hue Offensive in April 1972. By November, this number had dropped to 27,000. Source: Nguyen Duy Hinh, *Vietnamization and the Cease-Fire*, 27.
3. Ang Cheng Guan, *Ending the Vietnam War*, 102.
4. Nguyen, *Hanoi's War*, 260.
5. Ang Cheng Guan, *Ending the Vietnam War*, 111.
6. Note, Haig to Kissinger, 28 June 1972, *FRUS, 1969–1976*, 8:681.
7. Document No. 108, 28–29, VNDRN.
8. Document No. 109, 18, VNDRN.
9. Advisory Team 88, "Monthly Report for Kien Hoa Province, Period Ending 31 October, 1972," 3 November 1972, 1, RG 472, A1 495, Box 27, Folder: 1301–02, Province Reports, IV Corps, October 1972, NARA II.
10. Advisory Team 91, "Monthly Report for Binh Duong Province, Period Ending 31 October, 1972," 3 November 1972, 1, RG 472, A1 495, Box 27, Folder: 1301–02, Province Reports, III Corps, October 1972, NARA II.
11. Advisory Team 91, Monthly Report for Binh Duong Province, Period Ending 30 November 1972," 4 December 1972, 1, RG 472, A1 495, Box 27, Folder: Province Reports, III Corps, November 1972, NARA II.
12. Report, "MR 2 WIEU, Week Ending 8 November 1972," RG 472, A1 473, Box 10, Folder: 1601–11A, MR Overviews, June 1972—January 1973, NARA II.
13. Ibid.
14. Back-channel Message, Bunker to Kissinger, 26 July 1972, *FRUS, 1969–1976*, 8:767.
15. Message, Haig to Kennedy, 8 October 1972, *FRUS, 1969–1976*, 9:25.

16. Back-channel Message, Kissinger to Bunker, 12 October, 1972, *FRUS, 1969–1976*, 9:119.
17. Document No. 108, 27, VNDRN.
18. Editorial Note, *FRUS, 1969–1976*, 8:1076.
19. Kann, "Shape of Peace."
20. Back-channel Message, Bunker to Haig, 22 October 1972, *FRUS, 1969–1976*, 9:261.
21. Memo, Kissinger to Nixon, 20 December 1972, *FRUS, 1969–1976*, 9:774.
22. Thieu reportedly stated the following in October 1972: "Concerning South Vietnam our position is very unfortunate. We have been very faithful to the Americans and now feel that we are being sacrificed. . . . Now that you recognize the presence of North Vietnamese here, the South Vietnamese people will assume that we have been sold out by the U.S. and that North Vietnam has won the war." Ibid.
23. Nixon, *Memoirs*, 749–50.
24. Charles Corddry, "Viet Reds Said to Plan Final Grab," *Baltimore Sun*, 23 January 1973.
25. Document No. 110, 12, VNDRN.
26. "Agreement on Ending the War and Restoring the Peace in Vietnam" (copy), 24 January 197, RG 472, A1 473, Box 8, Folder: 228–08, Extract of Agreement to End War in Vietnam, January 1973, NARA II.
27. Ibid.
28. Parker, "Vietnam," 365.
29. "Agreement on Ending the War and Restoring the Peace in Vietnam."
30. Cao Van Vien, *Final Collapse*, 27.
31. Nguyen Duy Hinh, *Vietnamization and the Cease-Fire*, 139.
32. Herrington, *Peace with Honor?*, 57.
33. Cao Van Vien, *Final Collapse*, 37.
34. "Marxism, Communism and Western Society," January 1973, 278, Folder 04, Box 22, Pike Collection: Unit 02—Military Operations, TTUVA.
35. Tran Van Don, *Our Endless War*, 219.
36. Parker, "Vietnam," 353.
37. Peter Osnos, "On 9th Day of Ceasefire, Fight for Villages Goes On," *Washington Post*, 6 February 1973.
38. Document No. 115, "COSVN Directive 03," September 1973, 5, VNDRN.
39. Memorandum of Conversation, 20 December 1973, *FRUS, 1969–1976*, 10:492.
40. Brian Jenkins, "After the War," *Washington Post*, 1 April 1973.
41. Office of the Assistant Secretary of Defense (Systems Analysis) Regional Programs, "The Post-Ceasefire Period in Vietnam," April 1973, 1–2, RG 472, A1 210, Box 12, Folder: Assessment of the Situation Vietnam, Folder II, 1972–1973, NARA II.
42. Thomas Lippman, "Saigon Fights to Regain Land, People," *Washington Post*, 30 September 1973.
43. Senate Committee on Foreign Relations, *Vietnam: May 1974*, 1.
44. Hoang Van Thai, *Decisive Years*, 8–9.

45. Foreign Broadcast Information Service, Vietnam: "The Anti-U.S. Resistance War for National Salvation 1954–1975: Military Events, 3 June 1982," 161, Bell Collection, TTUVA.
46. Document No. 117, "Communist Guidance on 'New Phase' of the Revolution in South Vietnam," 12, VNDRN.
47. Isaacs, *Without Honor*, 86.
48. Defense Attaché Saigon, "RVNAF Quarterly Assessment, 4th Quarter FY 1973," 1.9, RG 472, A1 1727, Box 1, Folder: RVNAF Quarterly Assessment, 4th Quarter, FY 1973, NARA II.
49. Hosmer, Kellen, and Jenkins, *Fall of South Vietnam*, 50.
50. Lippman, "Saigon Fights to Regain Land, People."
51. "Viet Cong Reported to Attack Post 10 Miles from Saigon," *New York Times*, 3 September 1973; "Saigon Reports Heavy Fighting at Kontum and in the Hue Area," *New York Times*, 7 September 1973; "Fighting Increases in South Vietnam," *New York Times*, 9 September 1973.
52. Cao Van Vien, *Final Collapse*, 39.
53. National Intelligence Estimate, "Short-Term Prospects for Vietnam," 12 October 1973, 3, in Allen, Carver, and Elmore, *Estimative Products on Vietnam*, 600.
54. Senate Committee on Foreign Relations, *Vietnam: May 1974*, 1.
55. House Committee on Foreign Affairs, *Vietnam—a Changing Crucible*, 4.
56. Senate Committee on Foreign Relations, *Vietnam: May 1974*, 1, 4.
57. James M. Markham, "South Vietnam, a Year After the Truce, Is Still Racked by Indecisive War," *New York Times*, 27 January 1974.
58. Interagency Intelligence Memorandum, "South Vietnam: A New Military Assessment," 2 April 1974, B-2, in Allen, Carver, and Elmore, *Estimative Products on Vietnam*.
59. National Intelligence Estimate, "Short-Term Prospects for Vietnam," 23 December 1974, in Allen, Carver, and Elmore, *Estimative Products on Vietnam*, 633–43.
60. Defense Attaché Saigon, "RVNAF Quarterly Assessment, 3rd Quarter FY 1974," 9.12, RG 472, A1 1727, Box 2, Folder: RVNAF Quarterly Assessment, 3rd Quarter, FY 1974, NARA II.
61. Van Tien Dung, *Our Great Spring Victory*, 11–12.
62. Herrington, *Peace with Honor?*, 100.
63. House Committee on Foreign Affairs, *Vietnam—a Changing Crucible*, 8.
64. Parker, "Vietnam," 362.
65. Senate Committee on Foreign Relations, *Vietnam: May 1974*, 26.
66. Cao Van Vien and Dong Van Khuyen, *Reflections on the Vietnam War*, 114.
67. Herring, *America's Longest War*, 330–31.
68. Defense Attaché Saigon, "RVNAF Quarterly Assessment, 4th Quarter FY 1974," 2, RG 472, A1 1727, Box 2, Folder: RVNAF Quarterly Assessment, 3rd Quarter, FY 1974, NARA II.

69. Herrington, *Peace with Honor?*, 104–5.
70. Defense Attaché Saigon, "RVNAF Quarterly Assessment, 2nd Quarter FY 1974," 15.3, RG 472, A1 1727, Box 1, Folder: RVNAF Quarterly Assessment, 2nd Quarter, FY 1974, NARA II.
71. Nguyen Duy Hinh and Tran Dinh Tho, *South Vietnamese Society*, 42.
72. Ibid., 45.
73. Memo, Chambers to Colby, "A Personal Mid-Tour Report," 10 March 1970, RG 472, A1 456, Box 11, Folder: End of Tour Reports, 1601–11A, 1970, NARA II.
74. Tran Dinh Tho, *Pacification*, 1.
75. Memo, Laird to Nixon, 13 March 1969, *FRUS, 1969–1976*, 6:400.
76. Kissinger, *White House Years*, 273.
77. Peter Osnos, "The War: Uneasy Lull, Fading Optimism," *Washington Post*, 1 August 1971.
78. House Committee on Appropriations, *Department of Defense Appropriations for 1971*, 95.
79. Cosmas, *MACV*, 391–92.
80. Cao Van Vien and Dong Van Khuyen, *Reflections on the Vietnam War*, 113–14.
81. Dong Van Khuyen, *RVNAF*, 142.
82. Nguyen Duy Hinh and Tran Dinh Tho, *South Vietnamese Society*, 120.
83. Le Gro, *Vietnam from Ceasefire to Capitulation*, 69.
84. Goodman, "War without End?," 71–72.
85. Cao Van Vien and Dong Van Khuyen, *Reflections on the Vietnam War*, 93.
86. Kissinger, *White House Years*, 989.
87. House Committee on Appropriations, *Department of Defense Appropriations for 1971*, 437.
88. Cao Van Vien, *Final Collapse*, 39.
89. Hosmer, Kellen, and Jenkins, *Fall of South Vietnam*, 26.
90. Cao Van Vien and Dong Van Khuyen, *Reflections on the Vietnam War*, 124.
91. Ibid., 107.
92. Tran Van Don, *Our Endless War*, 230.
93. Nguyen Cong Luan, *Nationalist in the Viet Nam Wars*, 551.
94. CDEC, Captured Enemy Document: "Tan Uyen District PRP C.C. Anti Pacification Program," 12 December 1968, Folder 07, Box 12, Pike Collection: Unit 01—Assessment and Strategy, TTUVA.
95. Document No. 99, 11, VNDRN.
96. Michal Parks, "Hanoi Pursuing Popular Uprising," *Baltimore Sun* 15 April 1972.
97. James M. Markham, "Fighting in Vietnam," *New York Times*, 15 January 1975.
98. Hosmer, Kellen, and Jenkins, *Fall of South Vietnam*, 43.
99. Cao Van Vien and Dong Van Khuyen, *Reflections on the Vietnam War*, 114.
100. Cosmas, 392.

101. Document No. 118, "COSVN Directive 08/74 and the Binh Dinh Province Resolution: Communist Plans for 1975," 41, VNDRN.
102. Hoang Van Thai, *Decisive Years*, 66.
103. Le Gro, *Vietnam from Ceasefire to Capitulation*, 96–131.
104. "Viet Reds Now Reign Where GIs Conquered," *Los Angeles Times*, 5 August 1974.
105. Duiker, *Sacred War*, 243.
106. National Intelligence Estimate, "Short-Term Prospects for Vietnam," 23 December 1974, in Allen, Carver, and Elmore, *Estimative Products on Vietnam*, 639.
107. Kann, "Two Views in Vietnam."
108. Memo, Graham to Schlesinger, 24 January 1975, *FRUS, 1969–1976*, 10:611.
109. Memorandum of Conversation, 5 March 1975, *FRUS, 1969–1976*, 10:663.
110. Ang Cheng Guan, *Ending the Vietnam War*, 154–61.
111. Document No. 118, 40–41, VNDRN.
112. Telegram, Embassy to Department of State, 20 March 1972, *FRUS, 1969–1976*, 10:684–86.
113. Duiker, *Sacred War*, 244–46.

CONCLUSION

1. Robert L. Bartley, "Vietnam Pacification: Some Lessons," *Wall Street Journal*, 22 February 1967.
2. Charles Mohr, "In New Phase of Struggle in South Vietnam, Both Sides Have Assets," *New York Times*, 23 March 1973.
3. Sorley, *Better War*, 217.
4. Colby, *Lost Victory*, 310.
5. Palmer, *25-Year War*, 116.
6. Report, "The Revolutionary Development Program in Thua Thien Province," 5, RG 472, A1 462, Box 4, Folder: PSG Studies, 196, Book 1, Folder 1, 1968, NARA II.
7. Ithiel de Sola Pool, "Undercut VC Village Control," DepCORDS Files, CMH.
8. Nguyen Duy Hinh, *Vietnamization and the Cease-Fire*, 166.
9. Tran Van Tra, *Vietnam*, 5:73.
10. Tran Van Don, *Our Endless War*, 219.
11. William C. Westmoreland, "The Demise of South Vietnam," *New York Times*, 17 May 1975.
12. Andrade, *Trial by Fire*, 531.
13. Senate Committee on Foreign Relations, *Vietnam: May 1974*, 7.
14. Pribbenow, *Victory in Vietnam*, 444–45.
15. Le Gro, *Vietnam from Ceasefire to Capitulation*, 179.
16. Tran Van Tra, *Vietnam*, 38.
17. Cosmas, *MACV*, 403–4.
18. Parker, "Vietnam," 362.

BIBLIOGRAPHY

MANUSCRIPT AND ARCHIVAL SOURCES

Interviews with Ralliers and Prisoners of War. David W. P. Elliott Collection of Vietnamese Communist Documents, Cornell University, Ithaca, N.Y.

Military History Branch, Headquarters, United States Military Assistance Command, Vietnam.

Command History, 1968, vols. 1–2

Command History, 1969, vols. 1–2

Command History, 1970, vols. 1–2

Command History, 1971, vols. 1–2

Command History, 1972–1973, vols. 1–2

National Archives and Records Administration, College Park, Md.

Manuscript Collection

Record Group 472, Records of the U.S. Forces in Southeast Asia, 1950–1976

Central Intelligence Agency Records Search Tool (CREST)

U.S. Army Center of Military History, Washington, D.C.

Manuscript Collection

DepCORDS Files

U.S. Army Military History Institute, Carlisle, Pa.
Manuscript Collection
Robert M. Montague Papers
John Paul Vann Papers
U.S. Department of State. *Foreign Relations of the United States* (*FRUS*), Washington, D.C.
FRUS, 1969–1976, vol. 6: *Vietnam, January 1969–July 1970*
FRUS, 1969–1976, vol. 7: *Vietnam, July 1970–January 1972*
FRUS, 1969–1976, vol. 8: *Vietnam, January–October 1972*
FRUS, 1969–1976, vol. 9: *Vietnam, October 1972–January 1973*
FRUS, 1969–1976, vol. 10: *Vietnam, January 1973–July 1975*
U.S. Mission in Vietnam. *Vietnam Documents and Research Notes: Translation and Analysis of Significant Viet Cong/North Vietnamese Documents.*
The Vietnam Center and Archive, Texas Tech University, Lubbock.
Manuscript Collection
Douglas Pike Collection
John Donnell Collection
Vietnam Archive Collection

BOOKS AND ARTICLES

Advanced Research Projects Agency. *Village Defense Study—Vietnam*. Vol. 1, *Executive Report.* Santa Barbara, Calif.: General Research Corporation, 1969.

Allen, John K., John Carver, and Tom Elmore, eds. *Estimative Products on Vietnam, 1948–1975.* Pittsburgh, Pa.: U.S. Government Printing Office, 2005.

Anderson, David L. *The Columbia Guide to the Vietnam War.* New York: Columbia University Press, 2002.

Anderson, David L., and John Ernst, eds. *The War That Never Ends: New Perspectives on the Vietnam War.* Lexington: The University Press of Kentucky, 2007.

Anderson, M., M. Arnsten, and H. Averch. *Insurgent Organization and Operations: A Case Study of the Viet Cong in the Delta, 1964–1966.* Santa Monica, Calif.: RAND Corporation, 1967.

Andrade, Dale. *America's Last Vietnam Battle: Halting Hanoi's 1972 Easter Offensive.* Lawrence: University Press of Kansas, 2001.

———. *Ashes to Ashes: The Phoenix Program and the Vietnam War.* Lexington, MA: Lexington Books, 1990.

———. *Trial by Fire: The 1972 Easter Offensive, America's Last Vietnam Battle.* New York: Hippocrene Books, 1995.

———. "Westmoreland Was Right: Learning the Wrong Lessons from the Vietnam War." *Small Wars and Insurgencies* 19 (2008): 145–81.

Ang Cheng Guan. *Ending the Vietnam War: The Vietnamese Communists' Perspective.* London: Routledge, 2004.

———. *The Vietnam War from the Other Side: The Vietnamese Communists' Perspective.* London: Routledge, 2002.

Asselin, Pierre. *A Bitter Peace: Washington, Hanoi, and the Making of the Paris Agreement.* Chapel Hill: University of North Carolina Press, 2002.

———. *Hanoi's Road to the Vietnam War, 1954–1965.* Berkeley: University of California Press, 2013.

———. "Revisionism Triumphant: Hanoi's Diplomatic Strategy in the Nixon Era." *Journal of Cold War Studies* 13 (Fall 2011): 101–37.

Bergerud, Eric M. *The Dynamics of Defeat: The Vietnam War in Hau Nghia Province.* Boulder, Colo.: Westview Press, 1991.

Birtle, Andrew J. "Persuasion and Coercion in Counterinsurgency Warfare." *Military Review* (July–August 2008): 45–53.

———. *U.S. Army Counterinsurgency and Contingency Operations Doctrine, 1942–1976.* Washington, D.C.: U.S. Army Center of Military History, 2006.

Bole, Albert C., and Katsuji Kobata. *An Evaluation of the Measurement of the Hamlet Evaluation System.* Naval War College Center for Advanced Research Projects. Newport, R.I.: U.S. Naval War College, 1975.

Boylan, Kevin M. "The Red Queen's Race: Operation Washington Green and Pacification in Binh Dinh Province, 1969–1970." *Journal of Military History* 73 (October 2009): 1195–230.

Bradley, Mark Philip. *Vietnam at War.* Oxford: Oxford University Press, 2009.

Brigham, Erwin R. "Pacification Measurement." *Military Review* (May 1970): 47–55.

Brigham, Robert K. *Guerrilla Diplomacy: The NLF's Foreign Relations and the Vietnam War.* Ithaca, N.Y.: Cornell University Press, 1999.

Bullington, James R., and James D. Rosenthal. "The South Vietnamese Countryside: Non-Communist Political Perceptions." *Asian Survey* 10 (August 1970): 651–61.

Buttinger, Joseph. *The Smaller Dragon: A Political History of Vietnam.* New York: Praeger, 1958.

Cao Van Vien. *The Final Collapse.* Washington, D.C.: U.S. Army Center of Military History, 1983.

Cao Van Vien and Dong Van Khuyen. *Reflections on the Vietnam War.* Washington, D.C.: U.S. Army Center of Military History, 1980.

Carland, John M. *Combat Operations: Stemming the Tide, May 1965 to October 1966.* Washington, D.C.: U.S. Army Center of Military History, 2000.

Catton, Philip E. "Counterinsurgency and Nation Building: The Strategic Hamlet Program in South Vietnam, 1961–1963." *International History Review* 21 (December 1999): 918–40.

———. *Diem's Final Failure: Prelude to America's War in Vietnam.* Lawrence: University Press of Kansas, 2002.

Clarke, Jeffrey J. *Advice and Support: The Final Years, 1965–1973.* Washington, D.C.: U.S. Army Center of Military History, 1988.

Clifford, Clark M. "A Vietnam Reappraisal." *Foreign Affairs* 47 (July 1969): 601–22.

Colby, William E. *Lost Victory: A Firsthand Account of America's Sixteen-Year Involvement in Vietnam*. Chicago: Contemporary Books, 1989.

Collins, John M. "Vietnam Postmortem: A Senseless Strategy." *Parameters* (Winter 2010): 32–39.

Conley, Michael Charles. "Communist Thought and Viet Cong Tactics." *Asian Survey* 8 (March 1968): 206–22.

Cooper, Chester L., Judith E. Corson, Laurence J. Legere, David E. Lockwood, and Donald M. Weller. *The American Experience with Pacification in Vietnam*. 3 vols. Springfield, Va.: Institute for Defense Analyses, International and Social Studies Division, 1972.

Corson, William R. *The Betrayal*. New York: W. W. Norton, 1968.

Corum, James S. *Bad Strategies: How Major Power Fail in Counterinsurgency*. Minneapolis, Zenith Press, 2008.

Cosmas, Graham A. *MACV: The Joint Command in the Years of Withdrawal, 1968–1973*. Washington, D.C.: U.S. Army Center of Military History, 2006.

Cowen, Deborah, and Emily Gilbert. "The Politics of War, Citizenship, Territory." In *War, Citizenship, Territory*. New York: Routledge, 2008.

Daddis, Gregory A. *No Sure Victory: Measuring U.S. Army Effectiveness and Progress in the Vietnam War*. New York: Oxford University Press, 2011.

———. *Westmoreland's War: Reassessing American Strategy in Vietnam*. New York: Oxford University Press, 2014.

Davidson, Phillip B. *Secrets of the Vietnam War*. Novato, Calif.: Presidio Press, 1990.

———. *Vietnam at War: The History, 1946–1975*. New York: Oxford University Press, 1988.

Davison, W. P. *Some Observations on Viet Cong Operations in the Villages*. Santa Monica, Calif.: RAND Corporation, 1968.

Davison, W. P., and J. J. Zasloff. *A Profile of Viet Cong Cadres*. Santa Monica, Calif.: RAND Corporation, 1966.

Denton, Frank H. *Some Notes on the Consequences of Bombardment of Civilian Areas in South Vietnam*. Santa Monica, Calif.: RAND Corporation, 1967.

Dodd, Joseph W. "Faction and Failure in South Vietnam." *Asian Affairs* 2 (January–February 1975): 173–78.

Dong Van Khuyen. *The RVNAF*. Washington, D.C.: U.S. Army Center of Military History, 1978.

Donnell, John C. "Expanding Political Participation: The Long Haul from Villagism to Nationalism." *Asian Survey* 10 (August 1970): 688–704.

———. "Pacification Reassessed." *Asian Survey* 7 (August 1967): 567–76.

Donovan, David. *Once a Warrior King: Memories of an Officer in Vietnam*. New York: McGraw-Hill, 1985.

Duiker, William J. *The Communist Road to Power in Vietnam*. Boulder, Colo.: Westview Press, 1981.

———. *Sacred War: Nationalism and Revolution in a Divided Vietnam*. New York: McGraw-Hill, 1995.

Duncanson, Dennis J. "Pacification and Democracy in South Vietnam." *World Today* 23 (October 1967): 410–18.

Dunn, Jerry F. "New Look at Pacification." *Military Review* (January 1970): 84–87.

Elden, Stuart. *Understanding Henri Lefebvre: Theory and the Possible.* London: Continuum, 2004.

Elliott, David W. P. "Hanoi's Strategy in the Second Indochina War." In *The Vietnam War: Vietnamese and American Perspectives.* Edited by Jayne S.Werner and Luu Doan Huynh. Armonk, N.Y.: M. E. Sharpe, 1993.

———. *Pacification and the Viet Cong System in Dinh Tuong, 1966–1967.* Santa Monica, Calif.: RAND Corporation, 1969.

———. *The Vietnamese War: Revolution and Social Change in the Mekong Delta, 1930–1975.* Armonk, N.Y.: M. E. Sharpe, 2003.

Ellsberg, Daniel. *The Day Loc Tien Was Pacified.* Santa Monica, Calif.: RAND Corporation, 1968.

Embrey, James H. *Reorienting Pacification: The Accelerated Pacification Campaign of 1968.* Lexington: University Press of Kentucky, 1997.

Ewell, Julian J., and Ira A. Hunt. *Sharpening the Combat Edge: The Use of Analysis to Reinforce Military Judgment.* Washington, D.C.: Department of the Army, 1973.

Fall, Bernard. "South Vietnam at the Crossroads." *International Journal* 19 (Spring 1964): 139–54.

———. *The Two Vietnams: A Political and Military Analysis.* New York: Praeger, 1967.

Fitzgerald, Francis. *Fire in the Lake: The Vietnamese and the Americans in Vietnam.* Boston: Little, Brown, 1972.

Flint, Colin, ed. *The Geography of War and Peace: From Death Camps to Diplomats.* Oxford: Oxford University Press, 2005.

Foucault, Michel. *Discipline and Punish: The Birth of the Prison.* New York: Vintage, 2005.

———. "Space, Knowledge, and Power." In Rabinow, *Foucault Reader,* 239–56.

Galula, David. *Counterinsurgency Theory and Practice.* New York: Praeger, 1964.

———. *Pacification in Algeria, 1956–1958.* Santa Monica, Calif.: RAND Corporation, 2006.

Gentile, Gian. *Wrong Turn: America's Deadly Embrace of Counterinsurgency.* New York: New Press, 2013.

Goodman, Allan E. "The End of the War as a Setting for the Future Development of South Vietnam." *Asian Survey* 11 (April 1971): 341–51.

———. "South Vietnam: Neither War nor Peace." *Asian Survey* 10 (February 1970): 107–32.

———. "War without End?" *Asian Survey* 15 (January 1975): 70–84.

Grinter, Lawrence E. "How They Lost: Doctrines, Strategies, and Outcomes of the Vietnam War." *Asian Survey* 15 (December 1975): 1114–32.

Hammond, William M. *Reporting Vietnam: Media and Military at War.* Lawrence: University Press of Kansas, 1998.

Harmon, Russell S., Francis H. Dillon III, and John B. Garver Jr. "Perspectives on Military Geography: The Military Operating Environment." In *Studies in Military Geography and Geology*, edited by Douglass R. Caldwell, Judy Ehlen, and Russell S. Harmon. Dordrecht, Neth.: Kluwer Academic Publishers, 2004.

Hennessy, Michael A. *Strategy in Vietnam: The Marines and Revolutionary Warfare in I Corps, 1965–1975*. Westport, Conn.: Praeger, 1997.

Herring, George C. *America's Longest War: The United States and Vietnam, 1950–1975*. New York: John Wiley and Sons, 1979.

Herrington, Stuart A. *Peace with Honor? An American Reports on Vietnam, 1973–1975*. Novato, Calif.: Presidio Press, 1983.

———. *Silence Was a Weapon: The Vietnam War in the Villages*. Novato, Calif.: Presidio Press, 1982.

Hess, Gary R. *Vietnam: Explaining America's Lost War*. Malden, Mass.: Blackwell Publishing, 2009.

Hickey, Gerald C. *Accommodation and Coalition in South Vietnam*. Santa Monica, Calif.: RAND Corporation, 1970.

———. *U.S. Strategy in South Vietnam: Extrication and Equilibrium*. Santa Monica, Calif.: RAND Corporation, 1969.

———. *Village in Vietnam*. New Haven: Yale University Press, 1964.

Hoang Ngoc Lung. *The General Offensives of 1968–69*. Washington, D.C.: U.S. Army Center of Military History, 1981.

———. *Strategy and Tactics*. Washington, D.C.: U.S. Army Center of Military History, 1980.

Hoang Van Thai. *The Decisive Years: Memoirs of Vietnamese Senior General Hoang Van Thai*. Washington, D.C.: Foreign Broadcast Information Service, 1987.

———. *Some Aspects of Guerrilla Warfare in Vietnam*. Hanoi: Foreign Languages Publishing House, 1965.

Hosmer, Stephen T. *Viet Cong Repression and Its Implications for the Future*. Santa Monica, Calif.: RAND Corporation, 1970.

Hosmer, Stephen T., Konrad Kellen, and Brian M. Jenkins. *The Fall of South Vietnam: Statements by Vietnamese Military and Civilian Leaders*. Santa Monica, Calif.: RAND Corporation, 1978.

Hunt, David. *Vietnam's Southern Revolution: From Peasant Insurrection to Total War*. Amherst: University of Massachusetts Press, 2008.

Hunt, Ira A. *The 9th Infantry Division in Vietnam: Unparalleled and Unequaled*. Lexington: University Press of Kentucky, 2010.

Hunt, Richard A. *Pacification: The American Struggle for Vietnam's Hearts and Minds*. Boulder, Colo.: Westview Press, 1995.

———. "Strategies at War: Pacification and Attrition in Vietnam." In Hunt and Shultz, *Lessons from an Unconventional War*.

Hunt, Richard A., and Richard H. Schultz, eds. *Lessons from an Unconventional War: Reassessing U.S. Strategies for Future Conflicts*. New York: Pergamon Press, 1982.

Huntington, Samuel P. "The Bases of Accommodation." *Foreign Affairs* 46 (July 1968): 642–56.

Isaacs, Arnold R. *Without Honor: Defeat in Vietnam and Cambodia.* Baltimore: Johns Hopkins University Press, 1983.

Jamieson, Neil L. *Understanding Vietnam.* Berkeley: University of California Press, 1993.

Johnson, Chalmers. *Autopsy on People's War.* Berkeley: University of California Press, 1973.

———. "The Third Generation of Guerrilla Warfare." *Asian Survey* 8 (June 1968): 435–47.

Joiner, Charles. "South Vietnam: The Politics of Peace." *Asian Survey* 9 (February 1969): 138–55.

Kalyvas, Stathis N. *The Logic of Violence in Civil War.* Cambridge: Cambridge University Press, 2006.

Kalyvas, Stathis N., and Matthew Adam Kocher. "The Dynamics of Violence in Vietnam: An Analysis of the Hamlet Evaluation System (HES)." *Journal of Peace Research* 46 (May 2009): 335–55.

Katzenbach, E. L. "Time, Space, and Will: The Politico-Military Views of Mao Tse-tung." *Marine Corps Gazette* 40 (October 1956): 36–40.

Kellen, Konrad. *Conversations with Enemy Soldiers in Late 1968/Early 1969: A Study of Motivation and Morale.* Santa Monica, Calif.: RAND Corporation, 1970.

———. *A View of the VC: Elements of Cohesion in the Enemy Camp, 1966–1967.* Santa Monica, Calif.: RAND Corporation, 1969.

Kernig, C. D., ed. *Marxism, Communism, and Western Society: A Comparative Encyclopedia.* New York: Herder and Herder, 1972.

Kimball, Jeffrey. *Nixon's Vietnam War.* Lawrence: University Press of Kansas, 1998.

———. *The Vietnam War Files: Uncovering the Secret History of Nixon-Era Strategy.* Lawrence: University Press of Kansas, 2004.

Kinnard, Douglas. *The War Managers.* Hanover, N.H.: University Press of New England, 1977.

Kissinger, Henry A. "The Vietnam Negotiations." *Foreign Affairs* 47 (January 1969): 212.

———. *White House Years.* Boston: Little, Brown, 1979.

Kocher, Matthew Adam, Thomas B. Pepinsky, and Stathis N. Kalyvas. "Aerial Bombing and Counterinsurgency in the Vietnam War." *American Journal of Political Science* 55 (April 2011): 201–18.

Kolko, Gabriel. *Anatomy of a War: Vietnam, the United States, and the Modern Historical Experience.* New York: New Press, 1985.

Komer, Robert W. *Bureaucracy at War: U.S. Performance in the Vietnam Conflict.* Boulder, Colo.: Westview Press, 1986.

———. *Impact of Pacification on Insurgency in South Vietnam.* Santa Monica, Calif.: RAND Corporation, 1970.

———. *Organization and Management of the "New Model" Pacification Program, 1966–1969.* Santa Monica, Calif.: RAND Corporation, 1970.

———. *The Current Pacification Program in Vietnam*. Santa Monica, Calif.: RAND Corporation, 1970.

Krepinevich, Andrew F. *The Army and Vietnam*. Baltimore: Johns Hopkins University Press, 1986.

Lam Quang Thi. *The Twenty-five Year Century: A South Vietnamese General Remembers the Indochina War to the Fall of Saigon*. Denton: University of North Texas Press, 2001.

Le Duan. "Political and Military Forces in Revolutionary War." In *Guerrilla Warfare and Marxism*, edited by William J. Pomeroy. New York: International Publishers, 1968.

Le Gro, William E. *Vietnam from Ceasefire to Capitulation*. Washington, D.C.: U.S. Army Center of Military History, 1985.

Lefebvre, Henri. *The Production of Space*. Translated by Donald Nicholson-Smith. Oxford: Blackwell Publishing, 1974.

———. "Reflections on Politics of Space." *Antipode* 8 (May 1976): 30–37.

Leites, Nathan, and Charles Wolf. *Rebellion and Authority: An Analytic Essay on Insurgent Conflicts*. Santa Monica, Calif.: RAND Corporation, 1970.

Lewy, Guenter. *America in Vietnam*. New York: Oxford University Press, 1978.

Lin Biao. *Long Live the Victory of People's War*. Beijing: Foreign Language Press, 1966.

Lohman, Andrew D., and Colin Flint. "The Geography of Insurgency." *Geography Compass* 4 (2010): 1154–66.

Mallin, Jay. *Terror in Vietnam*. Princeton, N.J.: D. Van Nostrand, 1966.

Mao Tse-tung. *On Guerrilla Warfare*. Translated by Samuel B. Griffith. Urbana: University of Illinois Press, 2000.

———. *Selected Military Writings of Mao Tse-Tung*. Beijing: Foreign Language Press, 1966.

Matthews, Lloyd J., and Dale E. Brown. *Assessing the Vietnam War*. Washington, D.C.: Pergamon-Brassey's International Defense Publishers, 1987.

McAlister, John T. *Vietnam: The Origins of Revolution*. New York: Alfred A. Knopf, 1969.

McAlister, John T., and Paul Mus. *The Vietnamese and Their Revolution*. New York: Harper and Row, 1970.

McColl, Robert W. "The Insurgent State: Territorial Bases of Revolution." *Annals of the Association of American Geographers* 59 (December 1969): 613–31.

Milam, Ron. *Not a Gentleman's War: An Inside View of Junior Officers in the Vietnam War*. Chapel Hill: University of North Carolina Press, 2009.

Mitchell, Don. "The End of Public Space?: People's Park, Definitions of the Public, and Democracy." *Annals of the Association of American Geographers* 85 (1995): 108–33.

Moyar, Mark. *Phoenix and the Birds of Prey: The CIA's Secret Campaign to Destroy the Viet Cong*. Annapolis, Md.: Naval Institute Press, 1997.

Nagl, John A. *Learning to Eat Soup with A Knife: Counterinsurgency Lessons from Malaya and Vietnam*. Chicago: University of Chicago Press, 2005.

Nguyen, Lien-Hang T. *Hanoi's War: An International History of the War for Peace in Vietnam*. Chapel Hill: University of North Carolina Press, 2012.

Nguyen Cong Luan. *Nationalist in the Viet Nam Wars: Memoirs of a Victim Turned Soldier*. Bloomington: Indiana University Press, 2012.

Nguyen Duy Hinh. *Vietnamization and the Cease-Fire*. Washington, D.C.: U.S. Army Center of Military History, 1980.

Nguyen Duy Hinh and Tran Dinh Tho. *The South Vietnamese Society*. Washington, D.C.: U.S. Army Center of Military History, 1980.

Nixon, Richard M. *The Memoirs of Richard Nixon*. New York: Grosset & Dunlap, 1978.

———. *No More Vietnams*. New York: Arbor House, 1985.

Office of the Secretary of Defense, Comptroller. "Unclassified Statistics on Southeast Asia." In *The Air War in Southeast Asia*, edited by Rafael Littauer and Norman Uphoff. Boston: Beacon Press, 1972.

O'Neill, Bard E. *Insurgency and Terrorism: Inside Modern Revolutionary Warfare*. Washington, D.C.: Brassey's, 1990.

O'Sullivan, Patrick. "A Geographical Analysis of Guerrilla Warfare." *Political Geography Quarterly* 2 (April 1983): 139–50.

Palmer, Bruce. *Summons of the Trumpet: U.S.–Vietnam in Perspective*. Novato, Calif.: Presidio Press, 1978).

———. *The 25-Year War: America's Military Role in Vietnam*. Lexington: University Press of Kentucky, 1984.

Parker, Maynard. "Vietnam: The War That Won't End." *Foreign Affairs* 53 (January 1975): 352–74.

Pauker, Guy J. *An Essay on Vietnamization*. Santa Monica, Calif.: RAND Corporation, 1971.

Peterson, Michael E. *The Combined Action Platoons: The U.S. Marines' Other War in Vietnam*. New York: Praeger, 1989.

Phillips, Rufus. *Why Vietnam Matters*. Annapolis, Md.: Naval Institute Press, 2008.

Pike, Douglas, ed. *The Bunker Papers: Reports to the President from Vietnam, 1967–1973*. Vol. 2. Berkeley: Institute of East Asian Studies, University of California, 1990.

———. "Guerrilla Warfare in Vietnam." In *Guerrilla Warfare in Asia*. The Hague: International Documentation and Information Center, 1971.

———. *PAVN: People's Army of Vietnam*. Novato, Calif.: Presidio Press, 1986.

———. *Viet Cong: The Organization and Techniques of the National Liberation Front of South Vietnam*. Cambridge, Mass.: MIT Press, 1966.

———. *War, Peace, and the Viet Cong*. Cambridge, Mass.: MIT Press, 1969.

Popkin, Samuel L. "Pacification: Politics and the Village." *Asian Survey* 10 (August 1970): 662–71.

Porch, Douglas. *Counterinsurgency: Exposing the Myths of the New Way of War*. Cambridge: Cambridge University Press, 2013.

Porter, Gareth, ed. *Vietnam: The Definitive Documentation of Human Decisions*. Stanfordville, N.Y.: Earl M. Coleman Enterprises, 1979.

Pred, Allen. *Making Histories and Constructing Human Geographies: The Local Transformation of Practice, Power Relations, and Consciousness*. Boulder, Colo.: Westview Press, 1990.

Pribbenow, Merle L., trans. *Victory in Vietnam: The Official History of the People's Army of Vietnam, 1954–1975*. Lawrence: University Press of Kansas, 2002.

———. "General Vo Nguyen Giap and the Mysterious Evolution of the Plan for the 1968 Tet Offensive." *Journal of Vietnamese Studies* 3 (Summer 2008): 1–33.

Rabinow, Paul, ed. *The Foucault Reader*. New York: Pantheon Books, 1984.

Race, Jeffrey. "How They Won." *Asian Survey* 10 (August 1970): 628–50.

———. *War Comes to Long An: Revolutionary Conflict in a Vietnamese Province*. Berkeley: University of California Press, 1972.

Rottman, G. *Viet Cong and NVA Tunnels and Fortifications of the Vietnam War*. Oxford: Osprey Publishing, 2006.

Sack, Robert. *Human Territoriality: Its Theory and History*. Cambridge: Cambridge University Press, 1986.

Scales, Robert H. *Firepower in Limited War*. Washington, D.C.: National Defense University Press, 1990.

Schlight, John, ed. *Second Indochina War Symposium*. Washington, D.C.: U.S. Army Center of Military History, 1986.

Schultz, Richard H. "The Pacification-Vietnamization Strategy of 1969–1972: A Quantitative and Qualitative Reassessment." In Hunt and Schultz, *Lessons from an Unconventional War*.

Scott, James C. *Seeing Like a State: How Certain Schemes to Improve the Human Condition Have Failed*. New Haven: Yale University Press, 1998.

Sharp, Admiral Ulysses S. Grant, CINCPAC, and General William C. Westmoreland, COMUSMACV. *Report on the War in Vietnam (as of June 30, 1968)*. Washington, D.C.: U.S. Government Printing Office, 1968.

Silverman, Jerry Mark. "South Vietnam and the Elusive Peace." *Asian Survey* 13 (January 1973): 19–45.

———. "South Vietnam and the Return to Political Struggle." *Asian Survey* 14 (January 1974): 65–77.

Smith, Rupert. *The Utility of Force: The Art of War in the Modern World*. New York: Vintage Books, 2005.

Sorley, Lewis. *A Better War: The Unexamined Victories and Final Tragedy of America's Last Years in Vietnam*. New York: Harcourt Brace and Company, 1999.

———, ed. *Vietnam Chronicles: The Abrams Tapes, 1968–1972*. Lubbock: Texas Tech University Press, 2004.

Spector, Ronald H. *After Tet: The Bloodiest Year in Vietnam*. New York: Vintage, 1993.

Staniland, Paul. "States, Insurgents, and Wartime Political Orders." *Perspectives on Politics* 10 (June 2012): 243–64.

Taber, Robert. *War of the Flea: The Classic Study of Guerrilla Warfare*. Washington, D.C.: Potomac Books, 2002.

Tanham, George K. *Communist Revolutionary Warfare: From the Vietminh to the Viet Cong*. Westport, Conn.: Praeger Security International, 2006.

Taylor, K. W. *A History of the Vietnamese*. Cambridge: Cambridge University Press, 2013.

Taylor, Maxwell D. *Swords and Plowshares*. New York: W. W. Norton, 1972.

Thayer, Thomas C., ed. *A Systems Analysis View of the Vietnam War, 1965–1972*. 10 vols. Washington, D.C.: Office of the Assistant Secretary of Defense (Systems Analysis), 1975.

———. "Patterns of the French and American Experience in Vietnam." In Thompson and Frizell, *Lessons of Vietnam*.

———. "On Pacification." In Thompson and Frizell, *Lessons of Vietnam*.

———. *War without Fronts: The American Experience in Vietnam*. Boulder, Colo.: Westview Press, 1985.

Thompson, Robert. *Defeating Communist Insurgency: The Lessons of Malaya and Vietnam*. New York: Praeger, 1966.

———. *No Exit from Vietnam*. New York: David McKay, 1969.

———. "Squaring the Error." *Foreign Affairs* 46 (April 1968): 442–53.

Thompson, W. Scott, and Donaldson D. Frizell, eds. *The Lessons of Vietnam*. New York: Crane, Russak & Company, 1977.

Tran Dinh Tho. *Pacification*. Washington, D.C.: U.S. Army Center of Military History, 1980.

Tran Van Don. *Our Endless War: Inside Vietnam*. San Rafael, Calif.: Presidio Press, 1978.

Tran Van Tra. "Tet: The 1968 General Offensive and General Uprising." In Werner and Huynh, *Vietnam War*, 37–65.

———. *Vietnam: History of the Bulwark B2 Theatre*. Vol. 5, *Concluding the 30-Years War*. Springfield, Va.: Joint Publications Research Service, 1983.

Trullinger, James W. *Village at War: An Account of Revolution in Vietnam*. New York: Longman, 1980.

Truong Chinh. *Primer for Revolt: The Communist Takeover in Vietnam*. New York: Praeger, 1963.

Truong Ngo Quang. *The Easter Offensive*. Washington, D.C.: U.S. Army Center of Military History, 1980.

———. *Territorial Forces*. Washington, D.C.: U.S. Army Center of Military History, 1981.

Truong Nhu Tang, with David Chanoff and Doan Van Toai. *A Vietcong Memoir*. San Diego: Harcourt Brace Jovanovich, 1985.

Tucker, Richard P., and Edmund Russell, eds. *Natural Enemy, Natural Ally: Toward an Environmental History of War*. Corvallis: Oregon State University Press, 2004.

U.S. Congress. House. Committee on Appropriations. *Department of Defense Appropriations for 1971: Hearings before a Subcommittee of the Committee on Appropriations*. 91st Cong., 2nd Sess., 16 February 1970. Washington, D.C.: U.S. Government Printing Office, 1970.

———. *Supplemental Appropriation Bill, 1971: Hearings before Subcommittees of the Committee on Appropriations*. 91st Cong., 2nd Sess., 21 September 1970. Washington, D.C.: U.S. Government Printing Office, 1970.

U.S. Congress. House. Committee on Armed Services. *Progress of the Pacification Program*. 91st Cong., 2nd Sess., 1970. H. Doc. 91-38. Washington, D.C.: U.S. Government Printing Office, 1970.

U.S. Congress. House. Committee on Foreign Affairs. *Measuring Hamlet Security in Vietnam, Report of a Special Study Mission by John V. Tunney*. 91st Cong., 1st Sess., 1969. H. Doc. 91-25. Washington, D.C.: U.S. Government Printing Office, 1969.

———. *Vietnam—a Changing Crucible: Report of a Study Mission to South Vietnam*. 93rd Cong., 2nd Sess., 1974. Washington, D.C.: U.S. Government Printing Office, 1974.

U.S. Congress. House. Committee on Government Operations. *U.S. Assistance Programs in Vietnam: Hearings before Subcommittee of the Committee on Government Operations*. 92nd Cong., 1st Sess., 1971. Washington, D.C.: U.S. Government Printing Office, 1971.

U.S. Congress. House. Select Committee on United States Involvement in Southeast Asia. *United States Involvement in Southeast Asia*. 91st Cong., 2nd Sess., 1970. H. Doc. 91-1276. Washington, D.C.: U.S. Government Printing Office, 1970.

U.S. Congress. Senate. Committee on Foreign Relations. *Impact of the Vietnam War*. 92nd Cong., 1st Sess., 1971. Washington, D.C.: U.S. Government Printing Office, 1971.

———. *Vietnam Policy and Prospects, 1970: Civil Operations and Rural Development Support Program: Hearings before the Senate Committee on Foreign Relations*. 91st Cong., 2nd Sess., 17–20 February and 3–4, 17, and 19 March 1970. Washington, D.C.: U.S. Government Printing Office, 1970.

———. *Vietnam: May 1972*. 92nd Cong., 2nd Sess., 1972. Washington, D.C.: U.S. Government Printing Office, 1972.

———. *Vietnam: May 1974*. 93rd Cong., 2nd Sess., 1974. Washington, D.C.: U.S. Government Printing Office, 1974.

U.S. Department of the Army. *A Program for the Pacification and Long-Term Development of South Vietnam*. Washington, D.C., 1966.

Van Tien Dung. *Our Great Spring Victory: An Account of the Liberation of South Vietnam*. New York: Monthly Review Press, 1977.

Vo Nguyen Giap. *Banner of People's War, the Party's Military Line*. New York: Praeger, 1970.

———. *The Military Art of People's War: Selected Writings of General Vo Nguyen Giap*. Edited by Russell Stetler. New York: Monthly Review Press, 1971.

———. *People's War, People's Army: The Vietcong Insurrection Manual for Underdeveloped Countries*. Honolulu: University Press of the Pacific, 2001.

———. *The South Vietnamese People Will Win*. Hanoi: Foreign Languages Publishing House, 1965.

Werner, Jayne S., and Luu Doan Huynh, eds. *The Vietnam War: Vietnamese and American Perspectives*. Armonk, N.Y.: M. E. Sharpe, 1993.

West, F. J. *Area Security*. Santa Monica, Calif.: RAND Corporation, 1969.

———. *The Enclave: Some U.S. Military Efforts in Ly Tin District, Quang Tin Province, 1966–1968*. Santa Monica, Calif.: RAND Corporation, 1969.

———. *The Village*. New York: Harper amd Row, 1972.

Westmoreland, William C. *A Soldier Reports*. New York: Doubleday & Co., 1976.

Wiegersma, Nancy. *Vietnam: Patriarchy and Collectivity in the Rural Economy.* New York: St. Martin's Press, 1988.

Wiest, Andrew, ed. *Rolling Thunder in a Gentle Land: The Vietnam War Revisited.* New York: Osprey Publishing, 2006.

———. *Vietnam's Forgotten Army: Heroism and Betrayal in the ARVN.* New York: New York University Press, 2008.

Willbanks, James A. *Abandoning Vietnam: How America Left and South Vietnam Lost Its War.* Lawrence: University Press of Kansas, 2004.

———. *The Battle of An Loc.* Bloomington: Indiana University Press, 2005.

———. *The Tet Offensive: A Concise History.* New York: Columbia University Press, 2006.

NEWSPAPERS AND PERIODICALS

Baltimore Sun

Boston Globe

Chicago Tribune

Guardian

Los Angeles Times

Newsweek

New York Times

Wall Street Journal

Washington Post

INDEX